Pharmaceutical Calculations

version 1.0

Sean E. Parsons, CPhT

P³
Parsons Printing Press
328 Janice Drive
Pittsburgh, PA 15235
sean@pharmaceuticalcalculations.org

You may download this book in portable desktop format (PDF) format or in open document format from http://pharmaceuticalcalculations.org. Some additional materials, such as presentations and practice exams, may be found at the aforementioned website as well.

A link for purchasing this book may be found at http://pharmaceuticalcalculations.org or you may purchase directly from http://lulu.com and search for pharmaceutical calculations.

Please submit any changes and/or corrections to sean@pharmaceuticalcalculations.org

Financial contributions can be made at http://pharmaceuticalcalcutions.org or by mailing a check or money order to Sean Parsons, 328 Janice Drive, Pittsburgh, PA 15235

ISBN: 978-0-578-06373-7

PREFACE

The purpose of creating *Pharmaceutical Calculations* is to provide instructors, students, pharmacy technicians, or anyone else even interested in a math book that offers better flexibility and affordability than any other textbook you might find. By making it available in a modifiable electronic format with a license that guarantees your rights, you have the freedom to use this textbook for any purpose, the ability to adapt the book to your needs, the freedom to improve the book and release those improvements to the public, and the freedom to use this book to help your neighbor.

I do request a $15 donation for each person you provide this book to, but honestly that is up to you and your school to collect that donation. I will not show up at your front door demanding payment, I will not accuse you of cheating me, and I will not create a website where I bash your name. That request for a donation is simply to help me supplement all the time and effort that I have placed into writing and updating this book.

I honestly hope that this textbook becomes the best math book you ever teach/learn pharmacy math from, and with your help this book can achieve this lofty goal.

INTRODUCTION

I have broken this book up into four sections, *Basic Arithmetic*, *Basic Pharmacy Math*, *Community Pharmacy Math*, and *Institutional Pharmacy Math*.

- *Basic Arithmetic* is intended to reintroduce the students to some basic math concepts and get everyone on the same page. It includes things such as Roman numerals, decimals, fractions, percentages, 24-hour time, exponents, scientific notation, and basic problem solving methods.
- *Basic Pharmacy Math* provides an introduction to converting between different temperature scales, the household system, the metric system, the apothecary system, some basic terminology, some fairly simple work with providing 24 hour supplies of medication, drip rates, and even some percentage strength problems and dilutions.
- *Community Pharmacy Math* will teach some compounding math and how to bill for those compounds, days' supply, par levels, percent mark-up, and third party insurance billing and other concepts related to pharmacy business math.
- *Institutional Pharmacy Math* will teach basic parenteral dosage calculations, insulin dosing, how to calculate milliMoles, milliEquivalents, and how international units are derived. It will also teach drug reconstitution, percentage strength, milligram percents, ratio strength, parts, reducing and enlarging formulas, dosage calculations based on body weight, dosage calculations based on body surface area, carboplatin dosing, infusion rates and drip rates, dilutions and alligations, parenteral nutrition, aliquots and drug desensitization therapy.

Upon completion of this book, a student should be well prepared for the calculations expected of a pharmacy technician.

Acknowledgements

With any book, an author has to make a large sacrifice of time, and the people that suffer most dearly are his loved ones. In this particular case, I want to let everyone know how much I love and adore my wife Shannon and my daughter Sammantha, without either of which this book would have been impossible. I also want to thank my supervisor Connie Geiger, whom never seemed bothered by how much effort I placed into this book. I should also thank Bidwell Training Center for offering me a job to teach pharmacy math (if I had never been offered this job, I am pretty sure I wouldn't have written a math book).

My students also deserve a lot of *'props'* for all their initial encouragement that made me consider even writing a book, and their willingness to function as guinea pigs as I tested my hair brained ideas on how to teach pharmacy math. In that same vein, I also want to thank Barb Snyder and my other coworkers for their continued encouragement throughout this whole process.

I want to thank my editors Sharon Brendza, CPhT and Arshaya Hardy, CPhT whom provided innumerable corrections and suggestions. They are both former students that I now look upon as peers. Their attention to detail made them obvious candidates for the job.

Last, but certainly not least, I want to thank everyone else whom has ever sent me or ever will send me an edit or suggestion to improve this book. I believe that this book created along with the collaborative efforts of others is an example of the direction that the textbook industry will need to move to keep up with the rapid advancements of science and technology.

Sincerely,
Sean E. Parsons

CONTENTS

UNIT 1

BASIC ARITHMETIC

What is arithmetic?

Arithmetic or arithmetics (from the Greek word arithmētikē which literally means the art of counting) is the oldest and most elementary branch of mathematics, used by almost everyone, for tasks ranging from simple daily counting to advanced science and business calculations. In common usage, the word refers to a branch of (or the forerunner of) mathematics which records elementary properties of certain operations on numbers. Professional mathematicians sometimes use the term higher arithmetic as a synonym for number theory, but this should not be confused with elementary arithmetic.

What will be learned/reviewed in this unit?

- Roman numerals
- decimal places
- rounding
- significant figures
- addition, subtraction, multiplication and division of decimals
- parts of a fraction
- addition, subtraction, multiplication and division of fractions
- percentages
- 24 hour clocks
- exponents
- scientific notation
- ratios
- proportions
- dimensional analysis
- 5 step problem solving method

Math is usually taught as all scales and no music.

-- Persis Herold

Chapter 1

Numeral Systems Used In Pharmacy

There are two major number systems used in pharmacy :

- Arabic Numbers – these are the symbols we use every day for enumeration, such as 1, 16, and 1337.
- Roman Numerals – this is a numeral system originating in ancient Rome. Roman numerals are expressed by letters of the alphabet.

The following is a chart showing how to convert back and forth between Arabic numbers and Roman numerals:

i	1	xxi	21	xli	41	lxi	61	lxxxi	81
ii	2	xxii	22	xlii	42	lxii	62	lxxxii	82
iii	3	xxiii	23	xliii	43	lxiii	63	lxxxiii	83
iv	4	xxiv	24	xliv	44	lxiv	64	lxxxiv	84
v	5	xxv	25	xlv	45	lxv	65	lxxxv	85
vi	6	xxvi	26	xlvi	46	lxvi	66	lxxxvi	86
vii	7	xxvii	27	xlvii	47	lxvii	67	lxxxvii	87
viii	8	xxviii	28	xlviii	48	lxviii	68	lxxxviii	88
ix	9	xxix	29	xlix	49	lxix	69	lxxxix	89
x	10	xxx	30	l	50	lxx	70	xc	90
xi	11	xxxi	31	li	51	lxxi	71	xci	91
xii	12	xxxii	32	lii	52	lxxii	72	xcii	92
xiii	13	xxxiii	33	liii	53	lxxiii	73	xciii	93
xiv	14	xxxiv	34	liv	54	lxxiv	74	xciv	94
xv	15	xxxv	35	lv	55	lxxv	75	xcv	95
xvi	16	xxxvi	36	lvi	56	lxxvi	76	xcvi	96
xvii	17	xxxvii	37	lvii	57	lxxvii	77	xcvii	97
xviii	18	xxxviii	38	lviii	58	lxxviii	78	xcviii	98
xix	19	xxxix	39	lix	59	lxxix	79	xcix	99
xx	20	xl	40	lx	60	lxxx	80	c	100
								d	500
								m	1000

Most people that have previously learned Roman numerals learned them using all capital letters; but in health care you will usually see them written in lower case letters (although either way is acceptable). Roman numerals consist of a basic set of seven symbols:

I *or* i	1	C *or* c	100
V *or* v	5	D *or* d	500
X *or* x	10	M *or* m	1000
L *or* l	50		

Here is how to count from 1 to 10 using Roman numerals:

I *or* i for one
II *or* ii for two
III *or* iii for three
IV *or* iv for four*
V *or* v for five
VI *or* vi for six
VII *or* vii for seven
VIII *or* viii for eight
IX *or* ix for nine
X *or* x for ten

* four strokes seemed like too many to the Romans, so they limited letter repetition to three

The principles for reading Roman numerals are:

- A letter repeated once or twice repeats its value that many times (XXX = 30, CC = 200, etc.).
- One or more letters that are placed after another letter of greater value increases the greater value by the amount of the smaller (VI = 6, LXX = 70, MCC = 1200, etc).
- A letter placed before another letter of greater value decreases the greater value by the amount of the smaller (IV = 4, XC = 90, CM = 900, etc.).*

* Rules regarding Roman numerals often state that a symbol representing 10^x may not precede any symbol larger than 10^{x+1}. For example, C cannot be preceded by I or V, only by X (or, of course, by a symbol representing a value equal to or larger than C). Thus, one should represent the number "ninety-nine" as XCIX, not as the "shortcut" IC

Convert the following Roman numerals to Arabic numbers:

1) VII
2) xxii
3) XXVII
4) iv
5) XIX
6) xiv
7) XII
8) ii
9) XXIV
10) vi
11) IV
12) iii

Convert the following Arabic numbers to Roman numerals:

13) 28
14) 13
15) 17
16) 15
17) 9
18) 12
19) 50
20) 41
21) 89
22) 2007
23) 1776
24) 1337

1) 7 2) 22 3) 27 4) 4 5) 19 6) 14 7) 12 8) 2 9) 24 10) 6 11) 4 12) 3
13) xxviii 14) xiii 15) xvii 16) xv 17) ix 18) xii 19) l 20) xli 21) lxxxix
22) mmvii 23) mdcclxxvi 24) mcccxxxvii

Worksheet 1-1

Name:

Date:

Convert the following Arabic numerals to Roman numerals:

1) 1 = ____________
2) 2 = ____________
3) 3 = ____________
4) 4 = ____________
5) 5 = ____________
6) 6 = ____________
7) 7 = ____________
8) 8 = ____________
9) 9 = ____________
10) 10 = ____________
11) 20 = ____________
12) 40 = ____________
13) 45 = ____________
14) 100 = ____________
15) 400 = ____________
16) 1,000 = ____________
17) 900 = ____________
18) 69 = ____________
19) 24 = ____________
20) 1,999 = ____________

Convert the following Roman numerals to Arabic numerals:

21) ix = ____________
22) xviii = ____________
23) xxiv = ____________
24) xxxvi = ____________
25) iii = ____________
26) ccxl = ____________
27) lv = ____________
28) dlv = ____________
29) mcxi = ____________
30) cmxcix = ____________
31) iv = ____________
32) vii = ____________
33) xii = ____________
34) xvi = ____________
35) xxii = ____________
36) xlii = ____________

37) xxxi	=	____________	39) mmviii	=	____________
38) mcccxxxvii	=	____________	40) mdcclxxxiii	=	____________

Worksheet 1-2

Name:

Date:

Perform the indicated operations and record the results in Arabic numbers.

1) VII + XXII = ____________

2) xxvii – iv = ____________

3) XIX – XIV = ____________

4) XII × II = ____________

5) xxiv ÷ vi = ____________

6) IV × III = ____________

Perform the indicated operations and record the results in Roman numerals.

7) 5 × 4 = ____________

8) 18 + 12 = ____________

9) 16 ÷ 4 = ____________

10) 4 × 3 = ____________

11) 625 ÷ 125 = ____________

12) 17 + 14 – 11 + 4 = ____________

13) 6 + 3 = ____________

14) 20 – 16 + 3 = ____________

Fill in the following multiplication table using Roman numerals.

	I	**II**	**III**	**IV**	**V**	**VI**	**VII**	**VIII**	**IX**	**X**
I										
II										
III										
IV										
V										
VI										
VII										
VIII										
IX										
X										

XXIV

The following pages include a card game that can help students become more comfortable working rapidly with Roman numerals. If you've ever played *24 Game®*, the rules are very similar.

How to play:

1) First, someone will need to use a pair of scissors to cut out their XXIV cards.
2) Then, students will need to pair up in groups of three to six.
3) Then, all the students in a group will look at the card on the top of the pile and try to solve it. In order to solve it, you must use all four numbers on the card once and only once. You can use any combination of multiplying, dividing, adding, and subtracting to make the card total '24'.
4) After the card is solved, flip the card and work on the other side as a group. Once both sides of the card have been solved, discard it and work on the next card in the deck.

This game is not about who scores the most points, instead it is about working collaboratively to better everyones' understanding of Roman numerals.

example card:

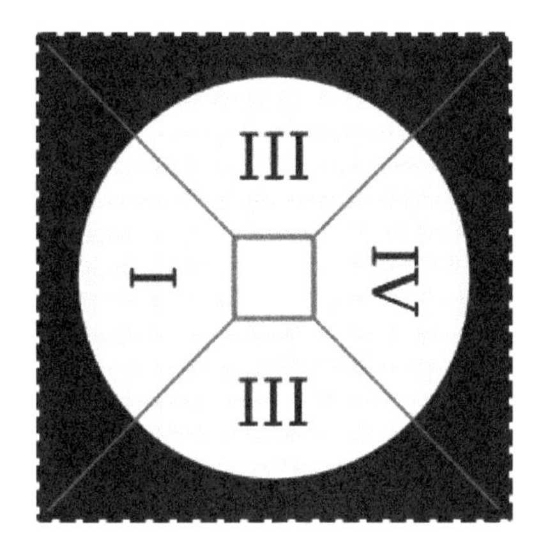

The Roman numerals on this card are III, IV, III, and I which translate into the Arabic numbers 3, 4, 3, and 1. I can solve this card any of the following ways:

$(3 \times 4) \times (3 - 1) = 24$

$[(3 - 1) \times 4] \times 3 = 24$

$(3 + 3) \times 4 \times 1 = 24$

Despite the fact that each card can usually be solved numerous ways, it only needs to be solved once.

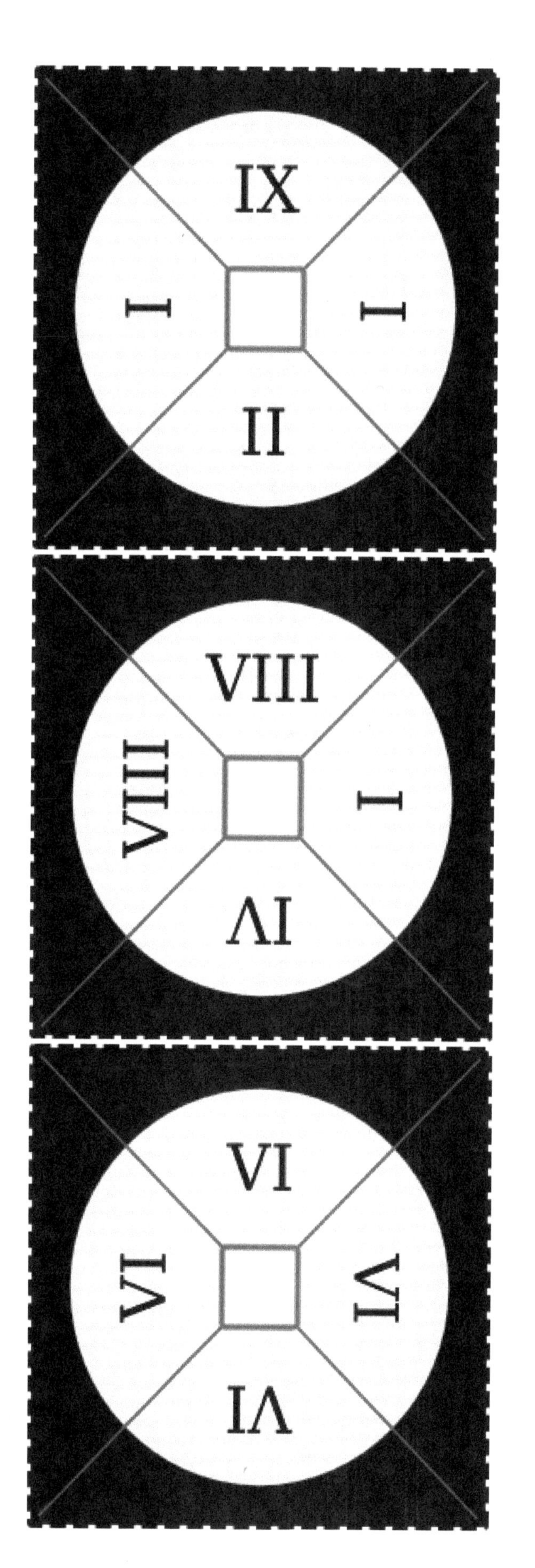

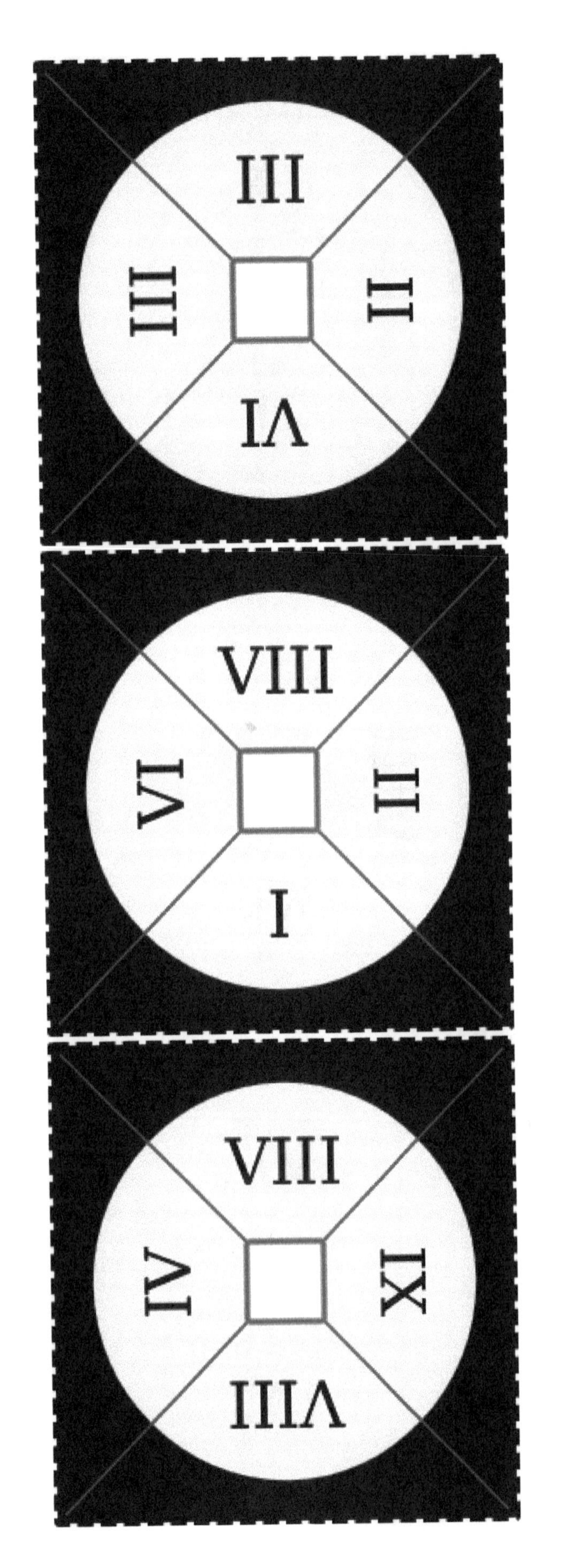
III
VIII
VIII

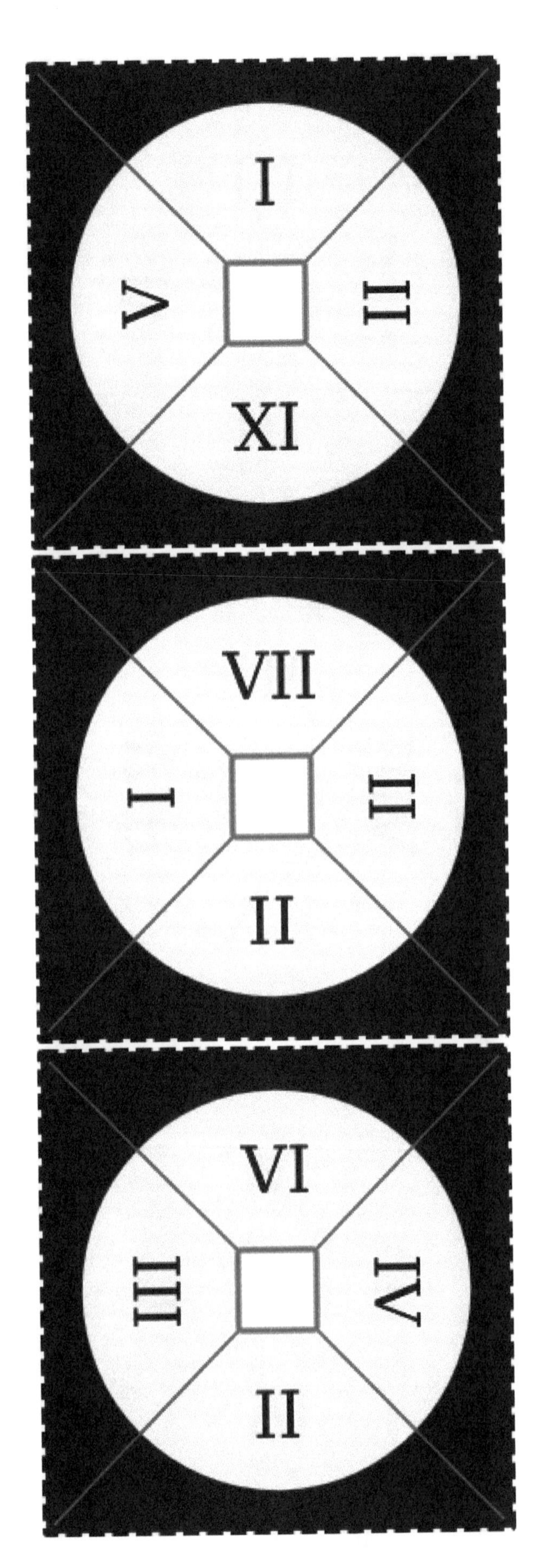
I
XI
VII
II
VI
II

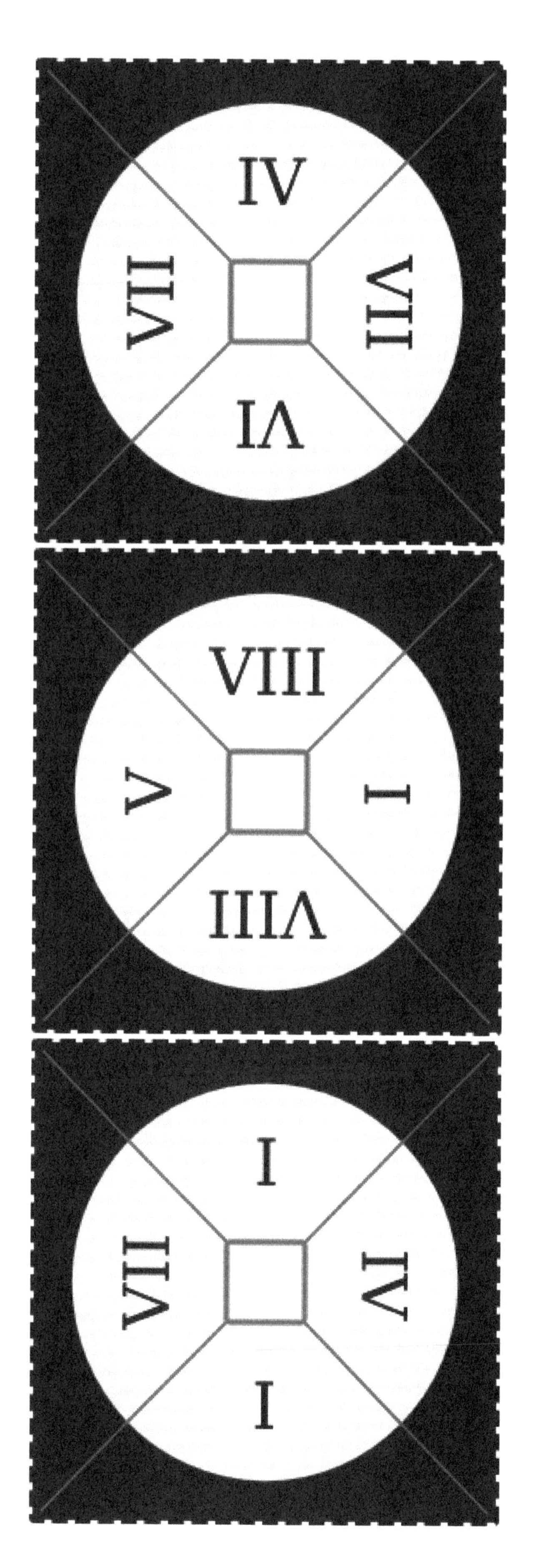
IV
VII
VII
IV
VIII
V
I
VIII
I
VII
IV
I

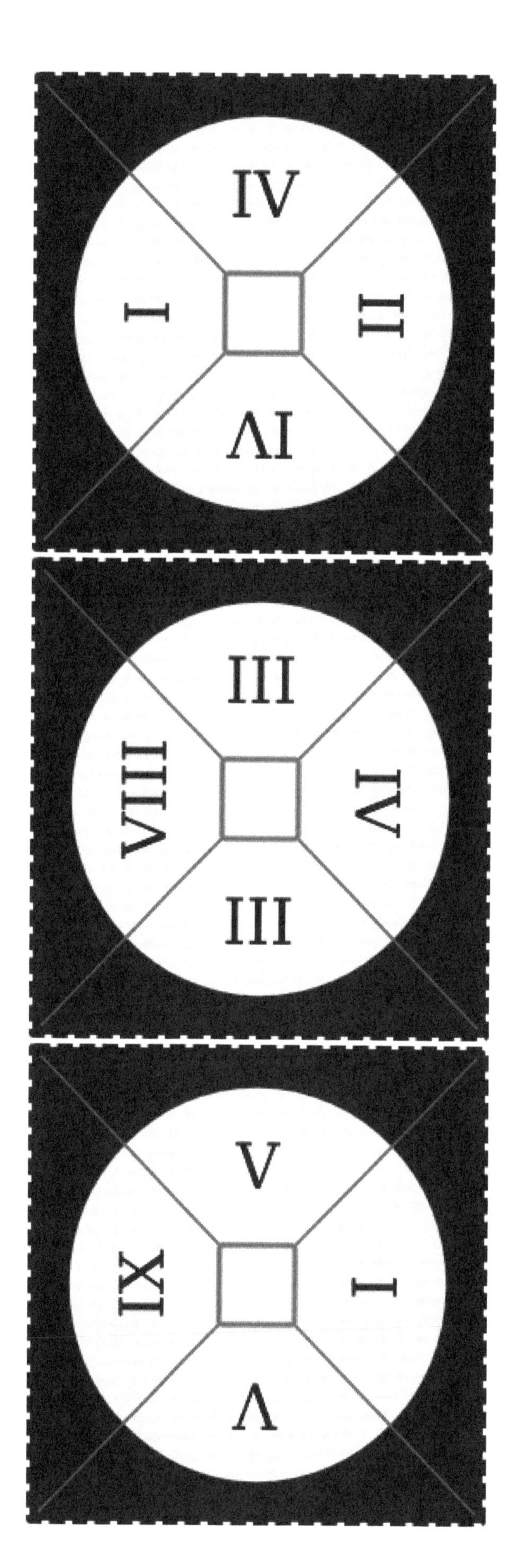
IV
I
II
IV
III
VIII
IV
III
V
IX
I
V

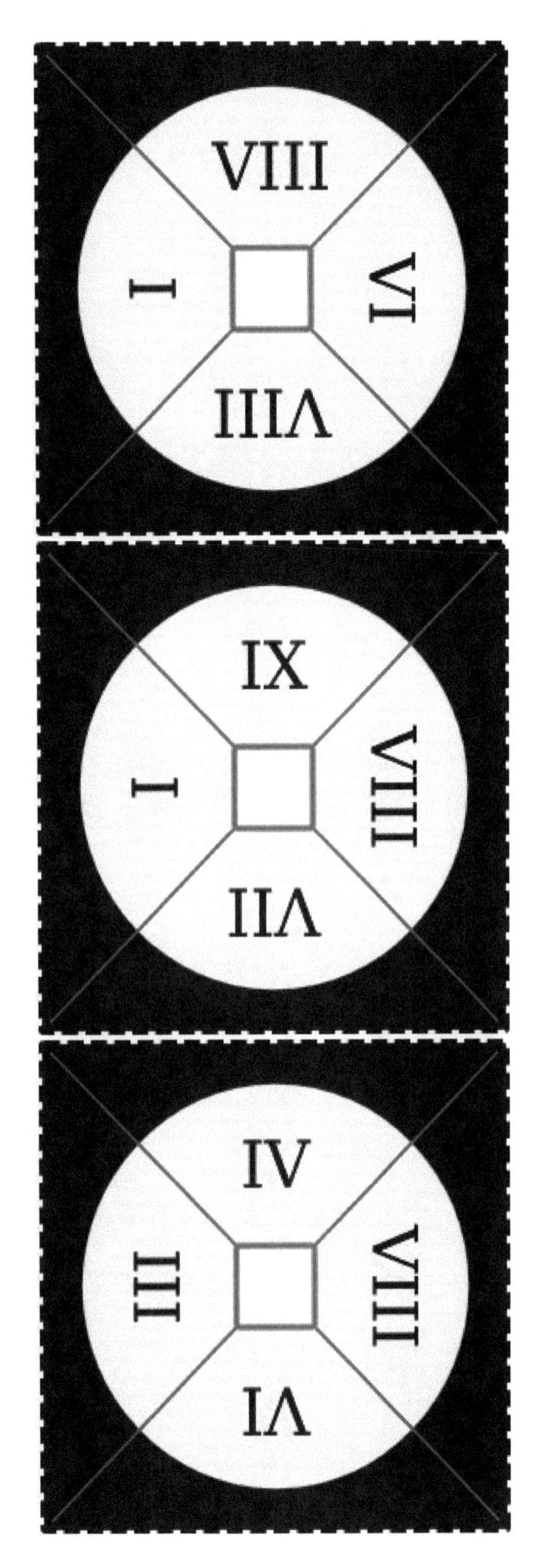
VIII
VI
VIII
I
IX
VIII
VII
I
IV
VIII
VI
III

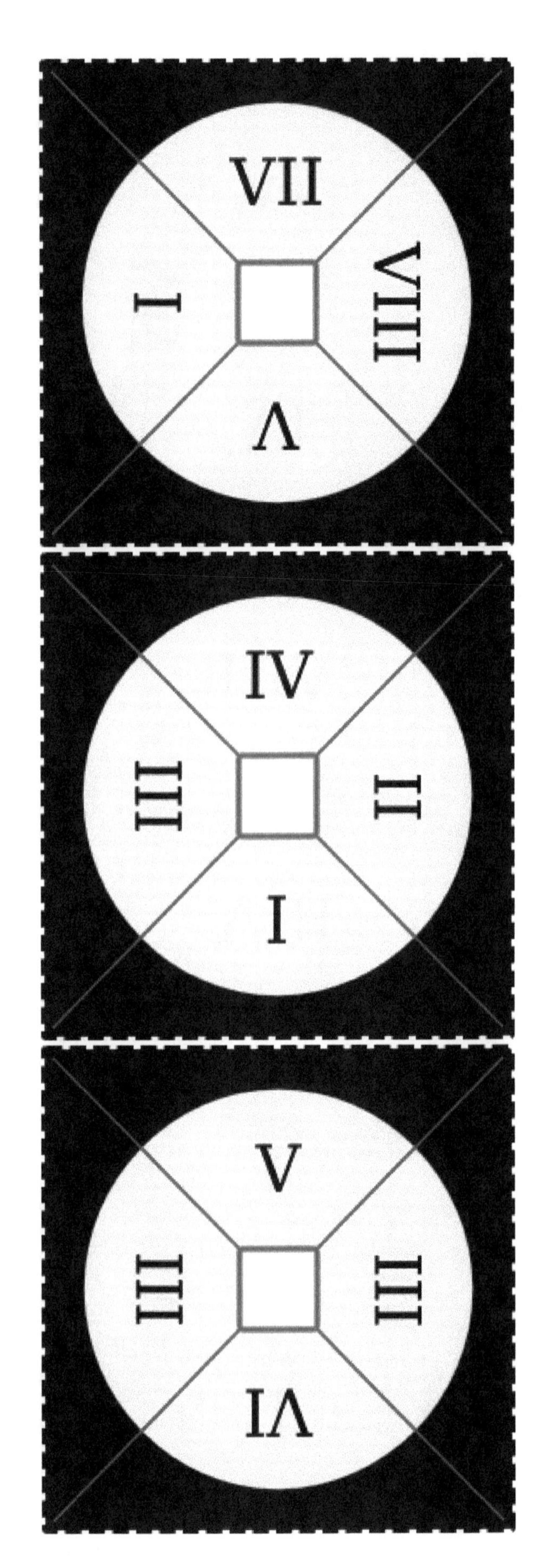
VII
VIII
V
I
IV
II
I
III
V
III
VI
III

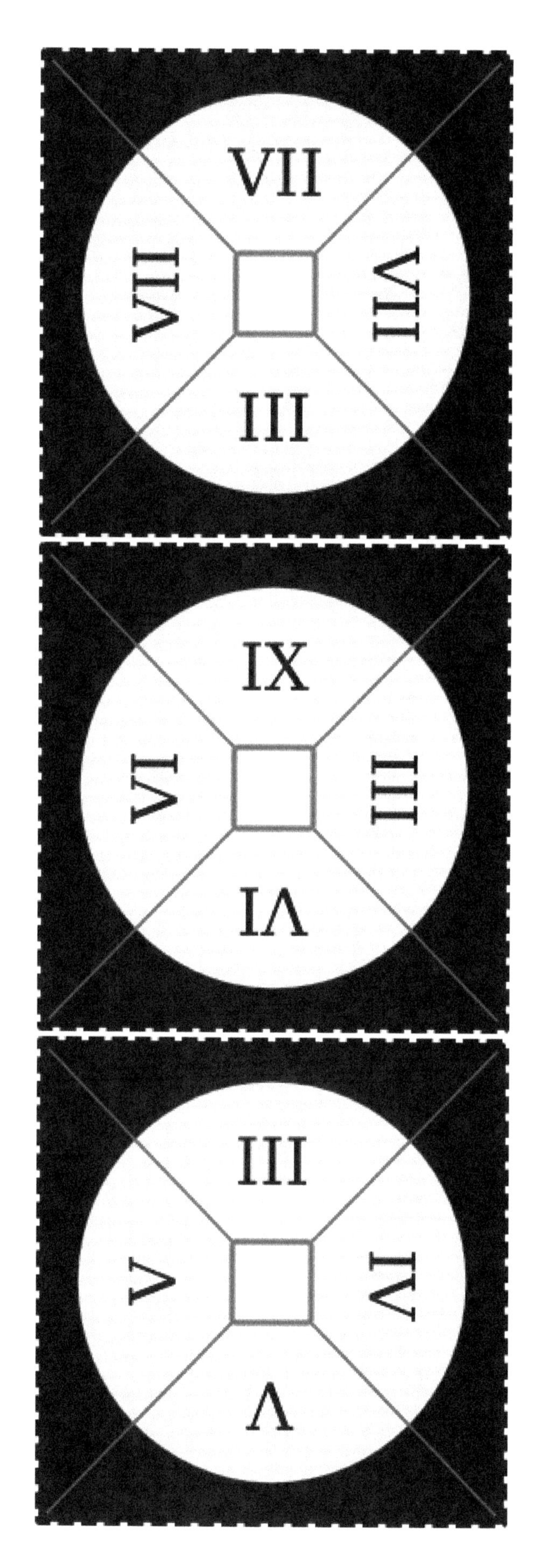

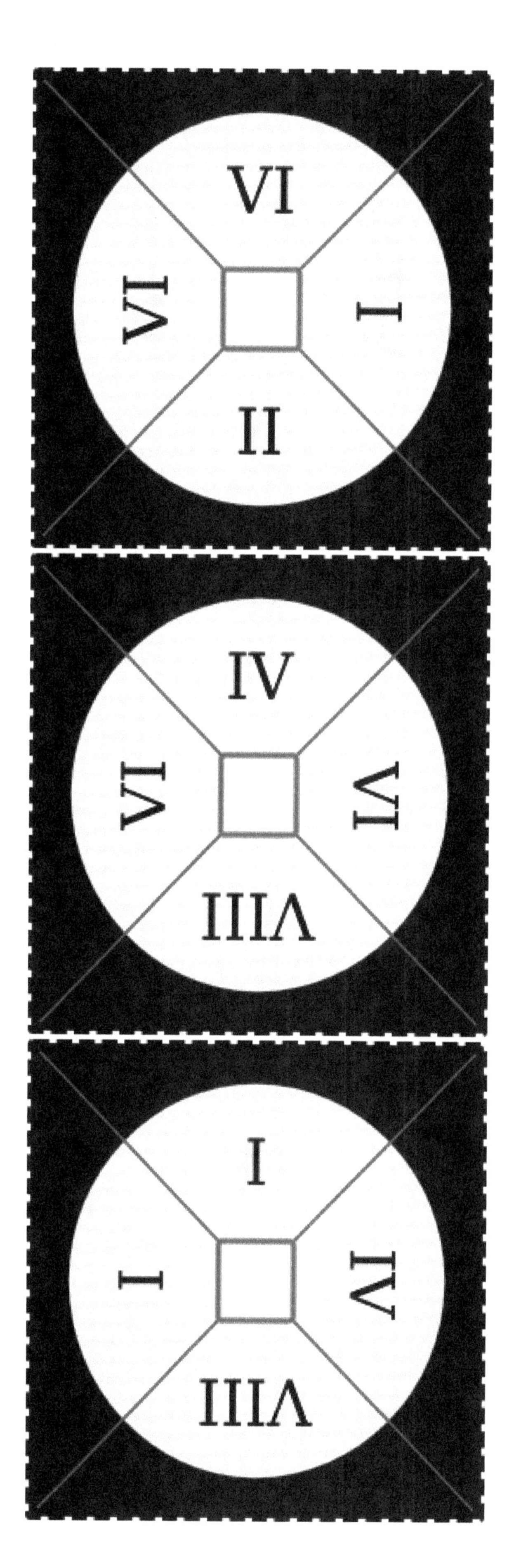

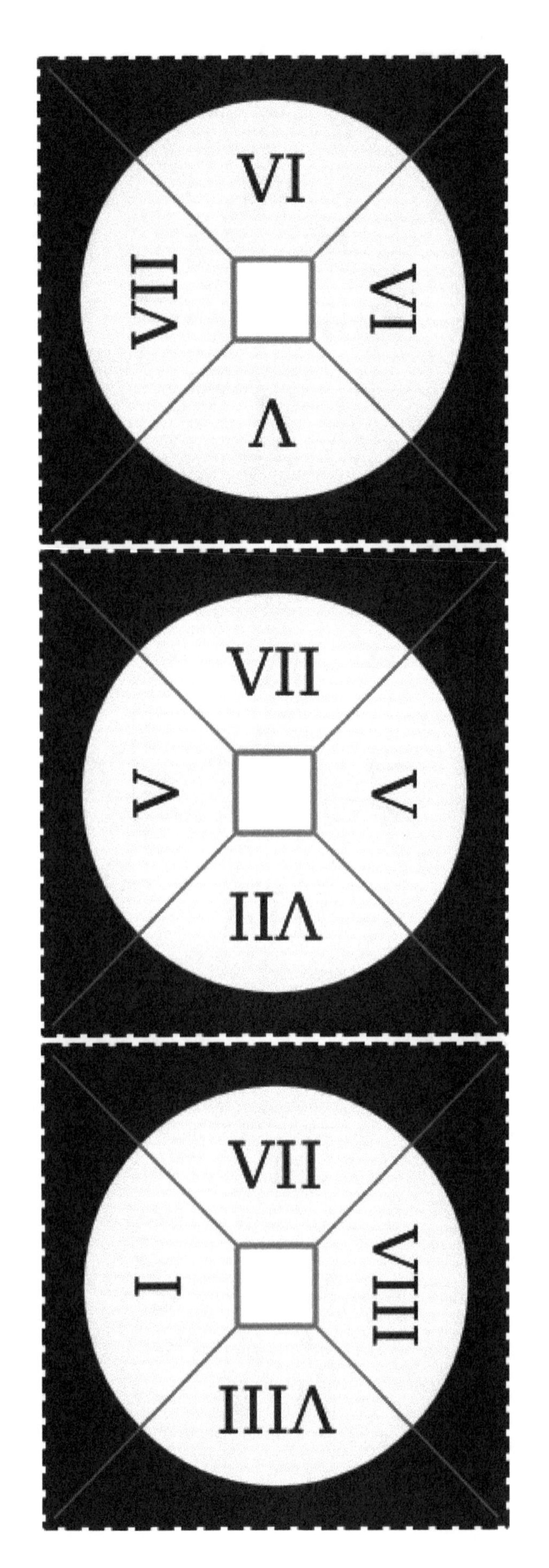
VI
VII
VI
V
VII
V
V
VII
VII
I
VIII
VIII

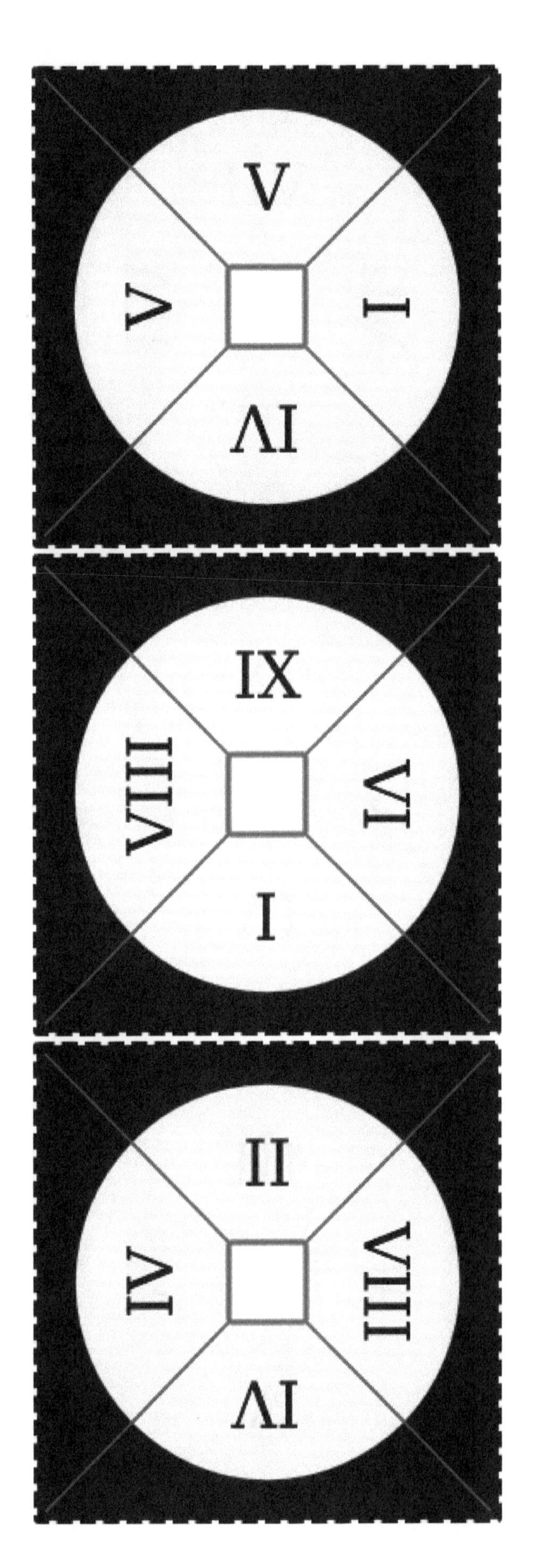
V
V
I
IV
IX
VIII
VI
I
II
IV
VIII
IV

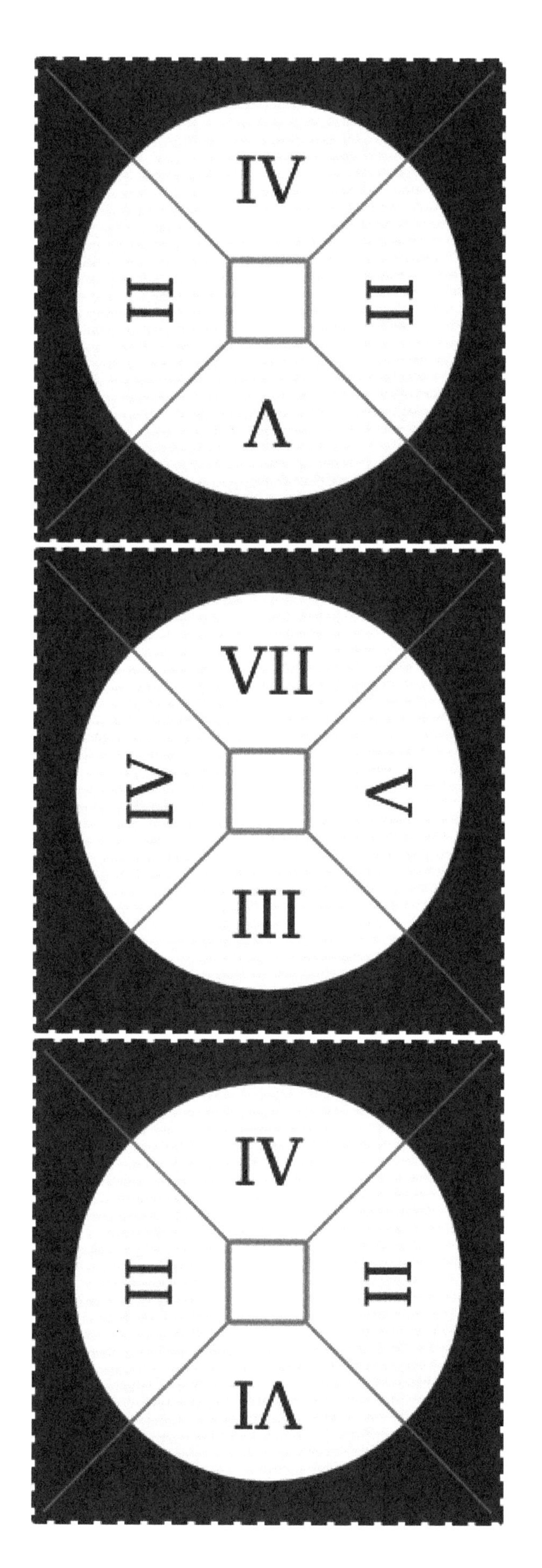

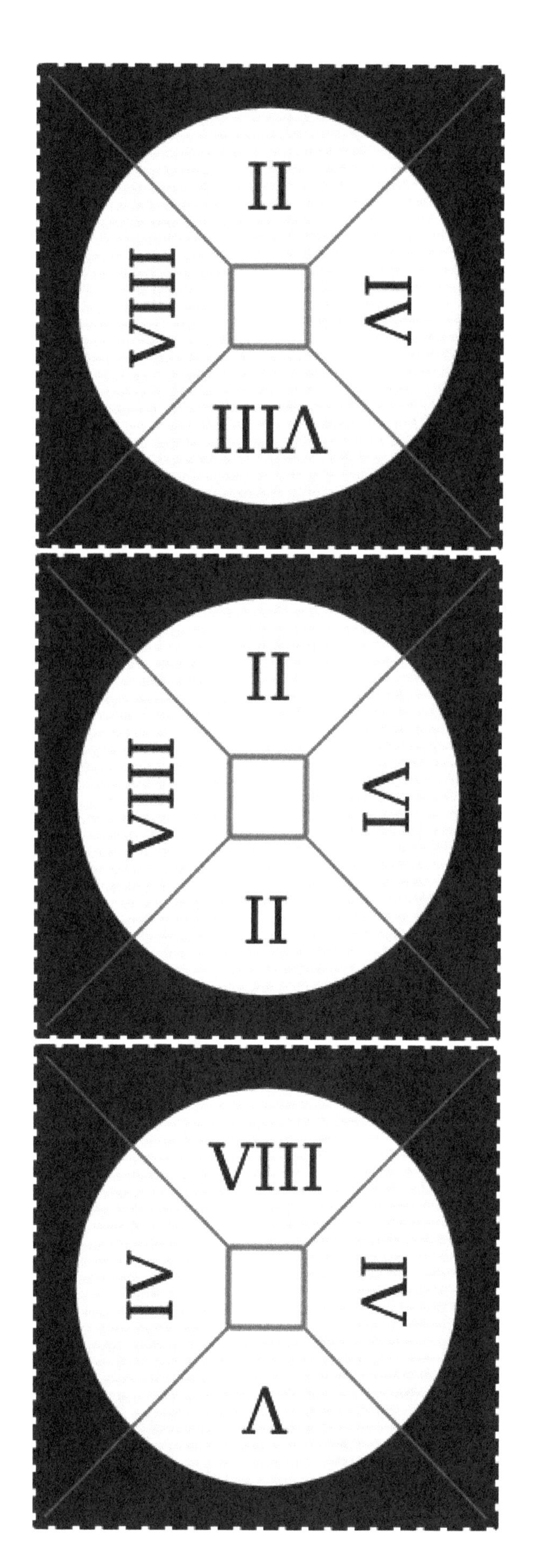

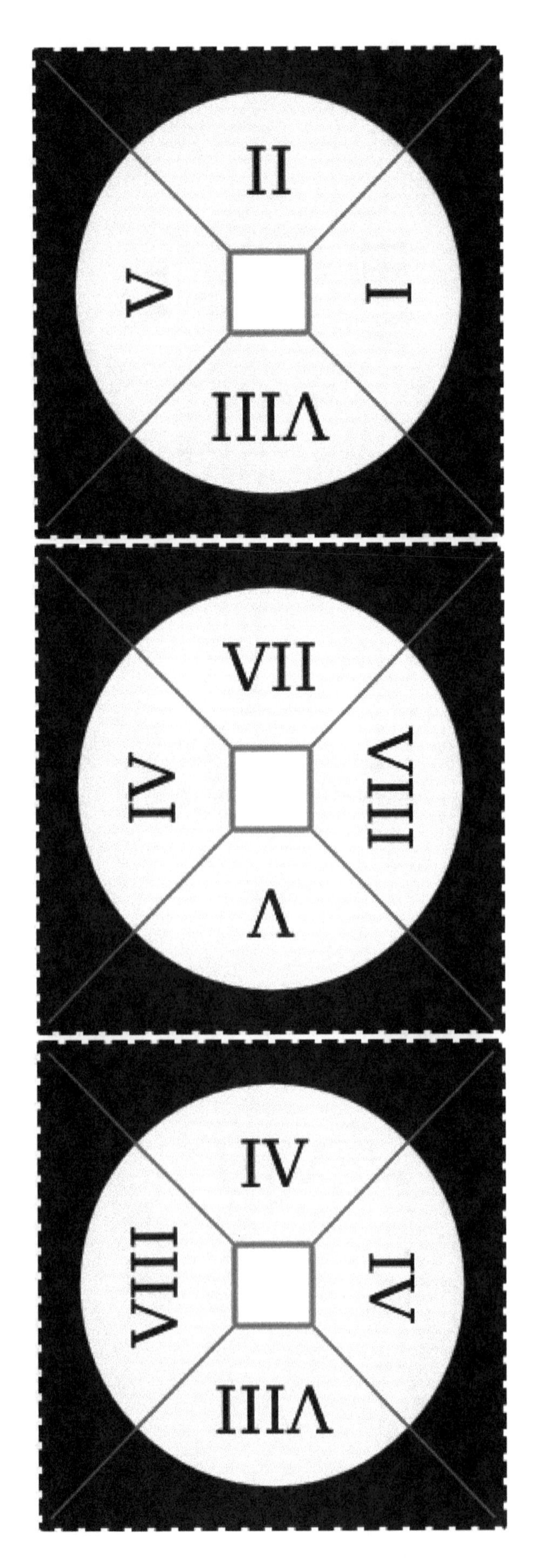

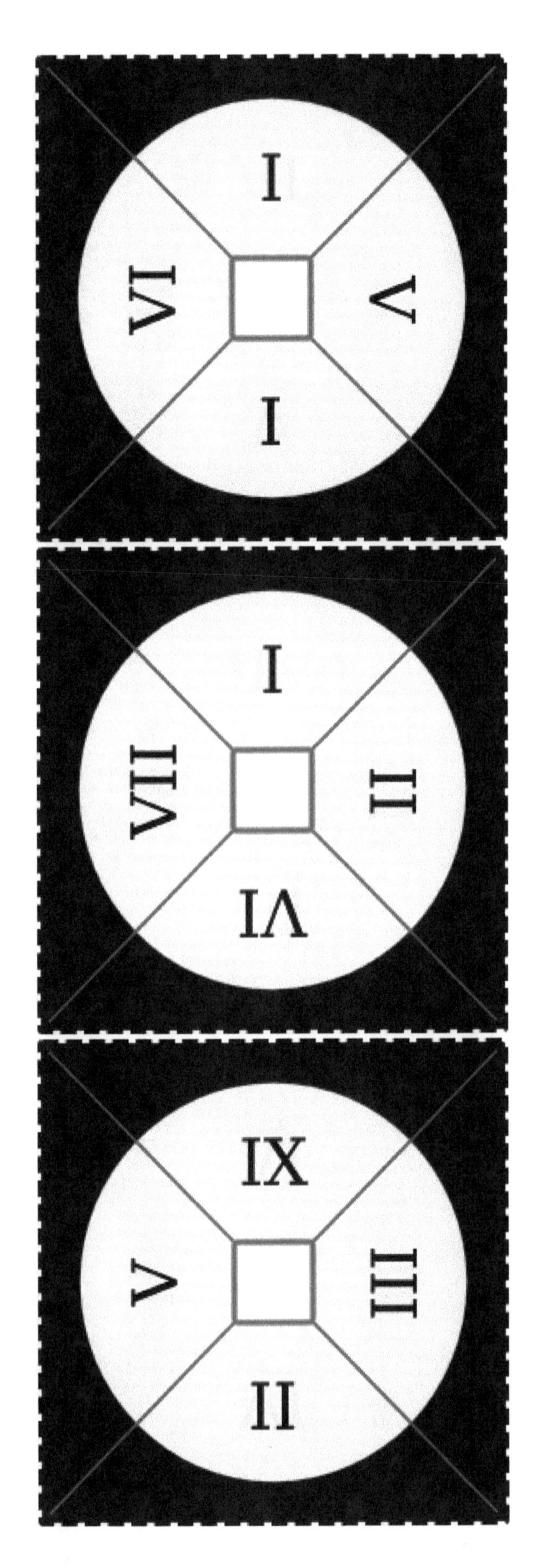

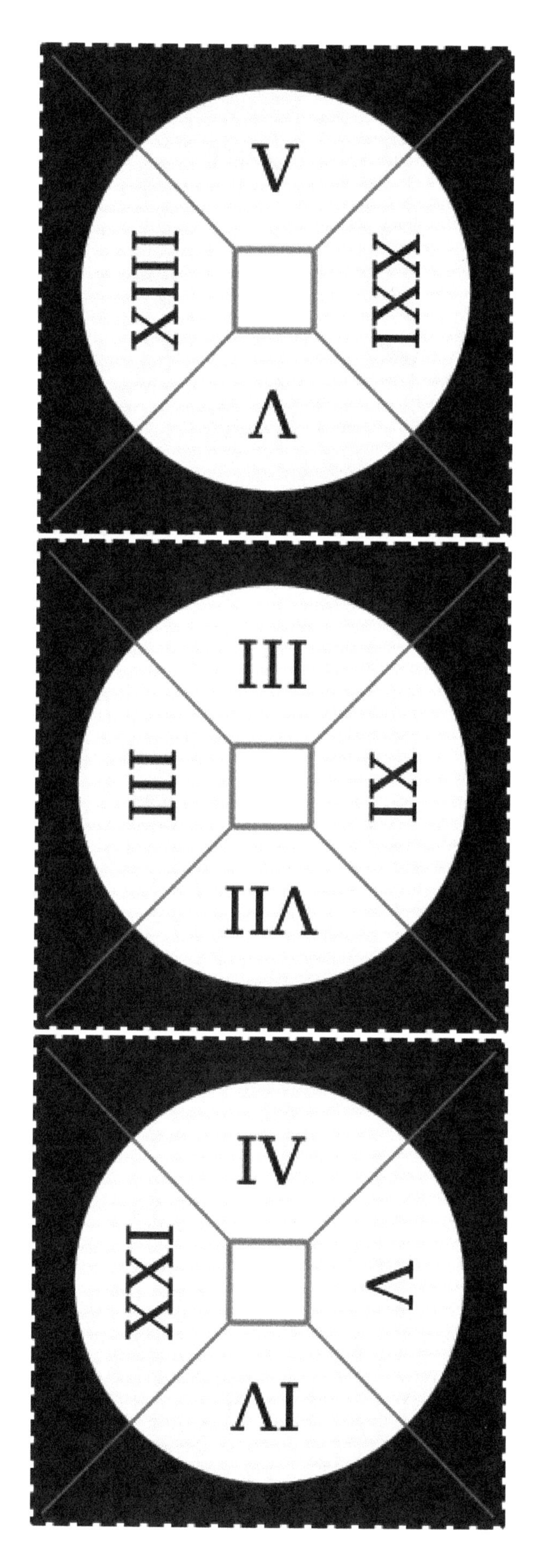
V
XIIII
XXI
V
III
III
XI
VII
IV
XXI
V
IV
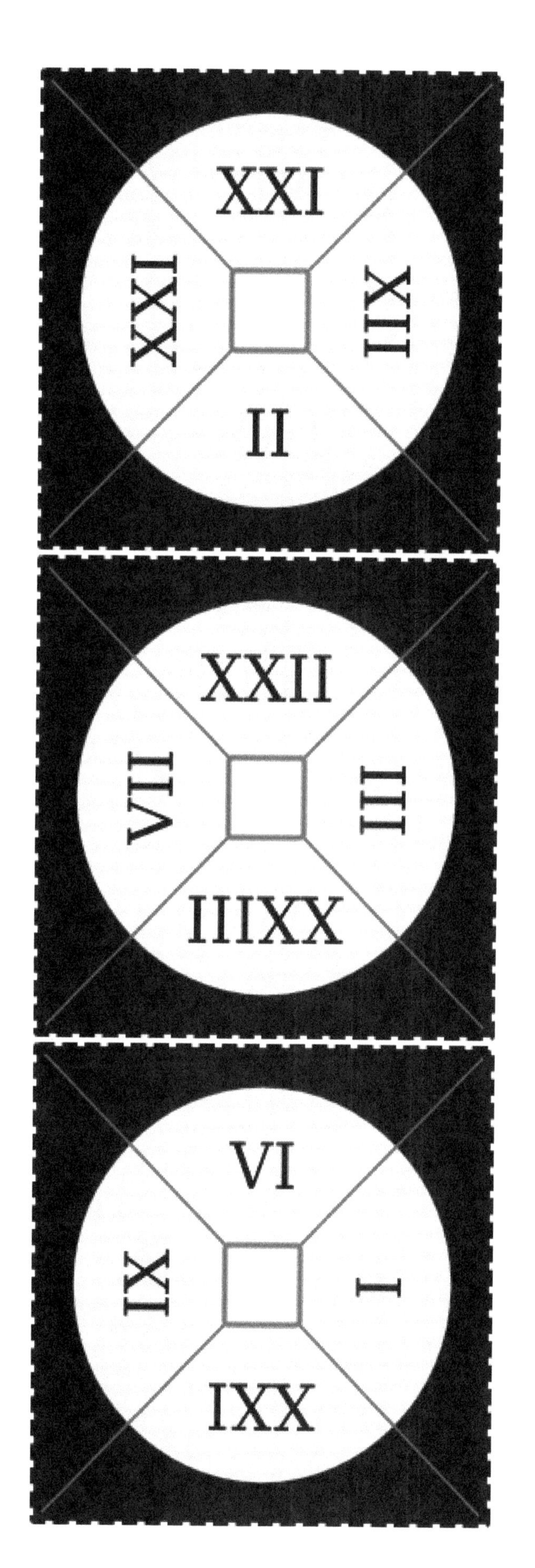
XXI
XXI
XII
II
XXII
VII
III
XXIII
VI
IX
I
XXI

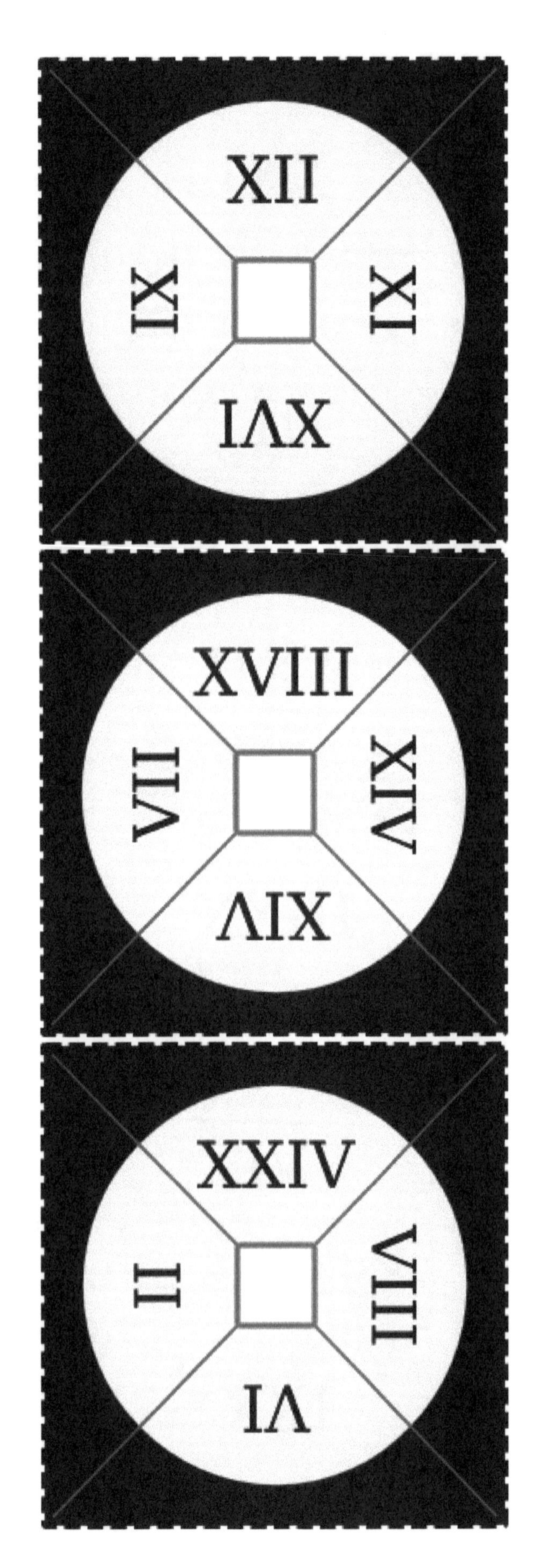
XII
XI
XVI
IX
XVIII
XIV
XIV
VIII
XXIV
VIII
VI
II

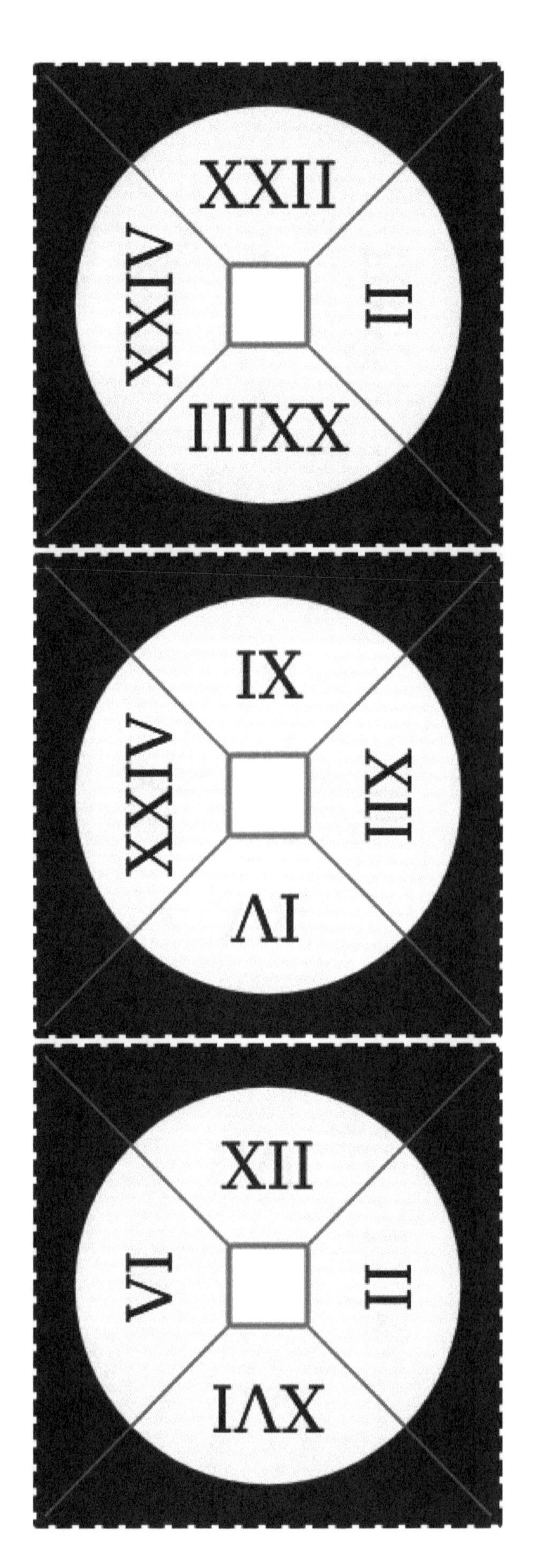
XXII
II
XXIII
XXIV
IX
XIII
IV
XXIV
XII
II
XVI
VI

XX
XXIV
XXII
II
I
XXIV
XXIV
XII
IX
XXII
X
XVII

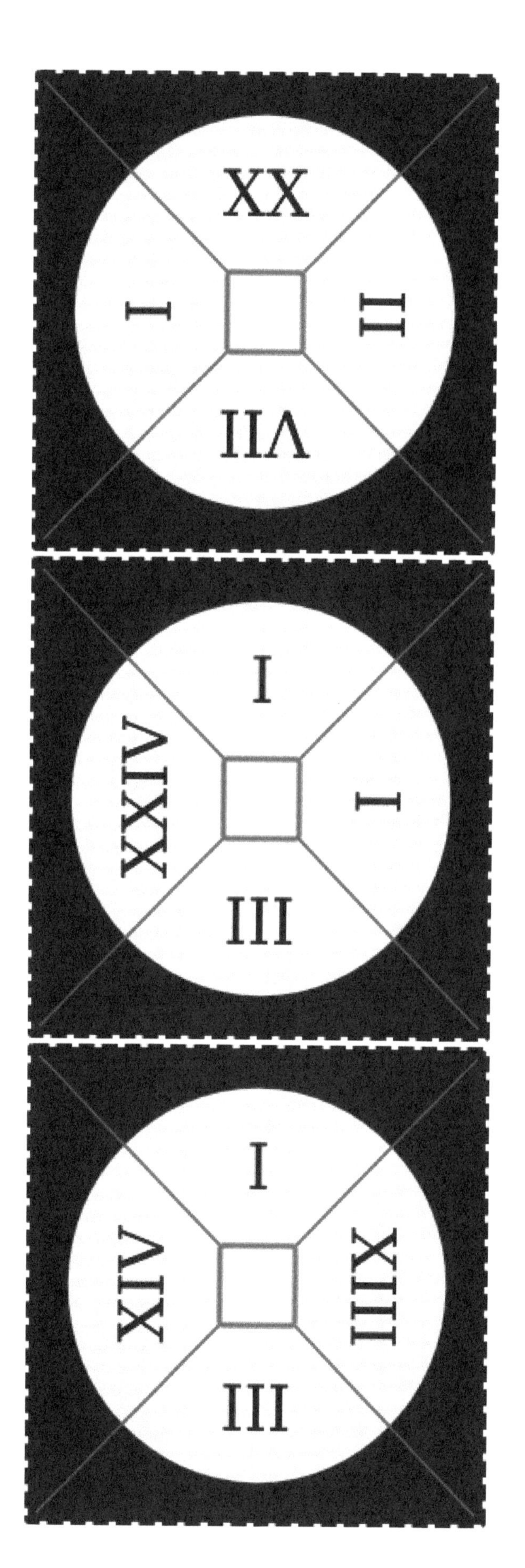
XX
II
VII
I
I
I
III
XXIV
I
XIII
III
XIV

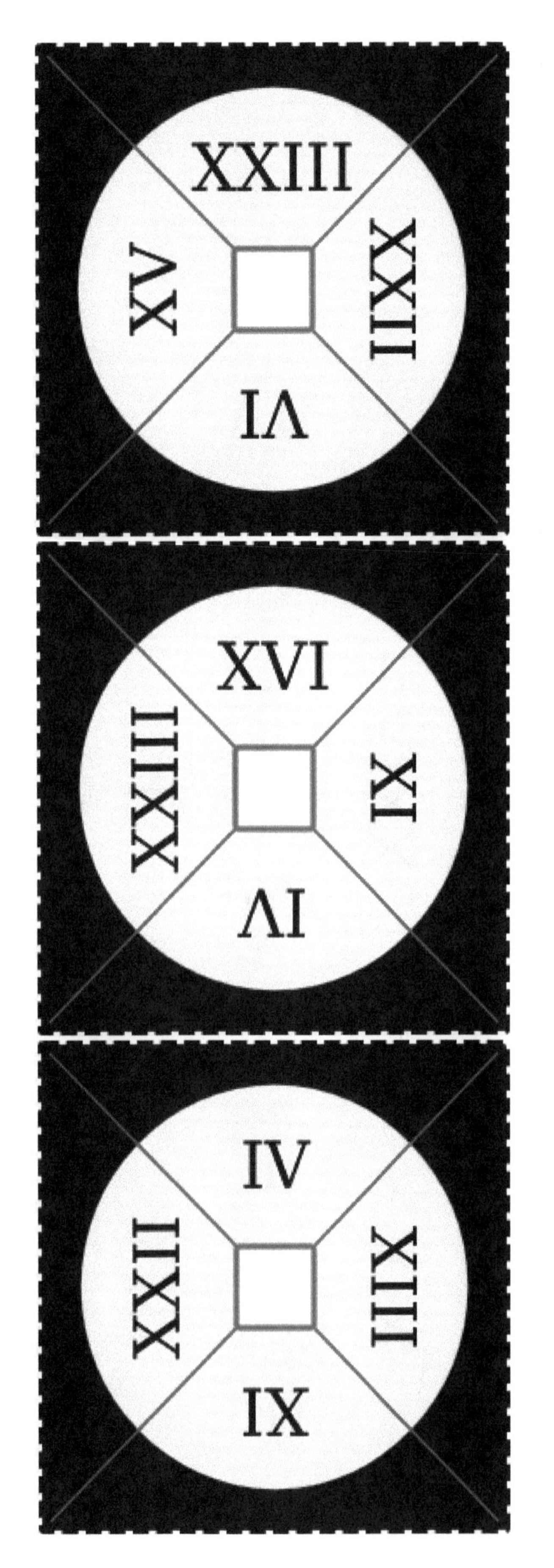
XXIII
XVI
IV

VII
XII
III

IX
I
I

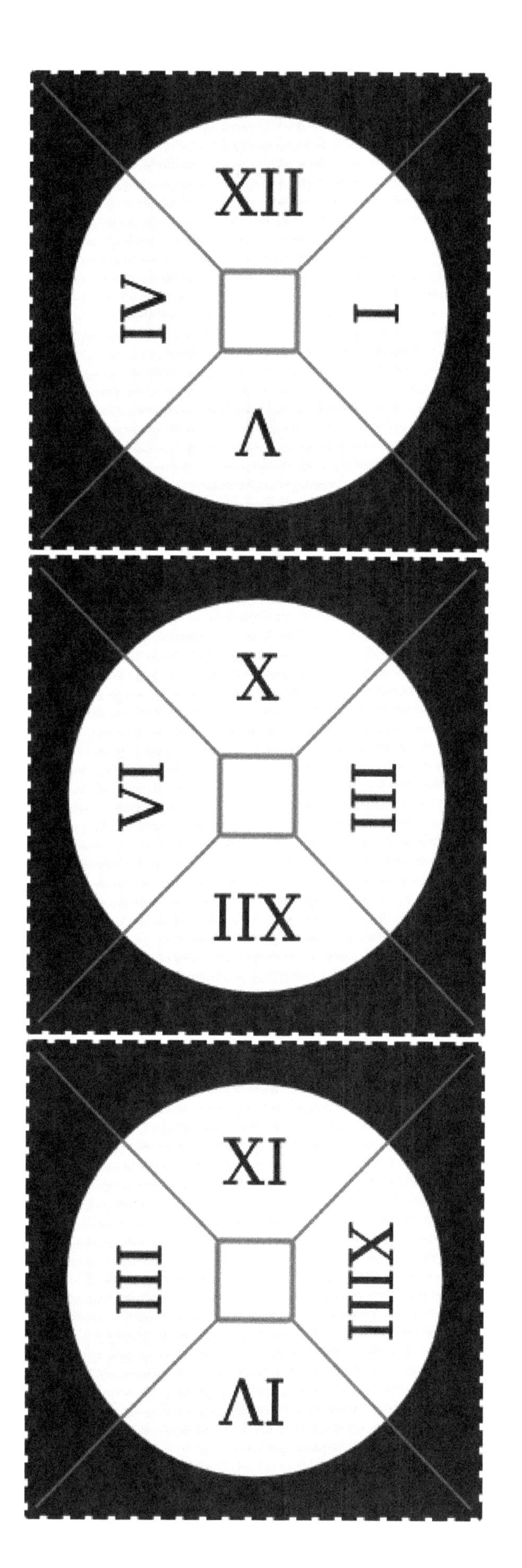
XII
X
XI

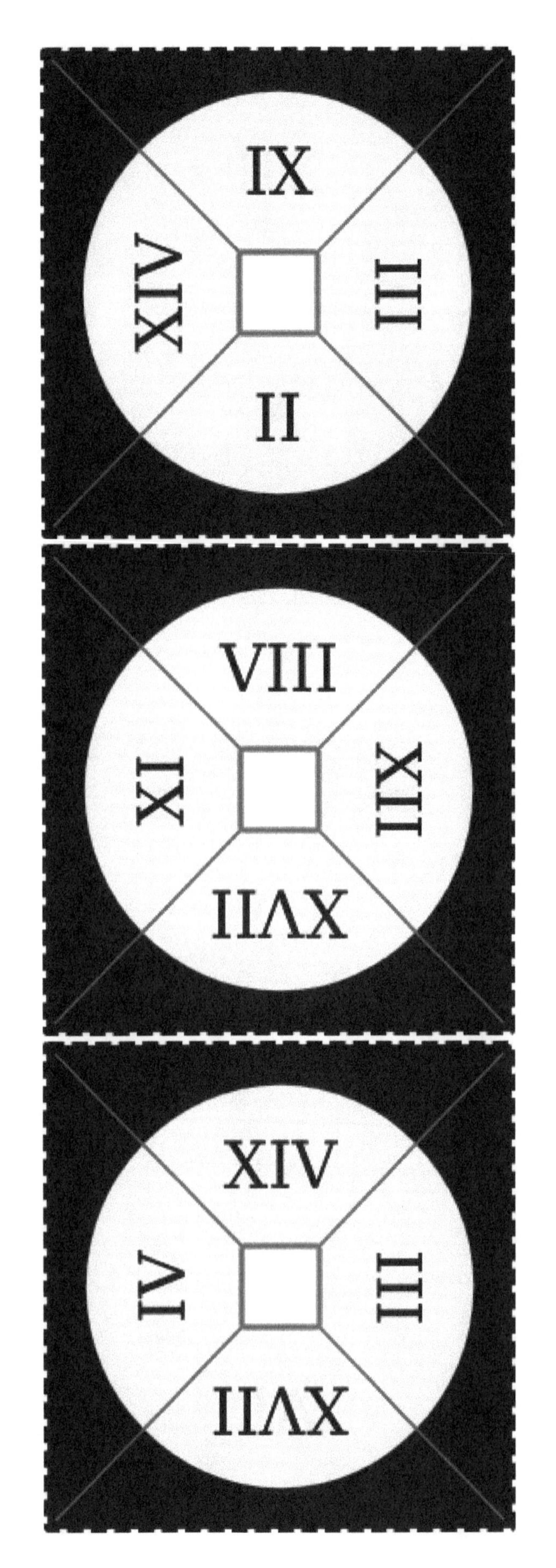
IX
XIV
III
II
VIII
XI
XIII
XVII
XIV
IV
III
XVII

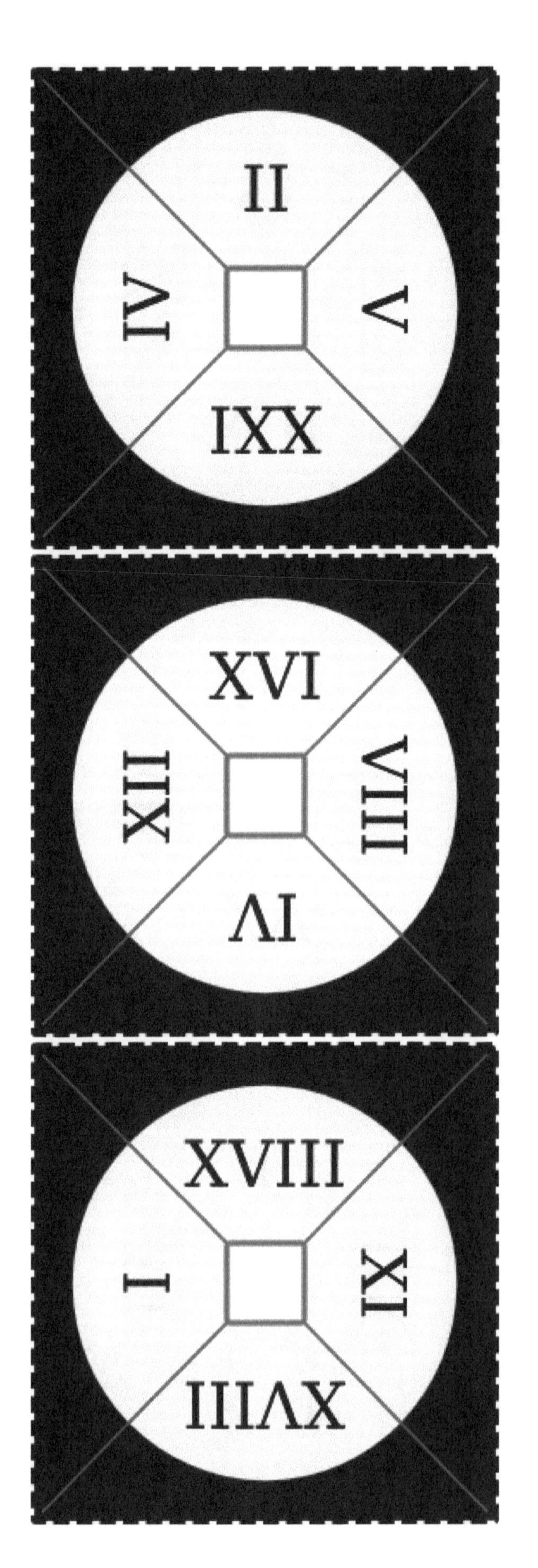
II
IV
V
XXI
XVI
XII
VIII
IV
XVIII
I
XI
XVIII

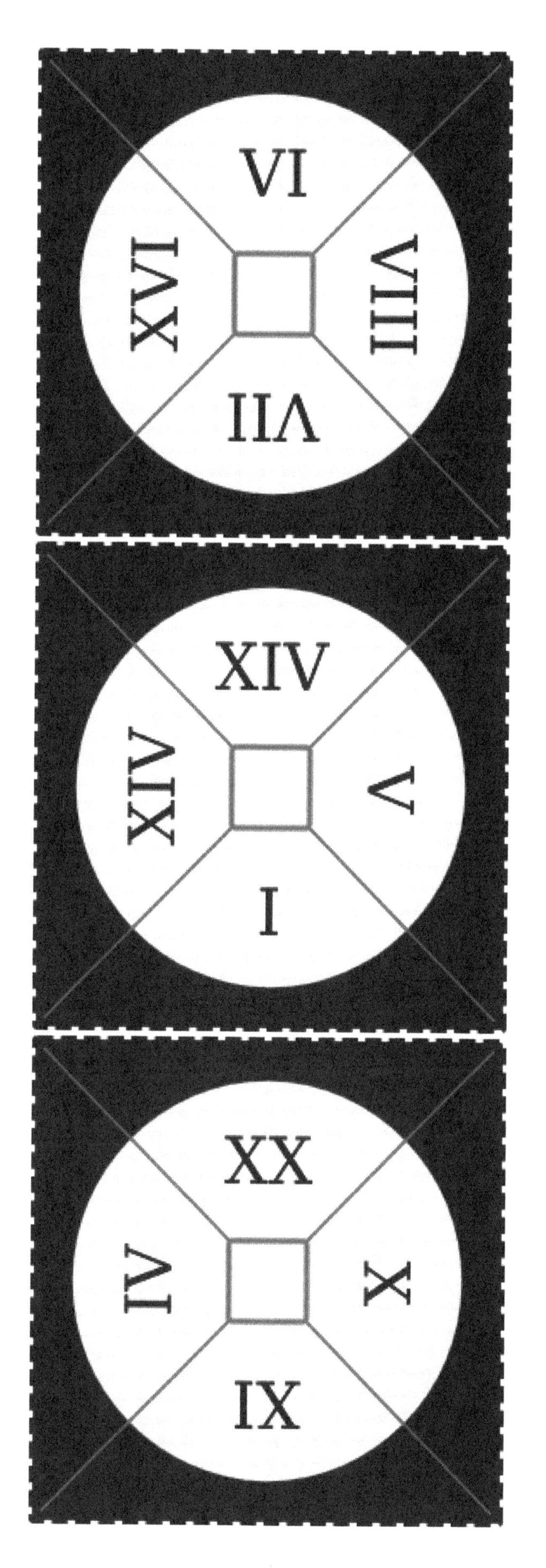
VI
XIV
I
XX

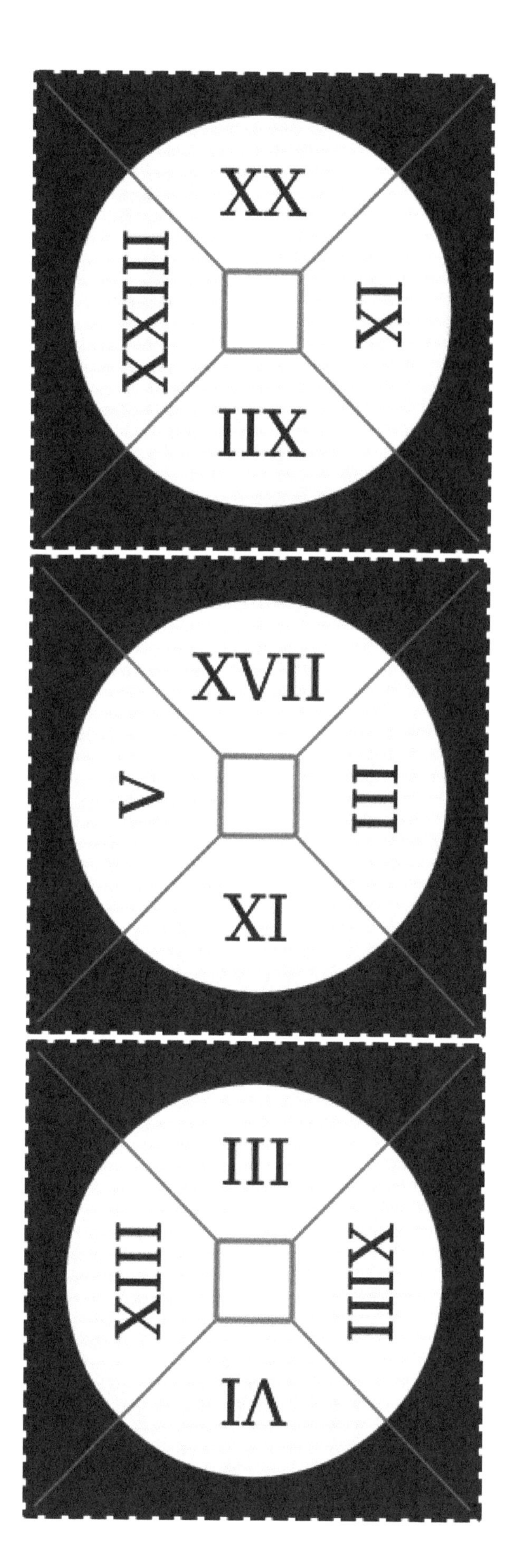
XX
XVII
XI
III

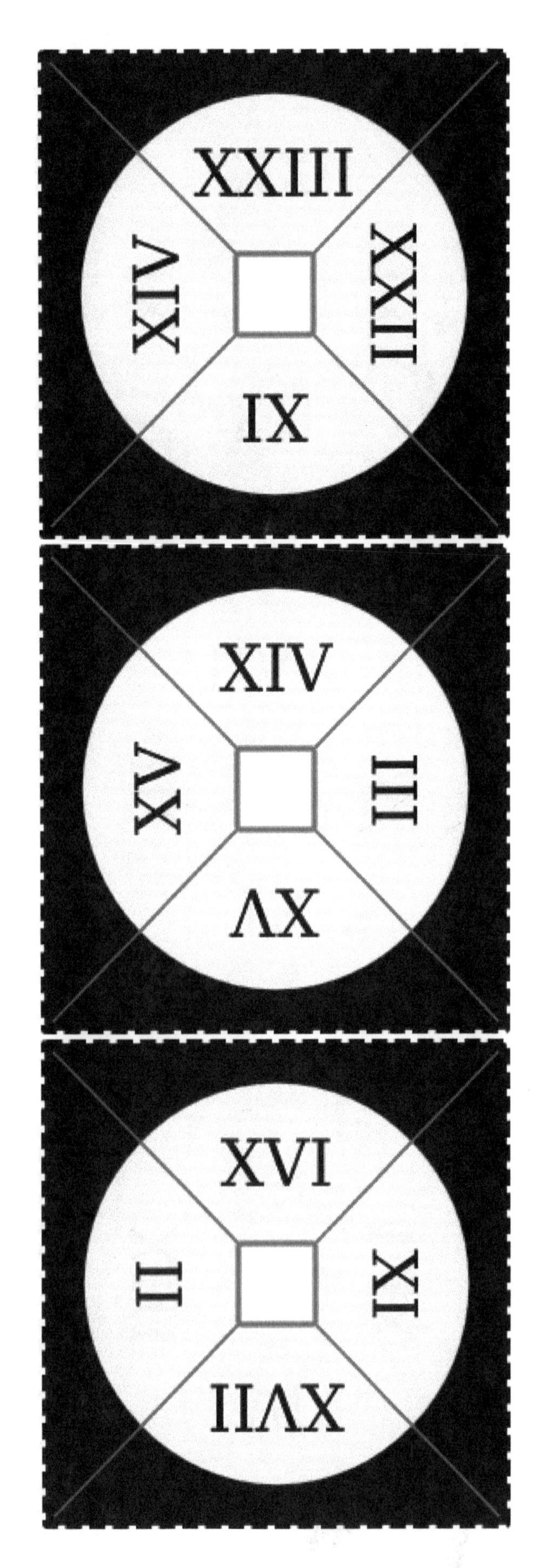
XXIII
XXII
IX
XIV
XIV
III
XV
XV
XVI
XI
XVII
II

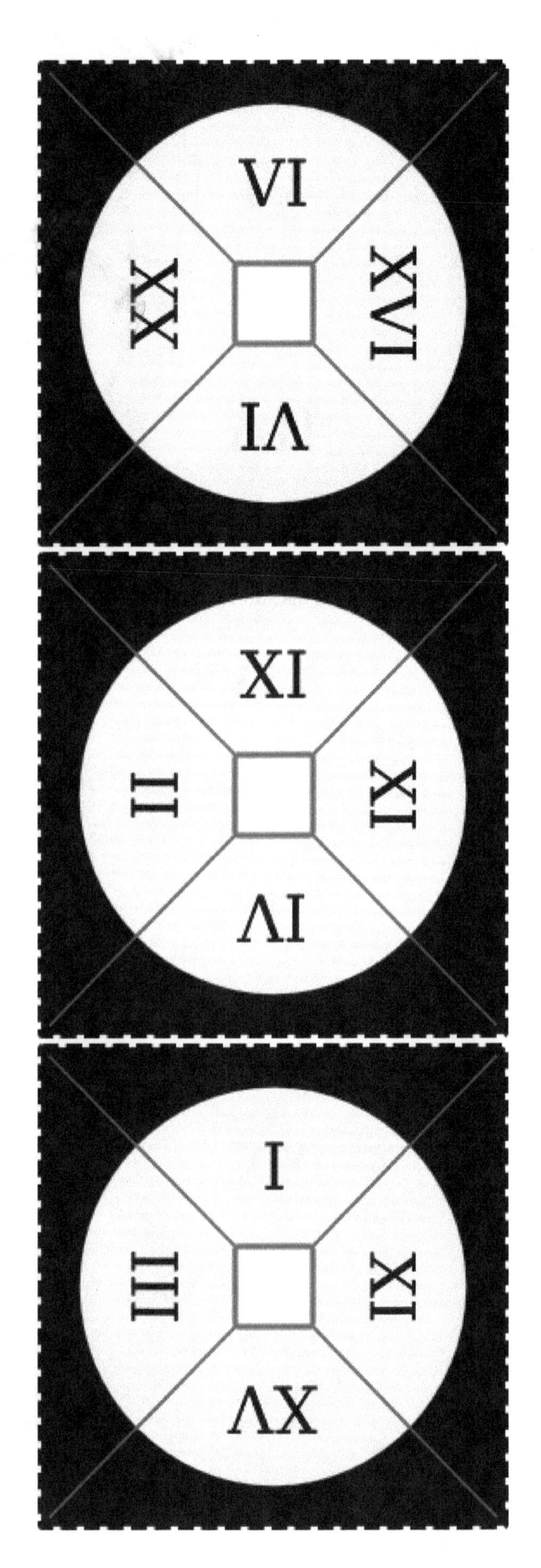
VI
XVI
IV
XX
XI
XI
IV
II
I
XI
XV
III

Decimal Places

Now that we've spent some time reviewing Roman numerals, we are going to spend some time reviewing Arabic numbers. With respect to this, I would like to start with working on writing fractions in decimal form. The following two lists represent whole number places and decimal places.

Whole Numbers

1 = one
10 = ten
100 = one hundred
1,000 = one thousand
10,000 = ten-thousand
100,000 = one hundred-thousand
1,000,000 = one million

Decimal Numbers

1/10 = 0.1 = one ten*th*
1/100 = 0.01 = one hundred*th*
1/1,000 = 0.001 = one thousand*th*
1/10,000 = 0.0001 = one ten-thousand*th*
1/100,000 = 0.00001 = one hundred-thousand*th*
1/1,000,000 = 0.000001 = one million*th*

So, based on the information above **9876.5432**

The **9** is in the thousands position
The **8** is in the hundreds position
The **7** is in the tens position
The **6** is in the ones position
The **5** is in the ten*th*s position
The **4** is in the hundred*th*s position
The **3** is in the thousand*th*s position
The **2** is in the ten-thousand*th*s position

If you were to write it as a fraction, it would be:

$$\mathbf{9876\,\frac{5432}{10{,}000}}$$

You could read it as **nine thousand eight hundred seventy six and five thousand four hundred thirty two ten-thousand*th*s**

Let's look at a decimal number and decide which position each number is in.

167.2543

a) 1 is in the ______________________ place.

b) 2 is in the ______________________ place.

c) 3 is in the ______________________ place.

d) 4 is in the ______________________ place.

e) 5 is in the ______________________ place.

f) 6 is in the ______________________ place.

g) 7 is in the ______________________ place.

a) hundreds b) tenths c) ten-thousandths d) thousandths e) hundredths f) tens g) ones

On the following page is a worksheet to practice writing fractions in decimal form.

Worksheet 1-3

Name:

Date:

What is the place name of the 7 in each of these numbers?

1) 743 __________

2) 5,763,482 __________

3) 2,178,000,013 __________

4) 26.74 __________

5) 0.0007 __________

6) 17.054 __________

Write the following numbers in decimal format.

7) twelve million four thousand twenty-five ____________________

8) thirty-one thousand three hundred thirty-seven ____________________

9) five million three hundred eighteen thousand eight ____________________

10) seven tenths ____________________

11) thirty-nine hundredths ____________________

12) fifteen and seven tenths ____________________

13) six and seven hundredths ____________________

14) seven hundred five and one hundred seven ten-thousandths____________________

Write the following fractions in decimal format.

15) $\frac{7}{10}$ __________

16) $95\ \frac{7}{100}$ __________

17) $\frac{57}{100}$ __________

18) $\frac{19}{10{,}000}$ __________

19) $406\ \frac{214}{10{,}000}$ __________

20) $507\ \frac{112}{10{,}000}$ __________

Change the following decimals into fractions (but do not reduce your fractions).

21) 0.6 _______________

22) 0.85 _______________

23) 0.006 ____________________

24) 0.0574 ____________________

25) 13.000013 ____________________

26) 80.08135 ____________________

The following are the answers to the odd problems.

1) hundreds place
3) ten-millions place
5) ten-thousandths place
7) 12,004,025
9) 5,318,008
11) 0.39
13) 6.07
15) 0.7
17) 0.57
19) 406.0214
21) $\frac{6}{10}$
23) $\frac{6}{1,000}$
25) 13 $\frac{13}{1,000,000}$

Rounding Decimals & Significant Figures

Sometimes when you are working with numbers, you will find them too large to be manageable. In these scenarios, you should round.

- Rounding makes numbers that are easier to work with in your head.
- Rounded numbers are only approximate.
- An exact answer generally can not be obtained using rounded numbers.
- Use rounding to get an answer that is close but that does not have to be exact.

How to round numbers

For whole numbers:

- Make the numbers that end in 1 through 4 into the next lower number that ends in 0. For example 74 rounded to the nearest ten would be 70.
- Numbers that end in a digit of 5 or more should be rounded up to the next even ten. The number 88 rounded to the nearest ten would be 90.

For decimal numbers:

- When looking to round a significant figure that is proceeded by a 1 through 4, you may stop your decimal there. For example 0.74 rounded to the nearest tenth would be 0.7
- When looking to round a significant figure that is proceeded by a 5 or more, it should be rounded up. The number 0.88 rounded to the nearest tenth would be 0.9

Significant Figures

A significant figure is one that is actually measured.

Rules for assigning significant figures:

- Digits other than zero are always significant. For example, 34.12 has four significant figures.
- Zeros used only to space the decimal are never significant. For example, 0.007 has one significant figure and 16,000 only has two significant figures.
- Final zeros after a decimal point are always significant. For example, 0.0070 has two significant figures.
- Zeros between two other significant digits are always significant. For example, 50.04 has four significant figures.

Practice Problem

Determine the number of significant figures in 0.20 ml of Lantus insulin.

0.20 ml of Lantus insulin has two significant figures

On the following page is a worksheet to attain some practice with determining significant figures and rounding numbers.

Worksheet 1-4

Name:

Date:

Round the following numbers to the nearest whole number.

1) 102.5 ___________

2) 1.0 ___________

3) 88.1 ___________

4) 99.7 ___________

5) 187.4999 ___________

6) 29,001.501 ___________

Round the following numbers to the nearest tenth.

7) 1.0 ___________

8) 99.7 ___________

9) 2.54 ___________

10) 29,001.501 ___________

11) 187.4999 ___________

12) 12.99 ___________

Round the following numbers to the nearest hundredth.

13) 187.4999 ___________

14) 12.99 ___________

15) 29,001.501 ___________

16) 1234567.8901 ___________

17) 66.666667 ___________

18) 5,454.5454 ___________

Determine the number of significant figures in each measurement.

19) 64.8 mg

______ significant figures

20) 1.609 km

______ significant figures

21) 0.25 mL

______ significant figures

22) 4.06 mEq

______ significant figures

23) 0.454 kg

______ significant figures

24) 0.06 mL

______ significant figures

25) 0.20 mcm

______ significant figures

26) 220 lb

______ significant figures

27) 0.001 kg

______ significant figures

Adding & Subtracting Decimal Numbers

Adding and subtracting decimals is the same as adding and subtracting whole numbers.

To add decimal numbers:

- Put the numbers in a vertical column, aligning the decimal points.
- Add each column of digits, starting on the right and working left.
- Place the decimal point in the answer directly below the decimal points in the terms.

$$\begin{array}{r} 9817.543 \\ +\quad 0.123 \\ \hline 9817.666 \end{array}$$

Practice Problems

Solve the following practice problems.

1) $\begin{array}{r} 324.5678 \\ +\quad 2.3456 \\ \hline \end{array}$

3) $\begin{array}{r} 32.255 \\ +\quad 1.0123 \\ \hline \end{array}$

5) $\begin{array}{r} 6.6663 \\ +\ 12.0007 \\ \hline \end{array}$

2) $\begin{array}{r} 0.6 \\ +\ 0.4 \\ +\ 1.5 \\ \hline \end{array}$

4) $\begin{array}{r} 4.0068 \\ +\quad 0.06 \\ +\ 43.667 \\ \hline \end{array}$

6) $\begin{array}{r} 5.004 \\ +\ 17 \\ +\ 12.02 \\ \hline \end{array}$

1) 326.9134 2) 2.5 3) 33.2673 4) 47.7338 5) 18.6670 6) 34.024

To subtract decimal numbers:

- Put the numbers in a vertical column, aligning the decimal points.
- Subtract each column, starting on the right and working left. If the digit being subtracted in a column is larger than the digit above it, "borrow" a digit from the next column to the left.
- Place the decimal point in the answer directly below the decimal points in the terms.

```
 9817.544
-   5.123
 --------
 9812.421
```

Practice Problems

Do the following practice problems.

1)	324.5678 - 2.3456	3)	32.255 - 1.0123	5)	6.6663 - 2.0007
2)	0.6 - 0.5	4)	4.0068 - 0.06	6)	5.004 - 2.02

1) 322.2222 2) 0.1 3) 31.2427 4) 3.9468 5) 4.6656 6) 2.984

Worksheet 1-5

Name:

Date:

Perform the following addition problems.

1) 28,291 + 27,595

2) 5.08 + 0.17

3) 25.09 + 18.7

4) 97.07 + 0.09

5) 31.05 + 4.7

6) 85.2 + 1.764

7) 31 + 24 + 18 = ________

8) 4951 + 3287 = ________

9) 35.07 + 19.1 = ________

10) 88.08 + 0.02 = ________

11) 1.28 + 31.05 + 4.7 = ________

12) 18 + 0.042 + 16.3 = ________

Perform the following subtraction problems.

13) 28,291 - 27,595

14) 5.08 - 0.17

15) 25.09 - 18.7

16) 97.07 - 0.09

17) 31.05 - 4.7

18) 85.2 - 1.764

19) 4951 – 3287 = ________

20) 35.07 – 19.1 = ________

21) 88.08 – 0.02 = ________

22) 0.97 – 0.012 = ________

23) 4.5 – 4.05 = ________

24) 0.951 – 0.112 = ________

Perform the following applied problems.

25) A neonate initially weighed 4.03 kg at birth. Three days later, he weighed 3.944 kg. How many kg did the neonate lose?

26) Normal body temperature is 37.0° C. What temperature is 0.4° above normal?

Multiplying Decimal Numbers

To multiply decimal numbers:

- Multiply the numbers just as if they were whole numbers.
- Line up the numbers on the right.
- Starting on the right, multiply each digit in the top number by each digit in the bottom number.
- Add the products.
- Place the decimal point in the answer by starting at the right and moving the point the number of places equal to the sum of the decimal places in both numbers that were multiplied.

$$\begin{array}{rcl} 36.3 & \leftarrow & \text{one decimal place} \\ \underline{\times\ 0.21} & \leftarrow & \underline{+\ \text{two decimal places}} \\ 363 & & \\ \underline{+\ 7260} & & \\ 7.623 & \leftarrow & \text{three decimal places} \end{array}$$

Let's try a few practice problems.

Practice Problems

1) $\begin{array}{r} 2020 \\ \underline{\times\ \ 1.1} \end{array}$ 2) $\begin{array}{r} 5.08 \\ \underline{\times\ 0.17} \end{array}$ 3) $\begin{array}{r} 25.09 \\ \underline{\times\ \ 18.7} \end{array}$ 4) $\begin{array}{r} 97.07 \\ \underline{\times\ \ 0.09} \end{array}$ 5) $\begin{array}{r} 31.05 \\ \underline{\times\ \ \ 4.7} \end{array}$ 6) $\begin{array}{r} 1.764 \\ \underline{\times\ \ 85.2} \end{array}$

1) 2222 2) 0.8636 3) 469.183 4) 8.7363 5) 145.935 6) 150.2928

Worksheet 1-6

Name:

Date:

Solve the following multiplication problems.

1) $\begin{array}{r} 28{,}291 \\ \times\ 27{,}595 \\ \hline \end{array}$ 2) $\begin{array}{r} 4.06 \\ \times\ 0.25 \\ \hline \end{array}$ 3) $\begin{array}{r} 22.33 \\ \times\ 16.4 \\ \hline \end{array}$ 4) $\begin{array}{r} 69.09 \\ \times\ 0.06 \\ \hline \end{array}$ 5) $\begin{array}{r} 50.13 \\ \times\ 7.4 \\ \hline \end{array}$ 6) $\begin{array}{r} 1.337 \\ \times\ 31.3 \\ \hline \end{array}$

7) $4951 \times 3287 =$ __________

9) $88.08 \times 0.02 =$ __________

11) $4.05 \times 4.5 =$ __________

8) $35.07 \times 19.1 =$ __________

10) $0.012 \times 0.97 =$ __________

12) $0.951 \times 0.112 =$ __________

Perform the following applied problems.

13) Sal is to take 1.25 grains of aspirin every day. Each grain weighs 64.8 mg, so how many mg of aspirin is Sal to take every day?

14) Nauseous Nancy bought 12 vials of ondansetron at a cost of $17.56 per vial. How much did she spend on ondansetron?

15) Theophylus Monk is to receive 81.6 mg of theophylline three times per day for the next three days. How many mg of theophylline should he receive in total?

Dividing Decimal Numbers

The statement "4 divided by 2" can be written several ways:

$2\overline{)4}$ $\qquad 4 \div 2 \qquad \dfrac{4}{2}$

Write the statement "48 divided by 6" three different ways (use the above example as a template if necessary):

The names of the numbers in a division problem are shown below:

$$\begin{array}{r} \mathbf{8} \\ \mathbf{6}\overline{)\mathbf{48}} \end{array}$$

- The **6** is called the **divisor**
- The **48** is called the **dividend**
- The **8** is called the **quotient**

The most common method of performing long division is called the long-division algorithm.

Example

Solve the following problem:

$$6\overline{)3108}$$

The pattern in the long division algorithm is:

- Divide
- Multiply back
- Subtract
- Bring down the next number

You can check how to solve the above example on the next two pages.

Solution to example on previous page:

```
    5
6)3108
```

Divide

6 can not go into 3, but it can go into 31 a total of 5 times.

```
    5
6)3108
  30
```

Multiply back

5 times 6 is 30.

```
    5
6)3108
 - 30
    1
```

Subtract

31 minus 30 equals 1

```
    5
6)3108
 - 30
   10
```

Bring down the next number

```
   51
6)3108
 - 30
   10
```

Divide

6 can go into 10 a total of 1 time.

```
   51
6)3108
 - 30
   10
    6
```

Multiply back

1 times 6 is 6.

```
   51
6)3108
 - 30
   10
  -  6
     4
```

Subtract

10 minus 6 equals 4

```
   51
6)3108
 - 30
   10
  -  6
    48
```

Bring down the next number

```
   518
6)3108
 - 30
   10
  -  6
    48
```

Divide

6 can go into 48 a total of 8 times.

```
   518
6)3108
 - 30
   10
  -  6
    48
    48
```

Multiply back

8 times 6 is 48.

```
   518
6)3108
 - 30
   10
  -  6
    48
   - 48
     0
```

Subtract

48 minus 48 equals 0.

There is nothing else to bring down; therefore, the final answer to “3,108 divided by 6” is 518.

Division of decimal numbers also uses long division.

- If the problem does not have a whole-number divisor, it is necessary to change the problem to an equivalent division with a whole number divisor.
- This is done by shifting the decimal point to the right in both the divisor and the dividend.
- The decimal point is shifted as many places as needed to make the divisor a whole number.
- The new place that the decimal is at in the dividend is directly below where the decimal point in the quotient should be.
- Now you may use the long division algorithm like you normally would.

Example

$8.2\overline{)73.8}$

Our initial problem is "73.8 divided by 8.2".
The **divisor** is **8.2**
The **dividend** is **73.8**

$8.2.\overline{)73.8.}$

The divisor must be a whole-number. To achieve this, shift the decimal one spot to the right.

Since you shifted the decimal in the divisor one spot, you should shift the decimal the same number of spots in the dividend.

Now, move the decimal straight up from the dividend to where it will be in the quotient.

$$\begin{array}{r} 9. \\ 82\overline{)738} \\ \underline{-738} \\ 0 \end{array}$$

Now, use the long division algorithm like you would normally to solve the problem.

Practice Problems

Solve the following practice problems.

1) $9\overline{)7.2}$ 2) $0.18\overline{)36}$ 3) $0.06\overline{)5.85}$

1) 0.8 2) 200 3) 97.5

Worksheet 1-7

Name:

Date:

Solve the following problems and round answers to the thousandths position when necessary.

1) $27\overline{)2,835}$ 2) $0.25\overline{)4.06}$ 3) $16.4\overline{)22.33}$ 4) $0.06\overline{)69.09}$ 5) $7.4\overline{)50.13}$ 6) $31.3\overline{)1.337}$

7) 49.51 ÷ 3287 =

____

9) 88.08 ÷ 0.2 =

____

11) 4.05 ÷ 4.5 =

____

8) 35.07 ÷ 19.1 =

____

10) 0.97 ÷ 0.012 =

____

12) 0.951 ÷ 0.112 =

____

Perform the following applied problems.

13) A dispensary has 256 ounces of rubbing alcohol on hand; how many 8 oz bottles can be filled with alcohol?

14) The average dose of a drug is 7.5 mg daily. If a vial contains 1,500 mg, how many days' supply are in the vial?

15) You receive a prescription for 6.25 mg capsules of metoprolol tartrate that you will have to compound. If you have 750 mg of metoprolol tartrate on hand, how many capsules will you be able to compound?

16) Four capsules of ampicillin are to be administered daily. How many days' supply would 32 capsules be?

Worksheet 1-8

Name:

Date:

Write the following Arabic numerals as Roman numerals.

1) 38 ____________ 2) 551 ____________ 3) 24 ____________

Write the following Roman numerals as Arabic numerals.

4) xxi ____________ 5) cd ____________ 6) xlviii ____________

What is the place name of the 3 in each of these numbers?

7) 2,432,484 ____________ 8) 24.032 ____________ 9) 107.3 ____________

Write the following numbers.

10) Two million fifteen thousand six hundred

____

11) Four thousand four ten-thousandths

____

12) Nine thousand eight hundred seventy six and five thousand four hundred thirty two ten-thousandths

____

Perform the following additions.

13) $\begin{array}{r} 19{,}867 \\ \underline{+\ 12{,}482} \end{array}$ 14) $\begin{array}{r} 82.07 \\ \underline{+\ 0.18} \end{array}$ 15) $\begin{array}{r} 23 \\ \underline{+\ 0.7} \end{array}$ 16) $\begin{array}{r} 38 \\ 0.042 \\ \underline{+\ 2.8} \end{array}$ 17) $\begin{array}{r} 107.3 \\ \underline{+\ 4.65} \end{array}$ 18) $\begin{array}{r} 81.9 \\ \underline{+\ 52.89} \end{array}$

Perform the following subtractions.

19) $\begin{array}{r} 19{,}867 \\ -\ 12{,}482 \\ \hline \end{array}$ 20) $\begin{array}{r} 82.07 \\ -\ 0.18 \\ \hline \end{array}$ 21) $\begin{array}{r} 23 \\ -\ 0.7 \\ \hline \end{array}$ 22) $\begin{array}{r} 93.8 \\ -\ 0.9 \\ \hline \end{array}$ 23) $\begin{array}{r} 107.3 \\ -\ 4.65 \\ \hline \end{array}$ 24) $\begin{array}{r} 81.9 \\ -\ 52.89 \\ \hline \end{array}$

Perform the following multiplications.

25) $\begin{array}{r} 48 \\ \times\ 5 \\ \hline \end{array}$ 26) $\begin{array}{r} 9.3 \\ \times\ 8 \\ \hline \end{array}$ 27) $\begin{array}{r} 0.41 \\ \times\ 0.27 \\ \hline \end{array}$ 28) $\begin{array}{r} 3.94 \\ \times\ 2.64 \\ \hline \end{array}$ 29) $\begin{array}{r} 2.8 \\ \times\ 0.042 \\ \hline \end{array}$ 30) $\begin{array}{r} 31.337 \\ \times\ 0.67 \\ \hline \end{array}$

Perform the following divisions and round your answers to the hundredths place.

31) $369 \div 6$ ________ 32) $1.5\overline{)156.5}$ 33) $16.4\overline{)22.33}$ 34) $2.04\overline{)84}$ 35) $33\overline{)6.6}$ 36) $12.3 \div 100$ ________

Solve the following word problems.

37) Find the total weight of two objects if one weighs 255.4 grams and the other weighs 198.6 grams.

38) If a patient weighs 75 kg one month and 72.7 kg the next month, how much weight did the patient lose?

39) What is the total cost of a dozen items at $2.49 each?

40) A patient is given 210 mL of medication to take. They are to take 15 mL of the medication daily till all of it is used. How many days should the medication last?

CHAPTER 2

FRACTIONS

I believe five out of four people have trouble with fractions.
--Steven Wright

Numerators, Denominators, and Reciprocals of Fractions

A fraction indicates a portion of a whole number. There are two types of fractions:

- common fractions such as ½ or ¾ ;
- decimal fractions such as 0.5 or 0.75 (we already covered these in Chapter 1).

A fraction is an expression of division, with one number placed over another number. The bottom number, or denominator, indicates the total number of parts into which the whole is divided. The top number, or numerator, indicates how many of those parts are considered. The fraction may be read as the "numerator divided by the denominator."

$$\frac{\textbf{Numerator}}{\textbf{Denominator}}$$

Examples

¼ = 1 part of 4 parts, or ¼ of the whole = 0.25

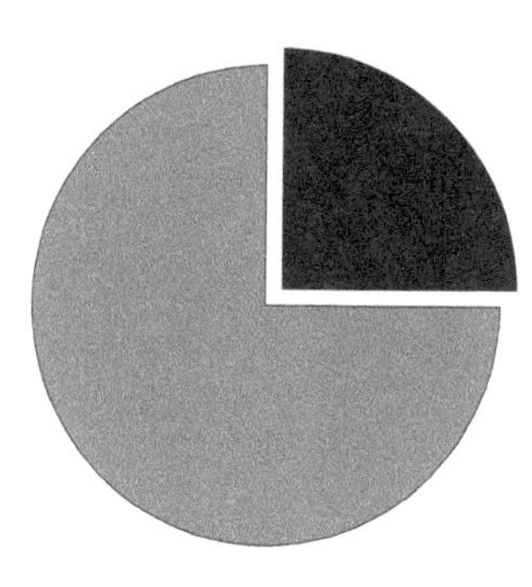

½ = 1 part of 2 parts, or ½ of the whole = 0.5

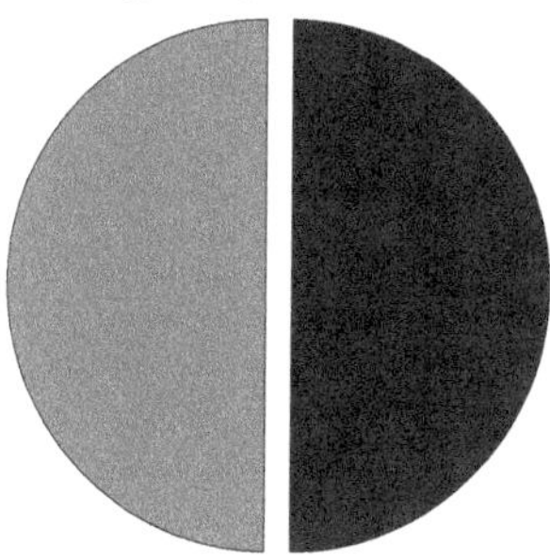

Reciprocals of fractions

To find the reciprocal of a fraction simply switch the numerator and denominator (flip it over).

- The reciprocal of $\frac{2}{3}$ is $\frac{3}{2}$
- The reciprocal of $\frac{4}{5}$ is $\frac{5}{4}$

A whole number could be considered to have a denominator of one, so the reciprocal of a whole number would be 1 over the original whole number.

- The reciprocal of 5 is $\frac{1}{5}$
- The reciprocal of 9 is $\frac{1}{9}$

The following page is a worksheet to help reinforce this information.

Worksheet 2-1

Name:

Date:

Convert the following fractions to decimals.

1)	$\frac{1}{5}$ = ______	8)	$\frac{2}{3}$ = ______	15)	$\frac{2}{22}$ = ______
2)	$\frac{1}{2}$ = ______	9)	$\frac{3}{4}$ = ______	16)	$\frac{5}{35}$ = ______
3)	$\frac{2}{4}$ = ______	10)	$\frac{4}{5}$ = ______	17)	$\frac{9}{55}$ = ______
4)	$\frac{1}{10}$ = ______	11)	$\frac{3}{8}$ = ______	18)	$\frac{13}{63}$ = ______
5)	$\frac{1}{12}$ = ______	12)	$\frac{5}{12}$ = ______	19)	$\frac{15}{71}$ = ______
6)	$\frac{1}{100}$ = ______	13)	$\frac{7}{18}$ = ______	20)	$\frac{23}{83}$ = ______
7)	$\frac{1}{1000}$ = ______	14)	$\frac{4}{11}$ = ______		

Determine the reciprocal of the following fractions.

21)	$\frac{1}{5}$ = ______	28)	$\frac{2}{3}$ = ______	35)	$\frac{2}{22}$ = ______
22)	$\frac{1}{2}$ = ______	29)	$\frac{3}{4}$ = ______	36)	$\frac{5}{35}$ = ______
23)	$\frac{2}{4}$ = ______	30)	$\frac{4}{5}$ = ______	37)	$\frac{9}{55}$ = ______
24)	$\frac{1}{10}$ = ______	31)	$\frac{3}{8}$ = ______	38)	$\frac{13}{63}$ = ______
25)	$\frac{1}{12}$ = ______	32)	$\frac{5}{12}$ = ______	39)	$\frac{15}{71}$ = ______
26)	$\frac{1}{100}$ = ______	33)	$\frac{7}{18}$ = ______	40)	$\frac{23}{83}$ = ______
27)	$\frac{1}{1000}$ = ______	34)	$\frac{4}{11}$ = ______		

Reducing fractions to lowest terms

In reducing a fraction to lowest terms (also called simplifying), you need to divide the numerator and denominator by their greatest common factor.

- The greatest common factor is the largest whole number that can divide into the numerator and the denominator.
- Some fractions are already in lowest terms if there is no factor common to the numerator and the denominator.

The steps to simplify a fraction:

- List the whole number factors of the numerator and the denominator.
- Find the factors common to both the numerator and denominator.
- Divide the numerator and denominator by the largest common factor.

Example $\frac{30}{70}$

List the factors of the numerator and the denominator:

- Numerator: 1,2,3,5,6,10,15,30
- Denominator: 1,2,5,7,10,14,35,70

The greatest common factor is 10

Divide the numerator and the denominator by 10

- Numerator: $30 \div 10 = 3$
- Denominator: $70 \div 10 = 7$

$\frac{3}{7}$ is the fraction when reduced to lowest terms.

Practice Problems

Reduce the fractions in the following problems.

1) $\frac{4}{8}$

2) $\frac{55}{99}$

3) $\frac{36}{144}$

1) 1/2 2) 5/9 3) 1/4

Worksheet 2-2

Name:

Date:

Reduce the following fractions to lowest terms.

1) $\frac{2}{4}$ = ______

2) $\frac{3}{9}$ = ______

3) $\frac{2}{16}$ = ______

4) $\frac{18}{54}$ = ______

5) $\frac{20}{240}$ = ______

6) $\frac{25}{125}$ = ______

7) $\frac{55}{75}$ = ______

8) $\frac{28}{35}$ = ______

9) $\frac{63}{90}$ = ______

10) $\frac{15}{75}$ = ______

11) $\frac{14}{36}$ = ______

12) $\frac{7}{11}$ = ______

13) $\frac{0}{18}$ = ______

14) $\frac{34}{51}$ = ______

15) $\frac{14}{63}$ = ______

16) $\frac{12}{35}$ = ______

17) $\frac{125}{500}$ = ______

18) $\frac{270}{2700}$ = ______

19) $\frac{65}{585}$ = ______

20) $\frac{82}{164}$ = ______

21) $\frac{79}{237}$ = ______

22) $\frac{17}{102}$ = ______

23) $\frac{19}{285}$ = ______

24) $\frac{18}{81}$ = ______

25) $\frac{121}{605}$ = ______

26) $\frac{63}{135}$ = ______

27) $\frac{42}{72}$ = ______

28) $\frac{33}{77}$ = ______

29) $\frac{108}{180}$ = ______

30) $\frac{110}{363}$ = ______

Adding and subtracting fractions

In order to add and subtract fractions, they must have a common denominator. Once they have a common denominator, you may add and subtract the numerators like you would in an ordinary addition or subtraction problem and then write the sum or the difference over top of the common denominator.

How to create equivalent fractions with a common denominator:

- Find a multiple for the denominator of both numbers.
- Rewrite the fractions as equivalent fractions with the common denominator.

Example $\frac{4}{5} - \frac{1}{3}$

You can multiply the denominators to find a common denominator:

$5 \times 3 = 15$

Now create equivalent fractions:

$\frac{4}{5} \times \frac{3}{3} = \frac{12}{15}$ $\frac{1}{3} \times \frac{5}{5} = \frac{5}{15}$

Now let's solve the problem with the equivalent fractions:

$$\frac{12}{15} - \frac{5}{15} = \mathbf{\frac{7}{15}}$$

Practice Problems

Solve the following practice problems.

1) $\frac{1}{3} - \frac{1}{5}$

2) $\frac{3}{8} + \frac{3}{5}$

3) $\frac{1}{19} + \frac{3}{38} + \frac{5}{76} + \frac{7}{152}$

1) 2/15 2) 39/40 3) 37/152

Worksheet 2-3

Name:

Date:

Solve the following problems.

1) $\frac{1}{3}+\frac{1}{3}$ = ______

2) $\frac{5}{8}-\frac{3}{8}$ = ______

3) $\frac{7}{8}-\frac{3}{8}$ = ______

4) $\frac{4}{25}+\frac{1}{5}$ = ______

5) $\frac{1}{8}+\frac{3}{16}$ = ______

6) $\frac{7}{8}-\frac{1}{4}$ = ______

7) $\frac{2}{7}+\frac{3}{7}+\frac{1}{7}$ = ______

8) $\frac{1}{8}+\frac{2}{8}+\frac{7}{8}$ = ______

9) $\frac{14}{38}-\frac{1}{19}$ = ______

10) $\frac{11}{12}-\frac{2}{3}$ = ______

11) $\frac{9}{24}-\frac{1}{6}$ = ______

12) $\frac{4}{5}-\frac{67}{100}$ = ______

13) $5+\frac{7}{10}+\frac{3}{1000}$ = ______

14) $\frac{5}{8}+\frac{4}{27}+\frac{1}{48}$ = ______

15) $1+\frac{7}{100}+\frac{3}{10}$ = ______

16) $\frac{1}{2}+\frac{1}{9}+\frac{1}{36}$ = ______

17) $\frac{1}{3}+\frac{1}{6}+\frac{1}{9}$ = ______

18) $\frac{11}{99}+\frac{4}{9}+\frac{11}{33}$ = ______

19) $\frac{1}{8}+\frac{1}{4}+\frac{1}{2}+\frac{3}{8}$ = ______

20) $\frac{12}{19}+\frac{14}{38}+\frac{1}{19}+\frac{1}{38}$ = ______

21) $\frac{1}{8}+\frac{2}{8}+\frac{7}{8}$ = ______

22) $\frac{10}{30}+\frac{13}{15}+\frac{1}{5}+\frac{17}{60}$ = ______

23) $\frac{1}{8}+\frac{2}{8}+\frac{7}{8}$ = ______

24) $\frac{1}{81}+\frac{2}{27}+\frac{1}{3}$ = ______

Multiplying fractions

- Reduce if possible.
- Multiply the numerators of the fractions to get the new numerator.
- Multiply the denominators of the fractions to get the new denominator.
- Simplify the resulting fraction if possible.

Example $\frac{2}{3} \times \frac{4}{7} = ?$

Reduce if possible

$\frac{2}{3} \times \frac{4}{7}$ no reduction is possible

Multiply the numerators

$2 \times 4 = 8$

Multiply the denominators

$3 \times 7 = 21$

Simplify the resulting fraction if possible

$\frac{8}{21}$ is already in simplest terms

$$\frac{2}{3} \times \frac{4}{7} = \mathbf{\frac{8}{21}}$$

Practice Problems

Multiply the following fractions.

1) $\frac{2}{5} \times \frac{1}{8} =$

2) $\frac{2}{3} \times \frac{2}{3} =$

3) $\frac{7}{8} \times \frac{9}{21} =$

1) 1/20 2) 4/9 3) 3/8

Multiplying mixed numbers

- Write mixed numbers as improper fractions.
- Reduce if possible.
- Multiply the numerators of the fractions to get the new numerator.
- Multiply the denominators of the fractions to get the new denominator.
- Simplify the resulting fraction if possible.

Example $4\frac{1}{8} \times 1\frac{5}{11} = ?$

Write mixed numbers as improper fractions.

$$\frac{33}{8} \times \frac{16}{11}$$

Reduce if possible

$$\frac{{}^{3}\cancel{33}}{{}_{1}\cancel{8}} \times \frac{{}^{2}\cancel{16}}{{}_{1}\cancel{11}}$$

Multiply the numerators

$$3 \times 2 = 6$$

Multiply the denominators

$$1 \times 1 = 1$$

Simplify the resulting fraction if possible

$$\frac{6}{1} = 6$$

$$4\frac{1}{8} \times 1\frac{5}{11} = \frac{{}^{3}\cancel{33}}{{}_{1}\cancel{8}} \times \frac{{}^{2}\cancel{16}}{{}_{1}\cancel{11}} = \frac{6}{1} = \mathbf{6}$$

Practice Problems

Solve the following practice problems by multiplying the fractions.

1) $13\frac{1}{3} \times 1\frac{4}{5} =$

2) $7\frac{1}{5} \times 3\frac{3}{4} =$

3) $20\frac{2}{3} \times 1\frac{4}{31} =$

1) 24 2) 27 3) 23 1/3

Worksheet 2-4

Name:

Date:

Solve the following problems.

1) $\frac{1}{3}$ of $\frac{1}{3}$ = _____

2) $\frac{1}{8}$ of $\frac{3}{8}$ = _____

3) $\frac{2}{3}$ of $\frac{1}{3}$ = _____

4) $\frac{5}{8}$ of $\frac{3}{8}$ = _____

5) $\frac{1}{5}$ of $\frac{3}{5}$ = _____

6) $\frac{7}{8}$ of $\frac{3}{8}$ = _____

7) $\frac{0}{4}$ of $\frac{5}{6}$ = _____

8) $\frac{1}{2}$ of $\frac{3}{5}$ = _____

9) $\frac{1}{9}$ of $\frac{1}{2}$ = _____

10) $\frac{2}{7}$ of $\frac{1}{5}$ = _____

11) $\frac{9}{10}$ of $\frac{3}{4}$ = _____

12) $\frac{3}{11}$ of $\frac{3}{4}$ = _____

Multiply the following fractions.

13) $\frac{3}{10} \times \frac{1}{4}$ = _____

14) $\frac{13}{15} \times \frac{7}{8}$ = _____

15) $\frac{21}{25} \times \frac{2}{5}$ = _____

16) $\frac{49}{50} \times \frac{1}{4}$ = _____

17) $\frac{6}{7} \times 21$ = _____

18) $25 \times \frac{3}{10}$ = _____

19) $\frac{8}{15} \times \frac{5}{6}$ = _____

20) $\frac{11}{14} \times \frac{2}{33}$ = _____

Multiply the following mixed numbers.

21) $4\frac{4}{5} \times 1\frac{1}{6}$ =

22) $7\frac{1}{2} \times 2\frac{2}{3}$ =

23) $1\frac{17}{18} \times 1\frac{1}{5}$ =

24) $1\frac{5}{7} \times 10\frac{1}{2}$ =

Dividing Fractions

Dividing fractions is just like multiplying fractions (with one exception), you need to "flip" (find the reciprocal of) the fraction you are dividing by.

The following are the steps to dividing fractions:

- Find the reciprocal of the fraction you are dividing by.
- Reduce if possible.
- Multiply the numerators of the fractions to get the new numerator.
- Multiply the denominators of the fractions to get the new denominator.
- Simplify the resulting fraction if possible.

A simpler way of remembering this is to simply think "Flip and Multiply"

Example $\frac{2}{3} \div \frac{4}{7}$

"Flip and Multiply"

$$\frac{{}^{1}\cancel{2}}{3} \times \frac{7}{{}_{2}\cancel{4}} = \frac{7}{6} = \mathbf{1\frac{1}{6}}$$

Practice Problems

Solve the following problems.

1) $\frac{2}{5} \div \frac{1}{8} \quad =$

2) $\frac{2}{3} \div \frac{2}{3} \quad =$

3) $\frac{7}{8} \div \frac{9}{21} \quad =$

1) 16/5 = 3 1/5 2) 1 3) 147/72 = 2 3/72 = 2 1/24

Worksheet 2-5

Name:

Date:

Divide the following fractions:

1) $\frac{1}{3} \div \frac{1}{3}$ = ______

2) $\frac{1}{8} \div \frac{3}{8}$ = ______

3) $\frac{1}{8} \div \frac{3}{16}$ = ______

4) $\frac{4}{25} \div \frac{1}{5}$ = ______

5) $\frac{2}{3} \div \frac{1}{2}$ = ______

6) $\frac{16}{27} \div \frac{8}{9}$ = ______

7) $\frac{5}{15} \div 5$ = ______

8) $\frac{9}{10} \div \frac{9}{10}$ = ______

9) $\frac{7}{5} \div \frac{5}{7}$ = ______

10) $12 \div \frac{3}{8}$ = ______

11) $3\frac{2}{3} \div 1\frac{1}{4}$ = ______

12) $1\frac{5}{6} \div 7\frac{1}{3}$ = ______

13) $15\frac{3}{4} \div 5\frac{1}{7}$ = ______

14) $15 \div 1\frac{1}{5}$ = ______

Solve the following word problems:

15) How many 6 ¼ milligram capsules of metoprolol tartrate can be made from 500 mg of metoprolol tartrate?

16) The OR likes to use phenylephrine syringes, each containing 4/5 of a milligram. How many phenylephrine syringes can I make if I have a 20 mg vial on hand?

Worksheet 2-6

Name:

Date:

Convert these fractions into equivalent decimals (you only need to solve them out to the thousandths position).

1) $\frac{7}{15}$ = ________

2) $\frac{13}{40}$ = ________

3) $\frac{5}{8}$ = ________

4) $\frac{13}{15}$ = ________

5) $\frac{17}{30}$ = ________

6) $\frac{7}{8}$ = ________

7) $\frac{9}{27}$ = ________

8) $\frac{17}{25}$ = ________

9) $\frac{13}{37}$ = ________

Determine the reciprocals of the following fractions.

10) $\frac{3}{4}$ = ________

11) $\frac{1}{5}$ = ________

12) $\frac{15}{24}$ = ________

13) $\frac{2}{8}$ = ________

14) $\frac{65}{585}$ = ________

15) 1 = ________

Reduce the following fractions.

16) $\frac{6}{8}$ = ________

17) $\frac{12}{48}$ = ________

18) $\frac{15}{24}$ = ________

19) $\frac{2}{8}$ = ________

20) $\frac{12}{36}$ = ________

21) $\frac{18}{24}$ = ________

22) $\frac{27}{45}$ = ________

23) $\frac{17}{31}$ = ________

24) $\frac{52}{148}$ = ________

Perform the following additions and subtractions.

25) $\frac{1}{6}+\frac{3}{6}$ = ________

26) $\frac{7}{6}+\frac{5}{24}$ = ________

27) $\frac{7}{8}-\frac{7}{12}$ = ________

28) $\frac{5}{6}-\frac{2}{3}$ = ________

29) $\frac{1}{8}+\frac{2}{3}$ = ________

30) $\frac{11}{16}-\frac{1}{2}$ = ________

31) $\frac{1}{6}+\frac{4}{7}$ = ________

32) $\frac{9}{11}-\frac{2}{5}$ = ________

Perform the following multiplications and divisions.

33) $\frac{1}{6}\times\frac{4}{7}$ = ________

34) $4\times\frac{7}{8}$ = ________

35) $\frac{3}{4}\div\frac{1}{6}$ = ________

36) $\frac{7}{16}\div 7$ = ________

37) $10\div\frac{1}{2}$ = ________

38) $\frac{7}{16}\times 7$ = ________

39) $\frac{3}{4}\times\frac{1}{3}$ = ________

40) $\frac{3}{4}\div\frac{1}{3}$ = ________

Solve the following word problems.

41) What is the total volume of a mixture of 1/8 milliliter of solution X mixed with 2 and 1/4 milliliters of solution Y?

42) A patient is to receive 3 liters of an intravenous solution. If 2 1/8 liters have already been administered, how much solution remains?

43) How many milligrams of drug are needed to make 20 tablets of 1/4 milligrams each?

44) A bottle of Children's Tylenol contains 20 teaspoons of liquid. If each dose for a 2 year-old child is 1/2 teaspoon, how many doses are available in this bottle?

45) A patient is on strict recording of fluid intake and output, including measurements of liquid medications. A nursing student gave the patient 1/4 ounce of medication at 8 AM and 1/3 ounce of medication at 12 noon. What is the total amount of medication to be recorded on the Intake and Output sheet?

Chapter 3

Percentages

Baseball is 90% mental and the other half is physical.
--Yogi Berra

What is a percentage?

The word **percent** means “per 100” or “out of 100”

The percent symbol (%) is an easy way to write a fraction with a common denominator of 100.

Example

27 out of 100 equals 27%

Example

19% = 19/100 = 0.19

Nineteen percent (19%) is the same as the fraction 19/100 and the decimal 0.19

Writing Decimals as Percents

2.61 = 261%
3 = 3.00 = 300%
0.5 = 0.50 = 50%
0.004 = 0.4%

Move the decimal points over two places to the right and add a percent sign. At times, you may need to insert zeros.

Try the following practice problems

1) 0.27
2) 0.13
3) 0.06
4) 0.2
5) 0.018
6) 1.5

1) 27% 2) 13% 3) 6% 4) 20% 5) 1.8% 6) 150%

Writing Percents as Decimals

Examples

49% = 0.49
180% = 1.8
3% = 0.03
0.7% = 0.007

Move the decimal point over two places to the left and omit the percent sign. You may need to add zeros.

Practice Problems

Turn the percentages into their decimal equivalents.

1) 100%
2) 77%
3) 1.8%
4) 50%
5) 9.3%
6) 0.2%

1) 1 2) 0.77 3) 0.018 4) 0.5 5) 0.093 6) 0.002

Writing Fractions as Percents

The following is an example of two different ways that you can turn a fraction into a percent.

- You can turn a fraction into an equivalent fraction so that the denominator becomes equal to 100.

$$\frac{1}{5}=\frac{1}{5}\times\frac{20}{20}=\frac{20}{100}=\mathbf{20\,\%}$$

- Or, you can turn it into a decimal and shift the decimal two places to the right.

$$\frac{1}{5}=1\div 5=0.2=\mathbf{20\,\%}$$

Practice Problems

Convert these fractions into percents.

1) $\frac{3}{10}$
2) $\frac{12}{25}$
3) $\frac{3}{8}$
4) $\frac{13}{40}$
5) $\frac{13}{20}$
6) $\frac{2}{3}$

1) 30% 2) 48% 3) 37.5% 4) 32.5% 5) 65% 6) 66.7%

Worksheet 3-1

Name:

Date:

Express the following decimals as percents.

1) 0.2444 = __________
2) 0.3 = __________
3) 0.5 = __________
4) 0.125 = __________
5) 0.75 = __________
6) 0.02 = __________
7) 0.09 = __________
8) 0.1 = __________
9) 0.8 = __________
10) 0.36 = __________
11) 0.52 = __________
12) 0.4 = __________
13) 0.65 = __________
14) 0.025 = __________
15) 0.035 = __________
16) 0.055 = __________
17) 0.004 = __________
18) 1.10 = __________
19) 1.75 = __________
20) 2 = __________

Express the following percents as decimals.

21) 33% = __________
22) 24% = __________
23) 33.3% = __________
24) 50.5% = __________
25) 20% = __________
26) 47% = __________
27) 93% = __________
28) 32.5% = __________
29) 75% = __________
30) 83.32% = __________
31) 66.66667% = __________
32) 18.5% = __________
33) 1.3% = __________
34) 0.25% = __________
35) 0.125% = __________

Write the following fractions as percents.

36) $\frac{4}{5}$ = __________

37) $\frac{2}{10}$ = __________

38) $\frac{7}{8}$ = __________

39) $\frac{15}{20}$ = __________

40) $\frac{30}{35}$ = __________

41) $\frac{85}{100}$ = __________

42) $\frac{5}{8}$ = __________

43) $\frac{3}{7}$ = __________

44) $\frac{6}{20}$ = __________

45) $\frac{12}{20}$ = __________

46) $\frac{35}{40}$ = __________

47) $\frac{40}{50}$ = __________

48) $\frac{55}{100}$ = __________

49) $\frac{80}{90}$ = __________

50) $\frac{32}{64}$ = __________

Writing Percents as Fractions

You can turn percents back into fractions, just remember that percent means out of 100.

Examples

$$68\% = \frac{68}{100} = \mathbf{\frac{17}{25}}$$

$$3\% = \mathbf{\frac{3}{100}}$$

Practice Problems

Try converting the following percentages to fractions.

1) 55%
2) 60%
3) 0.4%
4) 5%
5) 29%
6) 260%

1) 11/20 2) 3/5 3) 1/250 4) 1/20 5) 29/100 6) 2 3/5

Finding a Percentage of a Number

There are two common ways of finding a percentage of a number:

- you can turn your percent into a decimal and multiply it by the total value the percent is out of,
- or you can use the *'is over of'* method

Example

How much is 90% out of 20? *(this can be solved 2 ways)*

$n = 0.9 \times 20$

$n = \mathbf{18}$

or

$$\frac{is}{of} = \frac{\%}{100}$$

$$\frac{n}{20} = \frac{90}{100}$$

$$(n)(100)=(20)(90)$$

$$n = \mathbf{18}$$

Practice Problems

Complete the following practice problems.

1) How much is 75% of 200?
2) How much is 30% of 50?
3) How much is 25% of 16?
4) How much is 12% of 250?

1) 150 2) 15 3) 4 4) 30

Worksheet 3-2

Name:

Date:

Write the following percentages as fractions (reduce if possible).

1) 50% = ________
2) 75% = ________
3) 60% = ________
4) 35% = ________
5) 30% = ________
6) 100% = ________
7) 68% = ________
8) 41% = ________
9) 93% = ________
10) 1.2% = ________
11) 12.5% = ________
12) 67% = ________
13) 66.6% = ________
14) 36% = ________
15) 110% = ________

Calculate the following problems.

16) 25% of 600 = ________
17) 20% of 30 = ________
18) 15% of 20 = ________
19) 75% of 50 = ________
20) 12.5% of 24 = ________
21) 80% of 40 = ________
22) 90% of 100 = ________
23) 17% of 10 = ________
24) 110% of 5 = ________
25) 33% of 90 = ________
26) 5% of 50 = ________
27) 30% of 120 = ________
28) 40% of 50 = ________
29) 60% of 150 = ________
30) 70% of 400 = ________

Find the value of n:

31) If n is 20% of 50 n = ________
32) If n is 35% of 24 n = ________
33) If n is 60% of 8 n = ________
34) If n is 75% of 10 n = ________
35) If n is 80% of 15 n = ________
36) If n is 40% of 30 n = ________
37) If n is 50% of 100 n = ________
38) If n is 25% of 50 n = ________

39) If n is 30% of 90 n = ________

40) If n is 100% of 50 n = ________

Finding What Percent One Number is of Another

There are two common ways of finding what percentage one number is of another:

- you can divide your parts by the total it is out of to get a decimal and then turn your decimal into a percentage,
- or you can use the *'is over of'* method

Example

What percent of 40 is 15? *(this can be solved 2 ways)*

$n = 15 \div 40 = 0.375 = \mathbf{37.5\,\%}$

or

$\frac{is}{of} = \frac{\%}{100}$

$\frac{15}{40} = \frac{n}{100}$

$(n)(40) = (15)(100)$

$n = \mathbf{37.5\,\%}$

Practice Problems

Attempt the following practice problems.

1) What percent of 50 is 2?
2) 0.6 is what percent of 20?
3) What percent of 40 is 8?
4) 20 is what percent of 80?
5) What percent of 60 is 2.4?
6) 10 is what percent of 50?

1) 4% 2) 3% 3) 20% 4) 25% 5) 4% 6) 20%

Finding a Number When a Percent of it is Known

There are two common ways of finding a number when a percent of it is known:

- you can divide your parts by your percentage (as a decimal) and this will provide you with your total,
- or you can use the *'is over of'* method

Example

8 is 32% of what number? *(this can be solved 2 ways)*

$n = 8 \div 0.32 = \mathbf{25}$

or

$$\frac{is}{of} = \frac{\%}{100}$$

$$\frac{8}{n} = \frac{32}{100}$$

$$(n)(32) = (8)(100)$$

$n = \mathbf{25}$

Practice Problems

Attempt the following practice problems.

1) 12 is 60% of what number?
2) 95% of what number is 114?
3) 14 is 35% of what number?
4) 75% of what number is 60?
5) 35.7 is 42% of what number?
6) 3% of what number is 27?

1) 20 2) 120 3) 40 4) 80 5) 85 6) 900

Worksheet 3-3

Name:

Date:

Fill in the blank boxes with all boxes in the same row being equal.

Percent	*Decimal*	*Reduced Fraction*
20%		
	0.06	
		7/8
15%		
	0.11	
		4/5
96%		
	0.55	
		3/25
53%		
	1	
		3/40

Solve the following problems.

1) How much is 4% of 50?
2) What percent of 90 is 14.4?
3) 7 is 4% of what number?
4) 99% of 200 is how much?
5) 42 is what percent of 150?
6) 55% of what number is 1375?
7) How much is 65% of 500?
8) What percent of 48 is 12?
9) 20% of what number is 3?
10) 90% of 112 is how much?
11) 15 is what percent of 75?
12) 70% of what number is 14?
13) How much is 45% of 40?
14) What percent of 80 is 72?
15) 20 is 62.5% of what number?
16) 77.5% of 80 is how much?

17) 23 is what percent of 25?

18) 88% of what number is 440?

19) How much is 140% of 30?

20) What percent of 64 is 52?

Solve the following word problems.

21.)Smart Sally got 2 out of 25 wrong on her math quiz. What percent did she get correct?

22.)34.8% of the pharm tech class wanted the holiday party to be semi-formal. How many students are in the class if 8 wanted a semi-formal party?

23.)Of the 3,620 people polled, 85% enjoyed mixing IVs. How many people liked mixing IVs?

24.)Frugal Fred saved $78.25 by purchasing generic medications. He decided to splurge and buy a nice engraved mortar and pestle for himself. The mortar and pestle was $50.08. What percent of his savings was that?

25.)Alopecia Allen was having a problem with his hair falling out. He lost 112 hairs yesterday which was 4% of his total hair. How many hairs did he have at the beginning of the day?

26.)When asked, 270 out of the 400 patients at the Funny Farm said that they felt a sense of humor was the most important trait in a physician. What percent is this rounded to the nearest whole number?

Worksheet 3-4

Name:

Date:

Convert the following percents to fractions.

1) 16%
2) 45%
3) 6%

Convert the following percents to decimals

4) 33%
5) 4.5%
6) 2.94%

Convert the following fractions to percentages.

7) 12/48
8) 2/5
9) 7/10

Convert the following decimals to percentages.

10) 0.7
11) 0.467
12) 0.006

Solve the following problems involving percents.

13) What is 40% of 120?
14) What is 20% of 30?
15) What percent of 150 is 15?
16) What percent of 45 is 30?

17) 35 is 14% of what number?

18) 30 is 110% of what number?

19) What number is 75% of 280?

20) What percent of 140 is 7?

Solve the following applied problems.

21) At Bidwell Training Center Inc., 28.6% of the instructors in the medical department are male. If there are two male instructors in the medical department, then how many instructors are there in the medical department?

22) The recommended daily allowance (RDA) of a particular vitamin is 60 milligrams. If a multivitamin tablet claims to provide 45% of the RDA, how many milligrams of the particular vitamin would a patient receive from the multivitamin tablet?

23) A person weighed 130 pounds at his last doctor's visit. At this visit the patient lost 5% of his weight. How many pounds has the patient lost?

24) In a drug study, it was determined that 4% of the participants developed the *headache* side effect. If there were 600 participants in the study, how many developed headaches?

25) You are employed in a health clinic where each employee must work 25% of 8 major holidays. How many holidays will you be required to work?

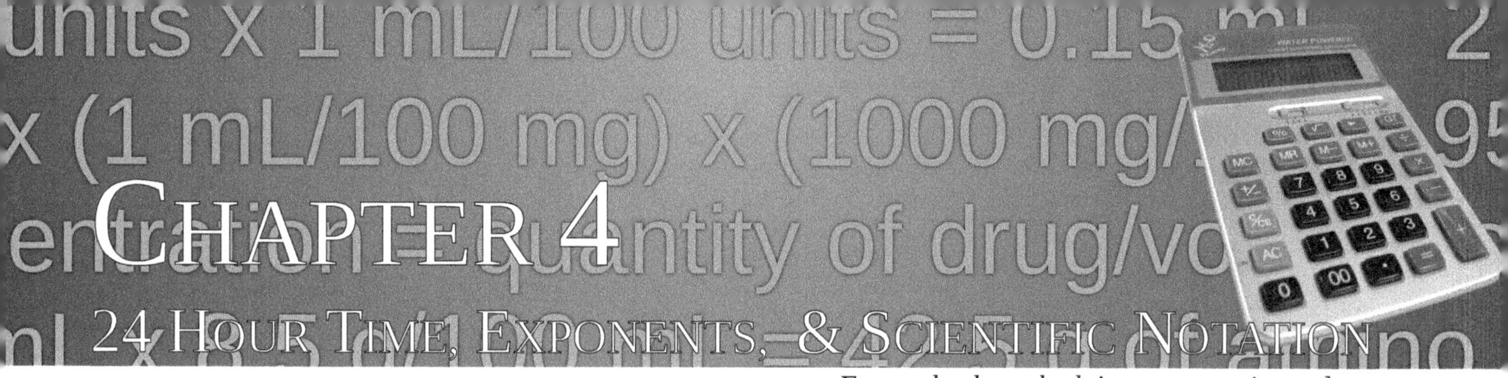

Chapter 4

24 Hour Time, Exponents, & Scientific Notation

Even a broken clock is correct twice a day.
--Polish Proverb

The 24 Hour-Clock

The 24-hour clock is increasingly used in medical facilities. This time system avoids confusion between *A.M.* and *P.M.* time. Four digits are used to express time in this system. The first two digits indicate the hour, and the last two the minutes.

Here are some examples of A.M. time on the 24-hour clock:

24-Hour Time	Reading	*A.M.* Time
0400	zero four hundred hours	4:00 A.M.
1130	eleven thirty hours	11:30 A.M.
0910	zero nine ten hours	9:10 A.M.

Notice several things:

1) A.M. times and 24 hour times *(excluding midnight to 12:59 A.M.)* possess the same digits except A.M. times drop the leading zero,
2) notice that when reading 24 hour time, you always say *'hours'* at the end,
3) military times that end in a double zero are read as *'hundred hours'*,
4) hours and minutes are read as groups,
5) if the number of hours and/or minutes is less than 9, then each digit is read separately *(example 0908 would be read as 'zero nine zero eight hours')*,
6) and if the number of hours and/or minutes is greater than 9, then each set is read as one number *(example 1132 would be read as 'eleven thirty two hours')*.

Based on the information you've been given so far, complete the following table:

24-Hour Time	Reading	*A.M.* Time
0230		
1045		
0100		
1150		
0600		
1030		
0235		
1120		

The following are the answers to the above table: 0230 - zero two thirty hours - 2:30 A.M.; 1045 -ten forty five hours - 10:45 A.M.; 0100 - zero one hundred hours - 1:00 A.M.; 1150 - eleven fifty hours - 11:50 A.M.; 0600 - zero six hundred hours - 6:00 A.M.; 1030 - ten thirty hours - 10:30 A.M.; 0235 - zero two thirty five hours - 2:35 A.M. - 1120 - eleven twenty hours - 11:20 A.M.

The following illustrates how P.M. time is indicated on the 24-hour clock:

24-Hour Time	Reading	*P.M.* Time
1300	thirteen hundred hours	1:00 P.M.
1830	eighteen thirty hours	6:30 P.M.
2150	twenty one fifty hours	9:50 P.M.

To obtain P.M. time from 24-hour time, subtract the number 1200 from the 24-hour time.

1300 – 1200 = 1:00 P.M.
1830 – 1200 = 6:30 P.M.
2150 – 1200 = 9:50 P.M.

Complete the following table:

24-Hour Time	Reading	*P.M.* Time
1800		
1945		
1415		
2320		
1500		
1730		
2145		
2350		

The following are the answers to the above table: 1800 – eighteen hundred hours – 6:00 P.M.; 1945 – nineteen forty five hours – 7:45 P.M.; 1415 – fourteen fifteen hours – 2:15 P.M.; 2320 – twenty three twenty hours – 11:20 P.M.; 1500 – fifteen hundred hours – 3:00 P.M.; 1730 – seventeen thirty hours – 5:30 P.M.; 2145 – twenty one forty five hours – 9:45 P.M.; 2350 – twenty three fifty hours – 11:50 P.M.

The time between 12:00 midnight and 1:00 A.M. is written and read as follows:

24-Hour Time	Reading	*A.M.* Time
0010	zero zero ten hours	12:10 A.M.
0045	zero zero forty five hours	12:45 A.M.

Complete the following table:

24-Hour Time	Reading	*A.M.* Time
0020		
0008		
0030		
0015		

The following are the answers to the above table: 0020 – zero zero twenty hours – 12:20 A.M.; 0008 – zero zero zero eight hours – 12:08 A.M.; 0030 – zero zero thirty hours – 12:30 A.M.; 0015 – zero zero fifteen hours – 12:15 A.M.

Technically, midnight is supposed to be *'zero zero zero zero hours'* but is often referred to as *'twenty four hundred hours'*. Expect to see and hear both done at some point when working in a hospital.

Worksheet 4-1

Name:

Date:

Complete the following table:

24-Hour Time	**Reading**	***A.M./P.M. Time***
0315	zero three fifteen hours	3:15 A.M.
1140		
1400		
2230		
0030		
1050		
		6:15 A.M.
		10:50 A.M.
		9:45 P.M.
		1:15 P.M.
		12:15 A.M.
		12:05 A.M.
	zero nine thirty hours	
	twenty-two thirty hours	
	zero zero twenty hours	
	fourteen twenty-five hours	
	zero zero zero seven hours	
	zero one hundred hours	

Exponents

Exponents show how many times a number is multiplied by itself. A number with an exponent is said to be "raised to the power" of that exponent.

Example

Three raised to the power of four would be written:

3^4

and it would equal:

$3^4 = 3 \times 3 \times 3 \times 3 = 81$

A couple of "powers" have their own names. If something is raised to the power of two, it may also be called "squared".

Example

Three squared = $3^2 = 3 \times 3 = 9$

If something is raised to the power of three, it may also be called "cubed".

Example

Three cubed = $3^3 = 3 \times 3 \times 3 = 27$

There are two additional rules to remember:

any number raised to the zero power (except 0) equals 1;

Example

$3^0 = 1$

any number raised to the power of one equals itself.

Example

$3^1 = 3$

Practice Problems

1) 2^3
2) 4^5
3) 6^2
4) 1^3
5) 3^5
6) 5^3
7) 7^4
8) 8^1
9) 9^0

1) 8 2) 1,024 3) 36 4) 1 5) 243 6) 125 7) 2,401 8) 8 9) 1

Scientific Notation

Scientific notation provides an easier way of writing very small and very large numbers and is sometimes referred to as "power-of-ten". A number written in scientific notation is written as a product of a number between 1 and 10 and multiplied by a power of 10 .

Here are some examples of large numbers:

$$100=1\times10\times10=1\times10^2$$

$$100{,}000=1\times10\times10\times10\times10\times10=1\times10^5$$

$$427{,}000=4.27\times10\times10\times10\times10\times10=4.27\times10^5$$

An easy shortcut to scientific notation is to move the decimal point so that there is only one digit to the left of it. Count the number of spaces the decimal has been moved, and this becomes the exponent on the base 10.

- If you move the decimal to the left, the exponent will be positive.
- If you move the decimal to the right, the exponent will be negative.

The mass of the Earth would be a very large number:

$$5{,}973{,}600{,}000{,}000{,}000{,}000{,}000{,}000 \text{ kg} = 5.9736 \times 10^{24} \text{ kg}$$

Here are some examples of small numbers:

$$0.008 = 8 \times 10^{-3}$$

$$0.000234 = 2.34 \times 10^{-4}$$

The mass of an electron would be a very small number:

$$0.00000000000000000000000000000091093826 \text{ kg} = 9.1093826 \times 10^{-31} \text{ kg}$$

Practice Problems

Write the following numbers in scientific notation:

1) 1,000,000
2) 1
3) 807,000
4) 0.1
5) 6,100,000
6) 0.0303
7) 0.0018
8) 0.00014

1) 1×10^6 2) 1×10^0 3) 8.07×10^5 4) 1×10^{-1} 5) 6.1×10^6 6) 3.03×10^{-2} 7) 1.8×10^{-3} 8) 1.4×10^{-4}

Worksheet 4-2

Name:

Date:

Solve for the values of the following exponential numbers.

1) 1^1 = ____________

2) 2^1 = ____________

3) 1^2 = ____________

4) 2^2 = ____________

5) 2^3 = ____________

6) 3^2 = ____________

7) 3^3 = ____________

8) 3^4 = ____________

9) 10^2 = ____________

10) 6^3 = ____________

11) 5^0 = ____________

12) 5^2 = ____________

13) 4^0 = ____________

14) 9^2 = ____________

15) 9^0 = ____________

16) 12^4 = ____________

17) 12^3 = ____________

18) 10^4 = ____________

19) 7^0 = ____________

20) 7^3 = ____________

Convert the following numbers to scientific notation.

21) 1,000 = ____________

22) 1 = ____________

23) 67,000,000 = ____________

24) 0.1 = ____________

25) 0.00306 = ____________

26) 1,000,000 = ____________

27) 909,000 = ____________

28) 0.001 = ____________

29) 0.00028 = ____________

30) 0.000000614 = ____________

Convert the following scientific notations to regular numbers.

31) 6.14×10^{-7} = ____________

32) 3.06×10^{-3} = ____________

33) 2.8×10^{-4} = ____________

34) 1×10^{-1} = ____________

35) 1×10^{-3} = ____________

36) 6.7×10^{7} = ____________

37) 9.09×10^5 = ____________

38) 1×10^0 = ____________

39) 1×10^6 = ____________

40) 1×10^3 = ____________

Worksheet 4-3

Name:

Date:

Complete the following table.

24-Hour Time	**Reading**	*A.M./P.M. Time*
0315	zero three fifteen hours	3:15 A.M.
1040		
2100		
1230		
		7:15 A.M.
		10:45 P.M.
	zero eight thirty hours	
	twenty-two fifteen hours	
1020		
2200		
1130		
		7:15 P.M.
		10:45 A.M.
	twelve thirty hours	
	twenty-two ten hours	

Solve for the values of the following exponential numbers.

1) 6^3 = ____________

2) 5^2 = ____________

3) 4^7 = ____________

4) 3^4 = ____________

5) 10^2 = ____________

6) 6^2 = ____________

7) 5^3 = ____________

8) 4^5 = ____________

9) 3^3 = ___________ 10) 10^4 = ___________

Express the following numbers in scientific notation.

11) 10,000,000 = ___________

12) 351 = ___________

13) 7,100 = ___________

14) 0.037 = ___________

15) 0.3750 = ___________

16) 0.0064 = ___________

17) 1,000,000 = ___________

18) 35 = ___________

19) 6,100 = ___________

20) 0.37 = ___________

21) 0.375 = ___________

22) 0.0056 = ___________

Worksheet 4-4 (tougher problems)

Name:

Date:

Solve the following problems to further your knowledge on 24 hour time, exponents, and scientific notation.

1) While working in the pharmacy, you receive an order to give a medication every four hours around the clock. If the patient's nurse tells you that she wants to give the first dose at 1 p.m., what times will you schedule the medication? *(give your answers in 24-hour time)*

2) Avogadro's Number is very useful in pharmaceutical chemistry as it provides a means for accurately determining the number of atoms that exist in a mass of a specific substance. If you have 6.02×10^{23} atoms of a pure substance, it will have the same mass (in grams) as its atomic mass number found on the periodic table. Write out Avogadro's Number (6.02×10^{23}) without using scientific notation.

3) $27^2 \div 3^2 =$

4) Whenever two problems written in scientific notation are multiplied, you may multiply the initial numbers normally and add the exponents to get your answer. An example would be:

   $(2 \times 10^2)(3 \times 10^4) = 6 \times 10^6$

   With that in mind solve the following problem:

   $(1.2 \times 10^5)(5 \times 10^7) =$

5) Whenever two problems written in scientific notation are divided, you may divide the initial numbers normally and subtract the exponents to get your answer. An example would be:

   $(6 \times 10^6) \div (3 \times 10^4) = 2 \times 10^2$

   With that in mind solve the following problem:

   $(6 \times 10^{12}) \div (1.2 \times 10^5) =$

Chapter 5

Problems Solving Methods

The essence of mathematics is not to make simple things complicated, but to make complicated things simple.
--Stanley Gudder

Most calculations in pharmacy can be solved using either ratios and proportions or dimensional analysis (*a.k.a.*, factor labels).

Ratios

We use ratios to make comparisons between two things. When we express ratios in words, we use the word "to" -- we say "the ratio of something to something else"

Example :

The ratio of squares to triangles in the illustration below. Ratios can be written in several different ways:

as a fraction	$\frac{3}{4}$
using the word "to"	3 to 4
using a colon	3:4

Using the above images, make a comparison of triangles to all shapes written the following ways:

1) as a fraction

2) using the word "to"

3) using a colon

1) 4/7 2) 4 to 7 3) 4:7

Multiplying or dividing each term by the same nonzero number will give an equal ratio.

Example

The ratio 2:4 is equal to the ratio 1:2. To tell if two ratios are equal, use a calculator and divide. If the division gives the same answer for both ratios, then they are equal.

$$1:2 = 2:4 = 4:8 = 6:12 \quad \textit{or} \quad \frac{1}{2}=\frac{2}{4}=\frac{4}{8}=\frac{6}{12}$$

Are 3:9, 1:3, and 9:27 all equal?

Yes, 3:9 = 1:3 = 9:27

Example

Janine has a bag with 3 flash drives, 4 marbles, 7 books, and 1 orange.

1) What is the ratio of books to marbles?
    - Expressed as a fraction, with the numerator equal to the first quantity and the denominator equal to the second, the answer would be 7/4.
    - Two other ways of writing the ratio are 7 to 4, and 7:4.

2) What is the ratio of flash drives to the total number of items in the bag?
    - There are 3 flash drives, and 3 + 4 + 7 + 1 = 15 items total.
    - The answer can be expressed as 3/15, 3 to 15, or 3:15.

Complete the following practice problems.

1) What is the ratio of squares to total?

2) What is the ratio of circles to triangles?

3) What is the ratio of triangles to squares?
What is the ratio of circles to all?
What is the ratio of triangles to squares to circles?

1) 7:8 2) 7:2 3) 3:2 ; 4:9 ; 3:2:4

Attempt the following practice word problems.

Practice Problems

1) Write a ratio comparing the oxygen tension of arterial blood (100 milliliters) to that of venous blood (40 milliliters).

2) If 2 oz of boric-acid are added to 598 oz of water, how many ounces are in the total solution? What is the ratio of ounces of boric acid to ounces of solution?

3) If 35 oz of a chemical are combined with 65 oz of water, how many oz are in the solution? What is the ratio of oz of chemical to oz of solution?

1) 100:40 would be an acceptable answer, I would also accept it if someone reduced it down to 5:2
2) There are 600 ounces in the total solution and the ratio of boric acid to solution is 2:600 which could be reduced to 1:300
3) There are 100 ounces in the solution and the ratio of chemical to solution is 35:100 which could be reduced to 7:20

Worksheet 5-1

Name:

Date:

Express the following as ratios reduced to lowest terms.

1) 1 to 5
2) 2 to 11
3) 2 grams to 9 grams
4) 15 cm to 23 cm
5) 3 feet to 27 feet
6) 25 liters to 5 liters
7) 36 cm to 6 cm
8) 24 inches to 9 inches
9) 3 meters to 9 meters
10) 100 g to 1,000 g

Solve the following problems.

11) If 3 oz of boric-acid solution are added to 897 oz of water, how many ounces are in the total solution? What is the ratio of ounces of boric acid to ounces of solution?

12) If 30 oz of a chemical are combined with 60 oz of water, what is the ratio of chemical to ounces of solution?

13) If 20 oz of alcohol are added to 40 oz of water, how many ounces are in the total solution? What is the ratio of ounces of alcohol to ounces of total solution?

14) If 10 grams of salt are added to 40 grams of water, what is the total weight of the water-and-salt solution? What is the ratio of grams of salt to grams of solution? (*Ordinarily, you would not measure water by mass, but it is possible to do and therefore the problem could be reproduced.*)

15) If 3 mL of glycerin are added to 87 mL of water, what is the ratio of the mL of glycerin to the mL of solution?

16) If 2 grams of a drug are added to 38 grams of petrolatum, what is the ratio of the grams of drug to the grams of the mixture?

17) If 5 oz of a drug are added to 95 oz of water, how many ounces are in the total solution? What is the ratio of ounces of drug to ounces of solution?

Worksheet 5-2

Name:

Date:

Express the following as ratios reduced to lowest terms.

1) 1 to 4
2) 3 to 17
3) 3 grams to 15 grams
4) 2 cm to 7 cm
5) 5 feet to 19 feet
6) 23 liters to 25 liters
7) 3 cm to 18 cm
8) 35 inches to 7 inches
9) 1,000 m to 10 m
10) 12 grams to 48 grams

Solve the following problems.

11) To perform a test, a lab technician adds 5 oz of a test liquid to 395 oz of water. How many ounces are in the solution? What is the ratio of ounces of test liquid to ounces of solution?

12) If 2 oz of chemical are combined with 8 oz of water, how many ounces are in the total solution? What is the ratio of ounces of chemical to ounces of solution?

13) A marathon runner has a normal heart rate of 68 beats/min. After completing a marathon, her heart rate is 110 beats/min. Express the ratio of her normal heart rate to her heart rate after a marathon.

14) If 3 oz of boric-acid[1] solution are added to 741 oz of water, how many ounces are in the total solution? What is the ratio of boric acid to ounces of solution?

15) In 1964, the U.S. Public Health Service reported that the risk of developing lung cancer is 10 times greater for moderate smokers and 20 times greater for heavy smokers than for nonsmokers. Express ratio of risk of cancer for nonsmokers to moderate smokers to heavy smokers. *(Hint, even if you don't smoke you still have a chance of developing lung cancer.)*

---

1 Since it has been mentioned repeatedly for various problems, you may be asking, "What's boric-acid?". Boric-acid, also called boracic acid or orthoboric acid, is a mild acid often used as an antiseptic, insecticide, flame retardant, in nuclear power plants to control the fission rate of uranium, and as a precursor of other chemical compounds. It exists in the form of colorless crystals or a white powder and dissolves in water. It has the chemical formula $B(OH)_3$. When occurring as a mineral, it is called sassolite.

Proportions

A proportion is a statement of equality between two ratios.

Example

$$\frac{1}{2}=\frac{2}{4}$$

This proportion can also be expressed as "1:2 = 2:4" or "1 is to 2 as 2 is to 4".

Practice Problem

What would be two other ways to express the following:

$$\frac{a}{b}=\frac{c}{d}$$

This proportion can also be expressed as "a:b = c:d" or "a is to b as c is to d".

solving for a variable in a proportion

You have three options:

1) basic algebra
2) cross multiplying
3) "means" and "extremes"

Let's solve the following equation for *a* using all three methods

$$\frac{a}{b}=\frac{c}{d}$$

algebra

Solving for *a* using basic algebra. First, we need to isolate *a*. We can do this by multiplying both sides of the equation by *b*. Then you can cancel out the *b*s on the left hand side of the equation, and this will leave you with your final answer.

$\frac{a}{b}=\frac{c}{d}$ ➡ Multiply both sides by *b*. ➡ $\frac{b\times a}{b}=\frac{b\times c}{d}$ ➡ Now cancel out the *b*s on the left. ➡ $\frac{\not{b}\times a}{\not{b}}=\frac{b\times c}{d}$ ➡ Your final answer ➡ $\boldsymbol{a=\frac{b\times c}{d}}$

cross multiplying

Solving for *a* using cross multiplication. You can multiply diagonally and set both multiplication problems as equal. Then we will want to isolate the *a* by dividing both sides by *d*. Cancel out the *d*s on the appropriate side and you will be left with the final answer.

$\frac{a}{b}=\frac{c}{d}$ ➡ Cross multiply ➡ $\frac{a}{b}\times\frac{c}{d}$ ➡ $a\times d=b\times c$ ➡ Divide by *d*. ➡ $\frac{a\times d}{d}=\frac{b\times c}{d}$ ➡ Cancel ➡

$\frac{a\times \not{d}}{\not{d}}=\frac{b\times c}{d}$ ➡ Your final answer ➡ $\mathbf{a=\frac{b\times c}{d}}$

"means" and "extremes"

Solving for *a* using "means" and "extremes". To do this, you need to rewrite your proportion using colons. Next, you multiply your means (your inside numbers) and set them equal to the value of your extremes (outside numbers) when they are multiplied. Then we will want to isolate the *a* by dividing both sides by *d*. Cancel out the *d*s on the appropriate side and you will be left with the final answer.

$\frac{a}{b}=\frac{c}{d}$ ➡ Rewrite it using colons ➡ $a:b=c:d$ ➡ Multiply your "means" and "extremes" ➡ $a:b=c:d$ ➡ $a\times d=b\times c$ ➡ Divide by *d*.

➡ $\frac{a\times d}{d}=\frac{b\times c}{d}$ ➡ Cancel ➡ $\frac{a\times \not{d}}{\not{d}}=\frac{b\times c}{d}$ ➡ Your final answer ➡ $\mathbf{a=\frac{b\times c}{d}}$

Now that you've been shown three different ways to solve for a variable in a ratio proportion, let's try a practice problem. Solve for each of the variables in the statement "W is to X as Y is to Z".

1) The value of *W* is:

2) The value of *X* is:

3) The value of *Y* is:

4) The value of *Z* is:

1) $W=\frac{X\times Y}{Z}$ 2) $X=\frac{W\times Z}{Y}$ 3) $Y=\frac{W\times Z}{X}$ 4) $Z=\frac{X\times Y}{W}$

Practice Problems

Solve for the variable in each of the following practice problems.

1) $\frac{3.8\,L}{14\,L}=\frac{N}{25\,L}$

2) $\frac{x}{7.2\,g}=\frac{3.0\,g}{2.0\,g}$

3) $5:N=7:9$

4) $3:4=x:15$

1) 6.8 L 2) 10.8 g 3) 6.4 4) 11.25

Worksheet 5-3

Name:

Date:

Solve the following ratio proportion problems.

1) In the proportion 5:8 = 25:40,
   a) The extremes are _____ and _____.
   b) The means are _____ and _____.

Solve for N in each of these problems.

2) $\frac{2}{3}=\frac{4}{N}$ $N=$ _____

3) $\frac{5}{7}=\frac{15}{N}$ $N=$ _____

4) $\frac{3}{6}=\frac{1}{N}$ $N=$ _____

5) $\frac{5}{10}=\frac{7}{N}$ $N=$ _____

6) $\frac{2}{8}=\frac{3}{N}$ $N=$ _____

7) $\frac{N}{3}=\frac{4}{6}$ $N=$ _____

8) $\frac{N}{8}=\frac{3}{24}$ $N=$ _____

9) $\frac{N}{12}=\frac{5}{6}$ $N=$ _____

10) $\frac{N}{4}=\frac{6}{8}$ $N=$ _____

11) $\frac{N}{5}=\frac{10}{25}$ $N=$ _____

12) $\frac{N}{6}=\frac{10}{12}$ $N=$ _____

13) $\frac{N}{7}=\frac{6}{42}$ $N=$ _____

14) $\frac{N}{15}=\frac{2}{5}$ $N=$ _____

15) $\frac{N}{5}=\frac{4}{10}$ $N=$ _____

16) $\frac{3}{5}=\frac{N}{100}$ $N=$ _____

17) $\frac{125}{1{,}000}=\frac{N}{8}$ $N=$ _____

18) $\frac{3}{8}=\frac{375}{N}$ $N=$ _____

19) $\frac{2}{3}=\frac{N}{12}$ $N=$ _____

Worksheet 5-4

Name:

Date:

Write the following problems in fractional-equation form.

1) $3:x = 6:7$
2) $5:8 = N:10$
3) $K:8 = 4:16$
4) $7:9 = 14:N$
5) $2:3 = x:9$
6) $7:N = 14:28$

Solve the following proportions for *N*.

7) $N:7 = 6:14$
8) $8:10 = 4:N$
9) $3:N = 6:18$
10) $5:8 = N:24$
11) $N:9 = 6:18$
12) $8:N = 16:32$
13) $\frac{2.5}{16} = \frac{N}{8}$
14) $\frac{4.2}{N} = \frac{2.1}{100}$
15) $\frac{6.8}{50} = \frac{13.6}{N}$
16) $\frac{12.0\,g}{4.6g} = \frac{8.4\,g}{N}$

Solve the following word problems.

17) Given a boric-acid solution of 1:400, how many ounces are in 70 oz of the solution?

18) How much magnesium sulfate is needed for a preparation of 24 oz of a 1:4 mixture?

19) The ratio of salt to water in a very concentrated saline solution is 3:8. How much salt should be added to 360 g of water to prepare a solution with this ratio?

20) Given a boric-acid solution of 1:500, how many ounces of boric acid are in 80 oz of solution?

Worksheet 5-5

Name:

Date:

Write the following proportions in fractional-equation form.

1) $4:X = 7:9$

2) $3:5 = N:8$

3) $N:72 = 4.8:12.0$

4) $15.0:N = 31.0:1$

5) $5:17.1 = N:2$

6) $315:32 = N:35$

Solve the following proportions for N.

7) $N:2 = 15:3$

8) $4:5 = 8:N$

9) $5:N = 10:17$

10) $5:9 = N:10$

11) $\frac{3.5}{N} = \frac{70}{100}$

12) $\frac{N}{1.4} = \frac{12}{28}$

13) $\frac{N}{2.2} = \frac{1.7}{1}$

14) $\frac{14.0\,g}{70\,mL} = \frac{12.0\,g}{N}$

15) $\frac{2.1\,L}{8\,L} = \frac{N}{12.0\,L}$

16) $\frac{14.1\,oz}{5.8\,oz} = \frac{4\,oz}{N}$

17) $\frac{N}{\$712} = \frac{\$0.07}{\$1.00}$

Solve the following word problem.

18) Given a colloidal suspension of 3.5:100, how many mL of colloids are present in 946 mL of this colloidal suspension?

Dimensional Analysis *(a.k.a., Factor Label Method)*

Dimensional analysis is a conceptual tool often utilized in health care to understand physical situations involving a mix of different kinds of physical quantities. It is routinely used by pharmacists and pharmacy technicians to calculate things such as weight, volume, dose, dosage form, and time. To solve a problem using dimensional analysis, you need to first identify what information is provided by the problem as well as any conversion factors that you will need to solve the problem. Terms that are equal to each other may be written in the form of a fraction.

If 1 capsule is 250 mg, you could say that 1 capsule = 250 mg. To write that as a fraction, you could write:

$$\frac{1\,capsule}{250\,mg} \quad \text{or} \quad \frac{250\,mg}{1\,capsule}$$

Example

How many tablets will be taken in seven days if a prescription order reads zafirlukast 20mg/tablet, one tablet twice a day?

QUESTION:
How many tablets?

DATA:

7 days	20mg/tablet
1 tablet/dose	2 doses/day

MATHEMATICAL METHOD / FORMULA:
Dimensional Analysis

DO THE MATH

$$\frac{7\,days}{1} \times \frac{2\,doses}{day} \times \frac{1\,tablet}{dose}$$

cancel out dimensions where applicable

$$\frac{7\,\cancel{days}}{1} \times \frac{2\,\cancel{doses}}{\cancel{day}} \times \frac{1\,tablet}{\cancel{dose}} = \mathbf{14\,tablets}$$

DOES THE ANSWER MAKE SENSE ?
Yes

Notice in the above example how dimensions are used to cancel everything out until you are left with just what you need, in this case tablets.

5 Step Method

You may also notice the way the example was broken down with five questions. This is known as the "5 Step Method". The "5 Step Method" is simply a way to help you interpret the data in a problem and get the answer. While certainly the problems in this book can be solved without using the "5 Step Method", you will find that many example problems throughout the book are broken down this way. The five steps are: 1) QUESTION (identify what the question is asking); 2) DATA (identify the data in the problem and any necessary conversion factors); 3) MATHEMATICAL METHOD/FORMULA (identify the method or formula needed to solve the problem); 4) DO THE MATH (exactly what the statement says), and 5) DOES THE ANSWER MAKE SENSE? (sometimes an answer will not make sense, and that may mean that there is an error).

Now we should try a practice problem using dimensional analysis and the "5 Step Method".

How many tablets will you need to provide a 24 hour supply of metoprolol succinate 50 mg/dose, 1 dose/day, if you have 25 mg tablets available?

QUESTION

DATA

MATHEMATICAL METHOD / FORMULA

DO THE MATH

DOES THE ANSWER MAKE SENSE?

2 tablets

Worksheet 5-6

Name:

Date:

Solve the following problems using dimensional analysis.

1) How many tablets will be taken in three days if a prescription reads zafirlukast 20 mg/tablet, one tablet twice a day?

2) How many capsules will be taken in seven days if a prescription order reads tetracycline 250 mg/capsule, one capsule four times a day?

3) How many tablets will be taken in four days if a prescription order reads sucralfate 1 g/tablet, one tablet four times a day?

4) How many tablets will be taken in 10 days if a prescription order reads warfarin 5 mg/tablet, one tablet daily at bedtime?

5) How many tablets will be taken in 30 days if a prescription order reads metoprolol tartrate 50 mg/tablet, one tablet two times a day?

6) How many tablets will be taken in two days if a prescription reads famotidine 20 mg/tablet, one tablet three times a day before meals?

7) How many tablets are needed to fill a prescription for 34 days for albuterol 2 mg/tablet, four times a day?

8) How many capsules are needed to fill a 4 week supply of fluoxetine HCl controlled release 60 mg/capsule, one capsule every week?

9) How many tablets are needed to fill a prescription for 21 days for repaglinide 0.5 mg/tablet, one tablet three times a day?

10) How many tablets are needed to fill a 72 hour supply for dipyridamole 50 mg/tablet, one tablet four times a day?

Worksheet 5-7

Name:

Date:

Solve the following problems using dimensional analysis.

1) How many tablets will be taken in seven days if a prescription order reads furosemide 20 mg/tablet, one tablet twice a day?

2) How many tablets are needed to fill a prescription for seven days for alprazolam 0.5 mg/tablet, one tablet three times a day?

3) How many tablets are needed to fill a prescription for 90 days for dipyridamole 50 mg/tablet, one tablet four times a day?

4) How many capsules are needed to fill a prescription for 28 days for potassium chloride 10 mEq/capsule, one capsule four times a day?

5) How many tablets are needed to fill a four day prescription of azithromycin 500 mg/tablet, one tablet daily?

6) How many tablets will be taken in 30 days if a prescription order reads methylphenidate 10 mg/tablet, one tablet three times a day?

7) How many capsules are needed to fill a 90 day supply of zidovudine 100 mg/capsule, three capsules twice daily?

8) You receive the following order for 10 mg terazosin HCl capsules:

    Rx terazosin HCl 10 mg/capsules
    Dispense: 30 day supply
    Sig: Take 1 capsule by mouth at bedtime for 3 days,
    then take 2 capsules by mouth at bedtime for 5 days,
    then take 4 capsules by mouth at bedtime thereafter.

   How many capsules should you dispense? *(Hint ~ This problem will require several dimensional analysis equations to get your final answer.)*

Worksheet 5-8

Name:

Date:

Solve the following problems using dimensional analysis.

1) How many tablets will be needed to fill a prescription for three days for upsadasium 25 mg/tablet if it is ordered 50 mg of upsadasium three times a day?

2) How many capsules are needed to fill a 30 day prescription for 25 mg/capsule of downagain if it is ordered one capsule every day?

3) How many tablets will be taken in three days if a prescription order reads sucralfate 1 g/tablet, one tablet four times a day?

4) How many tablets will be taken in seven days if a prescription order reads Ambien 5 mg/tablet, one tablet daily at bedtime?

5) How many capsules are needed to fill a prescription for 90 days if a prescription order reads zidovudine 100 mg/capsules, three capsules twice daily?

6) How many capsules are needed to fill a prescription for 14 days for potassium chloride 10 mEq/capsule, one capsule four times a day?

7) How many capsules are needed to fill a prescription for 30 days for prazosin 1 mg/capsule, two capsules three times a day?

8) You dispense a prescription for furosemide 40 mg. The instructions read take one tablet by mouth twice a day for 10 days. What is the total number of tablets you should dispense?

9) An order is received for dexamethasone 10 mg twice a day for 5 days, 5 mg twice a day for 4 days, and 2.5 mg twice a day for 2 days. Your stock is 10 mg tablets scored in fourths. What would be the total amount of tablets for you to dispense? *(Hint ~ This problem will require several dimensional analysis equations to get your final answer.)*

Worksheet 5-9

Name:

Date:

Express the following as ratios reduced to lowest terms.

1) 1 to 9

2) 2 g to 12 g

3) 6 cm to 36 cm

Solve the following word problems.

4) What is the ratio of chemical to ounces of solution if 20 oz of chemical are combined with 80 oz of water?

5) How many ml of a drug are needed to prepare 2,500 ml of a 1:20 solution?

6) How much hydrocortisone is needed to prepare 120 grams of a 1:50 ointment?

Write the following proportions in fractional-equation form.

7) $5{:}x = 10{:}15$

8) $6{:}11 = Y{:}12$

9) $1{:}3 = N{:}18$

Solve the following ratio proportions for N.

10) $N{:}10 = 3{:}5$

11) $N{:}10 = 4{:}5$

12) $N{:}9 = 2{:}3$

13) $\frac{3.1\,g}{15.5\,g} = \frac{8.1\,g}{N}$

14) $\frac{4.2\,g}{16.8\,g} = \frac{7.1\,g}{N}$

15) $\frac{4.1\,g}{12.3\,g} = \frac{6.1\,g}{N}$

16) $\frac{4.4\,mL}{17.6\,mL} = \frac{N}{4\,mL}$

17) $\frac{5\,in.}{12\,in.} = \frac{N}{8\,in.}$

18) $\frac{50\,g}{100\,mL} = \frac{N}{4\,mL}$

Solve the following problems using dimensional analysis.

19) How many tablets will be taken in seven days if a prescription order reads zafirlukast 20 mg/tablet, one tablet twice a day?

20) How many tablets are needed to fill a prescription for seven days for cyproheptadine 4 mg/tablet, one tablet three times a day?

21) How many tablets are needed to fill a prescription for 30 days for dipyridamole 50 mg/tablet, one tablet four times a day?

22) How many capsules are needed to fill a prescription for 14 days for potassium chloride 10 mEq/capsule, one capsule four times a day?

23) How many capsules are needed to fill a 14 day prescription of ampicillin 500 mg/capsule, one capsule four times a day?

24) How many capsules will be taken in 90 days if a prescription order reads atomoxetine 10 mg/capsule, one capsule three times a day?

25) How many capsules are needed to fill a three day supply of zidovudine 100 mg/capsule, three capsules twice daily?

26) How many tablets will be taken in two days if a prescription order reads promethazine 12.5 mg/tablet, one tablet three times a day?

27) Each tablet of TYLENOL WITH CODEINE contains 30 mg of codeine phosphate and 300 mg of acetaminophen. By taking two tablets daily for a week, how many milligrams of each drug would the patient take?

28) The biotechnology drug filgrastim (NEUPOGEN) is available in vials containing 480 micrograms (mcg) of filgrastim per 0.8 mL. How many micrograms of the drug would be administered by each 0.5 mL injection?

UNIT 2

BASIC PHARMACY MATH

What does basic pharmacy math consist of?

Before you start calculating dosages of antineoplastic infusions based on a patient's body surface area, you need to learn some basics. We need to be able to convert between various measurement systems, learn a little bit of terminology and how to manipulate data to find some simple dosages for oral medications, days' supply, and IV drip rates. That is what most would consider basic pharmacy math, and this unit is designed to provide a guide to learning this information.

What are the specific learning objectives in this unit?

- temperature conversions
- household measurements
- metric system
- apothecary system
- medication abbreviations
- calculating dosages when giving medications in tablet or capsule form
- calculating dosages when giving medications in liquid form
- preparing solutions
- diluting stock solutions
- determining the rate of intravenous medications
- dosages based on body weight
- dosages based on body surface area
- pediatric dosing

Chapter 6

Temperature Scale Conversions

It doesn't make a difference what temperature a room is, it's always room temperature.
--Steven Wright

The temperature for storing medication is extremely important for the stability—and therefore the effectiveness—of the medication. The two common temperature scales used in pharmacy are Celsius and Fahrenheit. Usually, storage requirements (including storage temperature) are listed in small print on the package label. The necessary temperature is usually given in both Celsius and Fahrenheit degrees, although not always. Furthermore, the Pharmacy Technician Certification Exam (a certification required by some of the higher paying technician jobs) always ask several questions involving temperatures. Hence, you need to be able to convert between Celsius and Fahrenheit.

Fahrenheit and Celsius are both linear scales, so let's gather some basic information on them and create a formula to compare them.

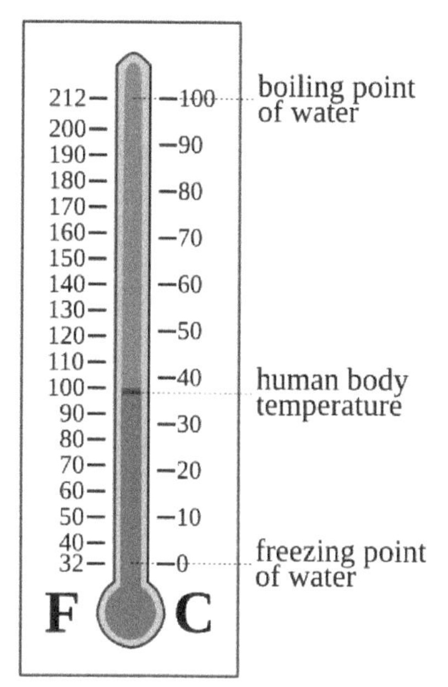

What is the temperature when water freezes in degrees Fahrenheit?

32° F

What is the temperature when water boils in degrees Fahrenheit?

212° F

What is the temperature when water freezes in degrees Celsius?

0° C

What is the temperature when water boils in degrees Celsius?

100° C

How many degrees does it take to get from the freezing point of water to the boiling point of water on the Fahrenheit scale?

180° F

How many degrees does it take to get from the freezing point of water to the boiling point of water on the Celsius scale?

100° C

Now lets create a ratio comparing the two ranges.

180° F : 100° C

That ratio can be reduced to....

9° F : 5° C

Now that we've established all that information, there is one other thing we need to look at before we can come up with a formula.

> *What is the difference between the freezing point of water in Fahrenheit and the freezing point of water in Celsius?*

32° F

So, based on this information I can create the following formula to convert from a Celsius temperature to a Fahrenheit temperature:

$$T^{\circ} C \times \frac{9^{\circ} F}{5^{\circ} C} + 32^{\circ} F = your\ equivalent\ degrees\ F$$

Often, you'll see this formula simplified to:

$$C \times \frac{9}{5} + 32 = F$$

From this point we can use some basic algebra to create a formula to convert from Fahrenheit temperature to Celsius temperature:

$$(F - 32) \times \frac{5}{9} = C$$

Practice Problem

Let's test our new formulas. You may already know that the ideal body temperature for a homo sapien is 98.6° F or 37° C.

Change 98.6° F to Celsius using the following equation:

$$(F - 32) \times \frac{5}{9} = C$$

Change 37° C to Fahrenheit using the following equation:

$$C \times \frac{9}{5} + 32 = F$$

Worksheet 6-1

Name:

Date:

Convert the following temperatures. Round all your answers to the tenths position.

1) 50° C = _____° F

2) 47° C = _____° F

3) 45° C = _____° F

4) 40° C = _____° F

5) 37° C = _____° F

6) 32° C = _____° F

7) 30° C = _____° F

8) 25° C = _____° F

9) 22° C = _____° F

10) 20° C = _____° F

11) 18° C = _____° F

12) 15° C = _____° F

13) 12° C = _____° F

14) 10° C = _____° F

15) 7° C = _____° F

16) 5° C = _____° F

17) 2° C = _____° F

18) 0° C = _____° F

19) -5° C = _____° F

20) -10° C = _____° F

21) -40° C = _____° F

22) 100° F = _____° C

23) 90° F = _____° C

24) 89° F = _____° C

25) 82° F = _____° C

26) 80° F = _____° C

27) 79° F = _____° C

28) 75° F = _____° C

29) 70° F = _____° C

30) 63° F = _____° C

31) 60° F = _____° C

32) 58° F = _____° C

33) 55° F = _____° C

34) 45° F = _____° C

35) 37° F = _____° C

36) 32° F = _____° C

37) 25° F = _____° C

38) 22° F = _____° C

39) 15° F = _____° C

40) 12° F = _____° C

41) 5° F = _____° C

42) 0° F = _____° C

43) -5° F = _____° C

44) -14° F = _____° C

45) -20° F = _____° C

Worksheet 6-2 (part I)

Name:

Date:

Specific storage conditions are required to be printed in product literature and on drug packaging and drug labels to ensure proper storage and product integrity. The conditions are defined by the following terms[1]:

Cold: any temperature not exceeding 8° C

Freezer: -25° to -10° C

Refrigerator: 2° to 8° C

Cool: 8° to 15° C

Room temperature: the temperature prevailing in a working area

Controlled room temperature: 15° to 30° C

Warm: 30° to 40° C

Excessive heat: any temperature above 40° C

Calculate all the above temperatures in Fahrenheit, round to the nearest whole number degree.

Cold: any temperature not exceeding ____ F

Freezer: ____ to ____ F

Refrigerator: ____ to ____ F

Cool: ____ to ____ F

Room temperature: the temperature prevailing in a working area

Controlled room temperature: ____ to ____ F

Warm: ____ to ____ F

Excessive heat: any temperature above ____ F

---

1 These important standards are contained in a combined publication that is recognized as the official compendium, the United States Pharmacopeia (USP) and the National Formulary (NF)

Worksheet 6-2 (part II)

Name:

Date:

Specific storage conditions are required to be printed in product literature and on drug packaging and drug labels to ensure proper storage and product integrity. The conditions are defined by the following terms[2]:

Cold: any temperature not exceeding 46° F

Freezer: -13° to 14° F

Refrigerator: 36° to 46° F

Cool: 46° to 59° F

Room temperature: the temperature prevailing in a working area

Controlled room temperature: 59° to 86° F

Warm: 86° to 104° F

Excessive heat: any temperature above 104° F

Calculate all the above temperatures in Celsius, round to the nearest whole number degree.

Cold: any temperature not exceeding ____ C

Freezer: ____ to ____ C

Refrigerator: ____ to ____ C

Cool: ____ to ____ C

Room temperature: the temperature prevailing in a working area

Controlled room temperature: ____ to ____ C

Warm: ____ to ____ C

Excessive heat: any temperature above ____ C

---

2 These important standards are contained in a combined publication that is recognized as the official compendium, the United States Pharmacopeia (USP) and the National Formulary (NF)

Worksheet 6-3

Name:

Date:

Solve the following temperature conversion problems.

1) When making a mixture, you are instructed to heat the mixture to 130° C. You have only a Fahrenheit thermometer. What is the equivalent temperature on the Fahrenheit scale?

2) Most of the drugs in a pharmacy need to be stored at controlled room temperature, which is defined by the USP/NF as 15° C to 30° C. The air conditioning at Bidwell Community Pharmacy breaks down on a warm day in August. The weather forecast says that it is supposed to get up to 90° F. Is that within the acceptable range?

3) The following sterile compounding request is sent to the hospital pharmacy:

Alteplase in a Syringe

Rx alteplase, 2mg/ml 50 mg
sterile water for injection 25 mL

1. Reconstitute the alteplase with SWFI.
2. Draw up 5 mL in 10 mL syringes.
3. Label syringes with contents, concentration, and date of preparation.
4. Place syringes in freezer. They should be frozen with premix piggybacks.

The syringes are stable for 45 days, at -25° C to -10° C

a) What is the Fahrenheit at which you should store this product?

b) What expiration should you put on this product if you made it today?

4) A prescription is sent to the pharmacy requesting a substance to be heated in a 300° F oven for 12-18 hours. At what Celsius temperature does the oven need to be set?

5) An autoclave is usually set to 250° F to sterilize medical instruments. What is the equivalent temperature in degrees Celsius?

6) Convert the following refrigerator temperatures and record them in the appropriate spaces on the log below. Note any temperatures out of the safe range (2° to 8° C).

Date	Degrees F	Degrees C
1/25	36.1	a.
1/26	37.7	b.
1/27	39.0	c.
1/28	34.5	d.
1/29	36.9	e.
1/30	36.7	f.
1/31	38.8	g.
2/1	43.8	h.
2/2	48.8	i.

Which days, if any, fell outside the safe range?

7) Convert the following freezer temperatures and record them in the appropriate spaces on the log below. Note any temperatures out of the safe range (-13° F to 14° F).

Date	Degrees C	Degrees F
1/25	-15.1	a.
1/26	-12.9	b.
1/27	-13.2	c.
1/28	-9.4	d.
1/29	-10.5	e.
1/30	-13.5	f.
1/31	-15.7	g.
2/1	-18.9	h.
2/2	-21.0	i.

Which days, if any, fell outside the safe range?

CHAPTER 7

UNITS OF MEASUREMENT

You can't control what you can't measure.
--Tom DeMarco

We will need to know three major systems of measurement thoroughly.

- The English/Household System
- The Metric System
- The Apothecary System

Health care has been combining science into its repertoire long before the majority of the world standardized on the metric system; therefore, many directions and recipes will still call for knowledge of all three systems, and how to convert between them.

The English/Household System

The household system (sometimes referred to as the English system) is the measurement system most commonly used in the United States today and is nearly the same as that brought by the colonists from England. These measures had their origins in a variety of cultures –Babylonian, Egyptian, Roman, Anglo-Saxon, and Norman French. The ancient "digit," "palm," "span" and "cubic" units of length slowly lost preference to the length units "inch," "foot," and "yard."

Because this is the system of measurement we are most familiar with, we tend to be capable of making conversions within this system very rapidly for length, volume, and weight (the term weight is more commonly used in health care than mass).

Some common conversions within this system include:

Length
12 inches = 1 foot
3 feet = 1 yard
5,280 feet = 1 mile

Weight
16 ounces = 1 pound

Volume
3 teaspoons = 1 tablespoon
2 tablespoons = 1 fluid ounce
8 fluid ounces = 1 cup
2 cups = 1 pint
2 pints = 1 quart
4 quarts = 1 gallon

Along with knowing their relationship to each other, you should also be aware of accepted abbreviations and symbols.:

inch ~ in., "
foot ~ ft., '
yard ~ yd.
ounce ~ oz.,
pound ~ lb., lbs, #
teaspoon ~ tsp , t
tablespoon ~ Tbsp , T
fluid ounce ~ fl. oz., $f ℥$
pint ~ pt.
quart ~ qt.
gallon ~ gal.

Example

Using the aforementioned conversions we can use dimensional analysis to figure out how many inches are in one mile.

$$\frac{1\,mile}{1}\times\frac{5{,}280\,ft}{1\,mile}\times\frac{12\,in.}{1\,ft.}=\mathbf{63{,}360}\,\boldsymbol{inches}$$

Practice Problem

How many teaspoons are in one gallon?

768 teaspoons

Review

1) The system of measurement most commonly used in the United States is the __________ or __________ system.

2) Quart and fluid ounce are household units that measure __________.

3) Pound and ounce are household units that measure __________.

4) Yard and inch are household units that measure __________.

Interpret these abbreviations

5) $f ℥$ = __________

6) tsp = __________

7) pt. = __________

8) lbs. = __________

9) ' = __________

10) oz. = __________

1) household *or* English 2) volume 3) weight 4) length 5) fluid ounce 6) teaspoon 7) pint 8) pound 9) foot 10) ounce

The Metric System

In 1790, in the midst of the French Revolution, the National Assembly of France requested the French Academy of Sciences to "deduce an invariable standard for all the measures and all the weights."

The Commission appointed by the Academy created a system that was, at once, simple and scientific. The unit of length was to be a portion of the Earth's circumference. Measures for capacity (volume) and mass were to be derived from the unit of length, thus relating the basic units of the system to each other and to nature. Furthermore, larger and smaller multiples of each unit were to be created by multiplying or dividing the basic units by 10 and its powers. This feature provided a great convenience to users of the system, by eliminating the need for such calculations as dividing by 16 (to convert ounces to pounds) or by 12 (to convert inches to feet). Similar calculations in the metric system could be performed simply by shifting the decimal point. Thus, the metric system is a "base-10" or "decimal" system.

The metric system is based on three basic units, meter(m) for length, gram(g) for weight, and liter(L) for volume. You can use the following prefixes to describe larger and smaller units of meters, grams, and liters:

Factor	***Name***	***Symbol***
10^9	giga	G
10^6	mega	M
10^3	kilo	k
10^2	hecto	h
10^1	deka	da
1	base unit	(m, g, L)
10^{-1}	deci	d
10^{-2}	centi	c
10^{-3}	milli	m
10^{-6}	micro	mc, µ
10^{-9}	nano	n
10^{-12}	pico	p

Based on the table above, 1.1 gigaliters would equal 1.1×10^9 liters or 1,100,000,000 liters.

$$\frac{1.1GL}{1} \times \frac{10^9 L}{1GL} = \mathbf{1{,}100{,}000{,}000}\,\boldsymbol{liters}$$

Some common conversions within the metric system include:

Length	*Weight*	*Volume*
10 mm = 1 cm	1,000 ng = 1 mcg	1 mL = 1 cc
1,000 mm = 1 m	1,000 mcg = 1 mg	1,000 mL = 1 L
100 cm = 1 m	1,000 mg = 1 g	1,000 cc = 1 L
1,000 m = 1 km	1,000 g = 1 kg	

Example

Using the aforementioned conversions we can use dimensional analysis to figure out how many centimeters are in one kilometer.

$$\frac{1\,km}{1} \times \frac{1{,}000\,m}{1\,km} \times \frac{100\,cm}{1\,m} = \mathbf{100{,}000\,\textit{centimeters}}$$

Practice Problem

How many micrograms are in one kilogram?

1,000,000,000 micrograms

Review

1) The measurement system most commonly used in prescribing and administering medication is the __________ system.

2) Liter and milliliter are metric units that measure __________.

3) Gram and milligram are metric units that measure __________.

4) Meter and millimeter are metric units that measure __________.

Interpret these abbreviations

5) mcg = __________

6) mL = __________

7) cc = __________

8) g = __________

9) mm = __________

10) kg = __________

1) metric 2) volume 3) weight 4) length 5) microgram 6) milliliter 7) cubic centimeter 8) gram 9) millimeter 10) kilogram

Converting between household units and metric units

One of the common challenges of health care is that we are often given units in the English (or household) system, but we will need to convert them to metric to do our jobs. The following are some of the more common conversions:

Length

1 in. = 2.54 cm
1 m = 1.09 yds.

Weight

1 lb. = 454 g
1 kg = 2.2 lbs
1 oz. = 28.4 g

Volume

1 tsp = 5 mL
1 Tbsp = 15 mL
1 fl. oz. = 29.6 mL (usually approximated to 30 mL)
1 pt. = 473 mL (usually approximated to 480 mL)

Practice Problem

Your instructor is 6' 1” and weighs 165 lbs. What is his height in meters, and his weight in kilograms?

Height: 1.85 m Weight: 75 kg

Worksheet 7-1

Name:

Date:

Make the following conversions.

1) 1 L = ________ ml
2) 10 cc = ________ ml
3) 697 ml = ________ L
4) 2.6 L = ________ ml
5) 0.25 L = ________ cc
6) 2.5 kg = ________ g
7) 415 g = ________ kg
8) 2,160 mg = ________ g
9) 8.9 mg = ________ mcg
10) 200 g = ________ lb
11) 4 liters = ________ qt
12) 50 miles = ________ km
13) 5 m = ________ in
14) 3 in = ________ cm
15) 4 lb = ________ kg
16) 50 g = ________ oz
17) 5 fl oz = ________ cc
18) 2 qt = ________ ml
19) 2 kg = ________ oz
20) 2.5 L = ________ fl oz
21) 20 oz = ________ kg
22) 5' 3" = ________ m

Solve the following word problems.

23) A neonatal patient needs to receive a medication based on their weight in grams. The patient weighs 4 lbs 6 oz. How many grams does the baby weigh?

24) If Richard Stallman weighs 220 lbs, how much does he weigh in kg?

25) A physician writes for 137 mcg of levothyroxine by mouth once a day. All the levothyroxine tablets on your shelves are measured in mg. How many mg will each tablet need to be?

Worksheet 7-2

Name:

Date:

Make the following conversions. Feel free to use the estimated numbers for milliliters to fluid ounces and pints.

1) 205 lb = ______ kg

2) 9,080 g = ______ lb.

3) 5' 10" = ______ inches = ______ cm = ______ m

4) 4 lb 8 oz. = ______ oz. = ______ g = ______ kg

5) 960 mL = ______ fl. oz. = ______ pt.

6) 1,920 mL = ______ qt.

7) 3 tsp = ______ mL = ______ Tbl = ______ f℥

8) 2 pt. = ______ cups = ______ oz. = ______ mL

9) 0.75 gal = ______ qt. = ______ pt. = ______ mL = ______ L

10) 7 lb 6 oz. = ______ kg

11) 6' 2" = ______ m

12) 1.5 cups = ______ cc

13) 5 L = ______ gal

14) 172 lb. = ______ kg

15) 1/3 tsp = ______ mL

Solve the following word problems.

16) If you purchased one-half ounce of papaverine hydrochloride and fill two prescriptions, one for 1.50 g and one for 0.125 oz., how many grams of papaverine hydrochloride do you have left in stock?

17) You're working in a pharmacy and receive the following prescription:

Rx Cefaclor 250 mg/5mL
Disp: 150 mL
Sig.: Take 375 mg by mouth two times a day for 10 days.

a) How many tsp should the patient take for each dose?

b) Did the physician write for you to dispense enough to cover a 10 day supply?

18) You're working in a pharmacy and receive the following prescription:

Rx Erythromycin 200 mg/5mL
Sig.: Take 200 mg by mouth three times a day for 5 days.

a) How many tsp should the patient take for each dose?

b) How many mL should you dispense to cover a 5 day supply?

The Apothecary System

Although fast becoming obsolete, the apothecary system for weighing and calculating pharmaceutical preparations is still used and must be taken into consideration. It has two divisions of measurement: weight and volume. In this system, the basic unit of weight is the grain, and the basic unit of volume is the minim. Some common conversions within the apothecary system are:

Weight

ounce (℥) = 8 drams = 480 grains
dram (ʒ) = 3 scruples = 60 grains
scruple (℈) = 20 grains
grain (gr) = 1 grain

Volume

fluid ounce (*f* ℥) = 8 fluid drams = 480 minims
fluid dram (*f* ʒ) = 3 fluid scruples = 60 minims
fluid scruple (*f* ℈) = 20 minims
minim (♍) = 1 minim
drop (gtt) = 1 minim

How apothecary units are written

In metric you write units of measurement in this order:

number of units + unit wanted

Examples

350 mg = three hundred fifty milligrams
1.8 m = one point eight meters

But in the apothecary system you write units of measurements in this order:

unit wanted + number of units (often written in Roman numerals)

Examples

gr ii = two grains
gtt xii = twelve drops

While this is the order in which these systems of measurement are usually written, you will see them in other orders.

Another item worth noting is that apothecary units frequently add in a Latin abbreviation with the Roman numerals to gain a subdivision between units. This Latin abbreviation is ss, which equals 0.5 and is often read as the fraction one-half.

Example

gr iiss = two and one-half grains

Practice Problems

Let's write the meaning or symbols of the given dosages:

1) f℥ iiiss
2) ♏ ix
3) five drams
4) three drops
5) three and one-half grain
6) two hundred fifty cubic centimeters

1) three and one-half fluid ounces 2) nine minims 3) ʒv 4) gtt iii 5) gr iiiss 6) 250 cc

Now, let's look at using dimensional analysis to make some conversions within the apothecary system.

Example

Convert ʒ iiiss to gr.

$$\frac{3.5\,drams}{1}\times\frac{60\,grains}{1\,dram}=\mathbf{210\,\mathit{grains}}\quad or\quad \mathbf{\mathit{gr\,ccx}}$$

Practice Problems

1) Convert fʒ xvi to f℥.

2) Convert 90 gr to ʒ.

1) f℥ ii 2) ʒ iss

Review

1) The measurement system dating back to the 11th century and still occasionally used in prescribing and administering medication is the __________ system.

2) Fluid dram and minim are apothecary units that measure __________.

3) Dram and grain are apothecary units that measure __________.

Provide the appropriate symbol for each unit.

4) ounce = __________
5) scruple = __________
6) grain = __________
7) fluid dram = __________
8) minim = __________
9) drop = __________

1) apothecary 2) volume 3) weight 4) ℥ 5) ℈ 6) gr 7) fʒ 8) ♏ 9) gtt

Conversions

Now for some tables with all the pertinent conversions needed for this chapter.

Apothecary	**Household**	**Metric**
Volume	Length	Length
1 ♏ = 1 gtt	12 in = 1 ft	100 cm = 1 m
60 ♏ = 1 *f* ʒ	Volume	Volume
8 *f* ʒ = 1 *f* ℥	3 tsp = 1 Tbs	1 cc = 1 ml
Weight	8 fl oz = 1 cup	1000 ml = 1 L
60 gr = 1 ʒ	16 fl oz = 1 pt	Weight
8 ʒ = 1 ℥	Weight	1000 ng = 1 mcg (µg)
	16 oz = 1 lb	1000 mcg (µg) = 1 mg
		1000 mg = 1g
		1000 g = 1 kg

	Apothecary	**Household**	**Metric**
Length		1 in.	2.54 cm
Volume	1 ♏ or 1 gtt		0.06 ml
	1 *f* ʒ		3.6 ml (often rounded all the way up to 5 mL)
		1 tsp	5 ml
		1 Tbs	15 ml
	1 *f* ℥		28.4 ml (usually rounded to 30 ml)
		1 *f* ℥	29.6 ml (usually rounded to 30 ml)
		1 pt	473 ml (usually rounded to 480 ml)
Weight	1 gr		64.8 mg (60 or 65 mg is frequently used)
		1 ℥	28.4 g (sometimes rounded to 30 g)
	1 ℥		31.1 g (sometimes rounded to 30 g)
		1 lb	454 g
		2.2 lb	1 kg

Something to pay careful attention to is how far many of the apothecary units are rounded. This can be difficult when trying to figure out how many milligrams of thyroid medication to dispense when a physician orders it in grains or how many milliliters to dispense of an antibiotic suspension when a physician doses it in fluid drams.

On the back of this page, we will look at some examples and practice problems.

Example

Convert ʒ iiiss to g.

$$\frac{3.5\,drams}{1} \times \frac{60\,grains}{1\,dram} \times \frac{64.8\,milligrams}{1\,grain} \times \frac{1\,gram}{1{,}000\,milligrams} = \mathbf{13.6\,grams}$$

or

$$\frac{3.5\,drams}{1} \times \frac{1\,apothecary\ ounce}{8\,drams} \times \frac{31.1\,grams}{1\,apothecary\ ounce} = \mathbf{13.6\,grams}$$

If you used rounded numbers you could have been as low as *12.6 grams* or as high as *13.7 grams*, which is something that can be difficult to keep track of when looking at the answers that others may achieve.

Practice Problems

1) If someone orders gr v of ferrous sulfate, may I give them a 300 mg tablet of ferrous sulfate?

2) If someone orders gr i of thyroid medication, may I give them a 65 mg tablet?

3) If I make 120 ml of GI Cocktail, will I be able to fit it in an amber vial marked 'iv *f*℥'?

1) Yes, but it requires you to do the calculation using 60 milligrams per grain. 2) Yes, but this time you need to do the calculation using 65 milligrams per grain. 3) Yes, remember that 1 fluid ounce is usually rounded to 30 milliliters.

Worksheet 7-3

Name:

Date:

Perform the appropriate conversions to solve the following problems and also write the appropriate symbol or abbreviation above each problem.

1) 1 ounce = _____ drams

2) 1 dram = _____ grains

3) 1 fluid ounce = _____ fluid drams

4) 1 fluid dram = _____ drops

5) 1 drop = _____ minim

6) 1 meter = _____ centimeters

7) 1 meter = _____ millimeters

8) 1 kilogram = _____ grams

9) 1 gram = _____ milligrams

10) 1 milligram = _____ micrograms

11) 1 microgram = _____ nanograms

12) 1 liter = _____ milliliters

13) 1 liter = _____ cubic centimeters

14) 1 cubic centimeter = _____ milliliter

15) 1 foot = _____ inches

16) 1 pound = _____ ounces

17) 1 gallon = _____ quarts

18) 1 quart = _____ pints

19) 1 pint = _____ cups

20) 1 pint = _____ fluid ounces

21) 1 cup = _____ ounces

22) 1 ounce = _____ tablespoons

23) 1 tablespoon = _____ teaspoons

24) 1 grain = _____ milligrams

25) 1 fluid ounce = _____ milliliters

26) 1 drop = _____ milliliters

27) 1 inch = _____ centimeters

28) 1 kilogram = _____ pounds

29) 1 pound = _____ grams

30) 1 ounce (household) = _____ grams

31) 1 pint = _____ milliliters

32) 1 tablespoon = _____ teaspoons

33) 1 teaspoon = _____ millilters

34) 3 teaspoons = _____ fluid ounces

35) 960 milliliters = _____ pints

36) 20 drops = _____ milliliters

37) 6 foot 1 inch = _____ centimeter

38) 165 pounds = _____ kilograms

39) 8 pounds 13 ounces = _____ kilograms

40) 7.5 milliliters = _____ teaspoons

41) 4 fluid drams = _____ milliliters

Worksheet 7-4

Name:

Date:

Solve the following problems, and know that when reviewing answers in class or with other students, some problems may have a small range of acceptable answers.

1) gr xvi = ʒ _____
2) 150 gr = ℥ _____
3) gr iii = _____ mg
4) fʒ viii = f℥ _____
5) f℥ ss = fʒ _____
6) 90 mL = f℥ _____
7) f℥ viii = _____ pt
8) fʒ ss = _____ gtt
9) fʒ iii = _____ mL
10) 120 mg = gr _____
11) ℥ ii = _____ g
12) 30 mL = f℥ _____
13) f℥ iv = _____ mL
14) 6 oz. = _____ g
15) 97.2 mg = gr _____
16) 60 mL = fʒ _____
17) 75 mL = f℥ _____
18) 15.5 g = ℥ _____
19) 195 mg = gr _____
20) 120 mL = f℥ _____
21) fʒ iv = _____ tsp
22) 125 mL = _____ tsp
23) 120 mL = _____ tsp
24) 240 mL = f ℥ _____
25) 15 mg = gr _____
26) gr xv = _____ mg
27) 10 mg = gr _____
28) 2.5 g = gr _____
29) 42 kg = _____ lb
30) 44 lb = _____ kg
31) f ℥ vi = _____ mL

Solve the following word problems.

32) During the total course of his treatment, a patient will receive 720 mL of medication. How many pints will he receive?

33) If an order calls for the patient to receive 2 tsp of cough syrup, how many milliliters of syrup should the patient receive?

34) A patient weighs 65 kg. How many pounds does she weigh?

35) A patient weighs 187 lbs. How many kilograms does he weigh?

36) A physician orders eight fluid drams of liquid medication per dose. How many tablespoons should the patient take?

37) A patient drinks 2 pt of liquid during the morning. How many milliliters did the patient drink?

38) An order is for gr iii of medication. How many milligrams should the patient be given?

39) A physician orders that a patient be given gr ss three times a day. How many milligrams will the patient receive in a day?

40) An order calls for 1.5 Tbs of medicated mouthwash. How many milliliters should the patient receive?

Worksheet 7-5

Name:

Date:

Complete the following equivalences.

1) 1 m = _____cm
2) 1 cm = _____mm
3) 1 L = _____ml
4) 1 m = _____mm
5) 1 cc = _____ml
6) 1 kg = _____g
7) f ℥ i = f ʒ_____
8) gtt i = ♍_____
9) 1 Tbs = _____tsp
10) 1 ft = ______in

Change the following measurements to the desired unit (feel free to use approximations for fluid ounces, as this is what you will commonly do in practice and when taking the Pharmacy Technician Certification Exam).

11) 3.6 m = ________mm
12) 47 cm = ________m
13) 9.4 cm = ________mm
14) 482 ml = ________L
15) 3.9 L = ________ml
16) 3.6 g = ________mg
17) 2 oz = ________g
18) 0.5 lb = ________g
19) 2 fl oz = ________cc
20) 5.08 cm = ________in
21) gr xc = ʒ________
22) gr xvss = ________g
23) 90 ml = ________Tbs
24) 5 tsp = ________ml
25) 3 tsp = ________Tbs
26) 120 lb = ________kg
27) 93 lb = ________kg
28) 3 lb 4.5 oz = ________g
29) 1 lb 5.2 oz = ________g
30) 5'11" = ________m
31) 36" = ________cm
32) ♍ iii = gtt________
33) ♍ xvii = ________ml

Select the best possible answer for the next two multiple choice problems.

34) 1/200 gr nitroglycerin SL = ________mg

You get a prescription for the above. Which product will you dispense?

a) Nitrostat 0.3 mg (nitroglycerin)
b) Nitrostat 0.4 mg (nitroglycerin)
c) Nitrostat 0.6 mg (nitroglycerin)
d) Nitrobid 2.5 mg capsules (nitroglycerin)

35) gr v ferrous gluconate = _______mg

A patient tells you her doctor wanted her to take one of these every day. Which product will you recommend?

a) ferrous sulfate 324 mg tablets
b) ferrous fumarate 324 mg tablets
c) ferrous gluconate 240 mg tablets
d) ferrous gluconate 325 mg tablets

Solve the following word problems.

36) A physician tells a patient to drink 2400 mL of fluid per day. How many cups of liquid should this patient drink?

37) Several months ago, a patient weighed 95 kg. When he comes in for his next appointment, he tells you that he lost 11 lbs. If he is correct, how many kilograms should he weigh?

Worksheet 7-6 (tougher problems)

Name:

Date:

Convert the following recipes to metric

1) The following is a historic formula for a cough ball to be used in a horse:

Rx	Ipecacuanha (powdered ipecac)	ʒ i
	Codeine sulfate	gr x
	Powdered digitalis	ʒ i
	Honey	qs to form 1 ball (which will weigh ℥ ss)

Hint: honey has a specific gravity of 1.425 g/mL according to the U.S.P.

2) The following recipe is similar to Vick's VapoRub

Rx	Camphor	ʒ $^{2}/_{3}$
	Eualyptol	gr x
	Menthol	ʒ ss
	Petrolatum	qs ℥ ii

3) The following is the real recipe for BC Powder

Rx	ASA	gr x
	Caffeine	gr ss
	Salicylamide	gr iii

4) The following is similar to the original Milk of Magnesia (MOM)

Rx		
	Magnesium Hydroxide Powder	ʒ x
	Sodium Hypochlorite Powder	ʒ i
	Purified Water	qs pt i

How many mg of magnesium hydroxide are in 2 tbs of this?

5) "BROMPTON'S COCKTAIL"

Rx		
	Morphine Powder	gr i
	Cocaine HCl	gr i
	Simple Syrup	*f* ℥ ss
	90% Ethanol	qs *f* ℥ ii

How many mg of morphine and how many mg of cocaine are in 1 tsp?

Worksheet 7-7

Name:

Date:

Complete the following equivalences.

1) 1 g = ________mg

2) 1 m = ________cm

3) 1 L = ________mL

Change the following measurements to the desired unit.

4) 5.8 m = ________cm

5) 92 cm = ________m

6) 3.7 L = ________mL

7) 247 mL = ________L

8) 4.6 L = ________mL

9) 1400 mg = ________g

10) 4.2 kg = ________g

11) 1.4 m = ________in

12) 5 oz = ________g

13) 0.6 lb = ________g

14) 5 qt = ________L

15) 3 fl oz = ________cc

16) 7.62 cm = ________in

17) 5 L = ________qt

18) 6.6 lb = ________kg

19) 450 g = ________oz

20) 150 fl oz = ________mL

21) 18 mL of water is ________cc of water.

Write the name or the abbreviation of the following apothecary and household units.

22) grain

23) ounce

24) fluid ounce

25) pint

26) tablespoon

27) tsp

28) gtt

29) qt

30) ʒ

31) ℥

32) ♏

33) f℥

Give the following relationships.

34) f℥ i = _______mL

35) ♏ i = ________mL

36) ℥ i = ________g

37) gtt i = ________mL

38) gr i = ________mg

39) gtt i = ♏________

40) 1 tsp = ________mL

41) 1 Tbs = ________tsp

Make these conversions.

42) gr xc = ʒ________

43) gr ss = ________mg

44) ʒ xvi = ℥ ________

45) f℥ iii = ʒ________

46) 150 ♏ = fʒ________

47) 0.5 pt = f℥ ________

48) 2 tsp = ________mL

49) gr iii = ________mg

50) 40 ♏ = ________mL

51) 10 g = gr ________

52) ʒ i = ________mg

CHAPTER 8

WORKING WITH PRESCRIPTIONS

A doctor is to give a speech at the local AMA dinner. He jots down notes for his speech. Unfortunately, when he stands in front of his colleagues later that night, he finds that he can't read his notes. So he asks, "Is there a pharmacist in the house?"

--Author Unknown

Have you ever seen a prescription and wondered, "What the heck does that mean?", and even thought, "That doesn't even look like English!". Now it is time to owe up to the truth...much of it is not English. Prescriptions have been obfuscated by a combination of Latin and English abbreviations (sometimes they even throw in Greek words). They are commonly used on prescriptions to communicate essential information on formulations, preparation, dosage regimens and administration of the medication. Our goal is to demystify this drug nomenclature. Our goals in this chapter include:

- learning common medical abbreviations,
- learning the parts of a prescription and how to incorporate medical abbreviations,
- and the additional prescription requirements and limitations when dealing with controlled substances.

Common Medical Abbreviations

In total there are nearly 20,000 medical abbreviations; instead of providing an exhaustive and meaningless list, we will present you with the most common medical abbreviations that are necessary for interpreting prescriptions and performing calculations.

There are several key things to point out about the tables on the next several pages.

categories – for ease of memorization, the abbreviations have been broken up into five categories: route, form, time, measurement, and other.

abbreviations – the abbreviations can often be written with or without the 'periods' and in upper or lower case letters (e.g., p.o. and PO both mean 'by mouth').

meaning – sometimes you will need to place an abbreviation in context to know its meaning (e.g., IV could mean a dosage form as in an 'IV bag', it could mean a route of administration as in 'to give a medication IV', or it could even be the roman numeral meaning 'four').

Latin root – not all the words on this list are derived from Latin words, nor is it necessary to know the Latin root words to be able to understand the abbreviations, but it is simply provided to help you understand how some of these abbreviations were derived.

Route

Abbreviation	*Meaning*	*Latin Root*
a.d.[1]	right ear	auris dexter
a.s.	left ear	auris sinister
a.u.	each ear	auris utro
IM	intramuscular	
IV	intravenous	
IVP	intravenous push	
IVPB	intravenous piggyback	
KVO	keep vein open	
n.g.t.	naso-gastric tube	
n.p.o.	nothing by mouth	nasquam per os
nare	nostril	
o.d.	right eye	oculus dexter
o.s.	left eye	oculus sinister
o.u.	each eye	oculus utro
per neb	by nebulizer	
p.o.	by mouth	per os
p.r.	rectally	per rectum
p.v.	vaginally	
SC, SQ	subcutaneously	
S.L.	sublingually (under the tongue)	
top.	topically	

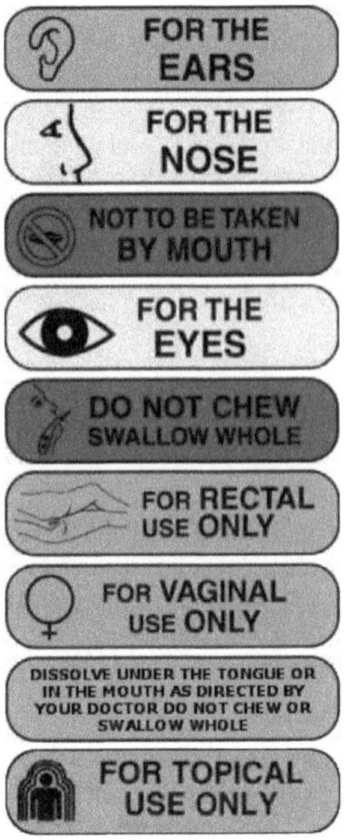

1 Always keep context in mind. In some prescriptions that require compounding, '*ad*' without the periods could mean *to* or *up to*.

Form

Abbreviation	*Meaning*	*Latin Root*
amp.	Ampule	
aq, aqua	water	aqua
caps	capsule	capsula
cm.[2]	cream	
elix.	elixir	
liq.	liquid	liquor
sol.	solution	
supp.	suppository	suppositorum
SR, XR, XL	slow/extended release	
syr.	syrup	syrupus
tab.	tablet	tabella
ung., oint	ointment	ungentum

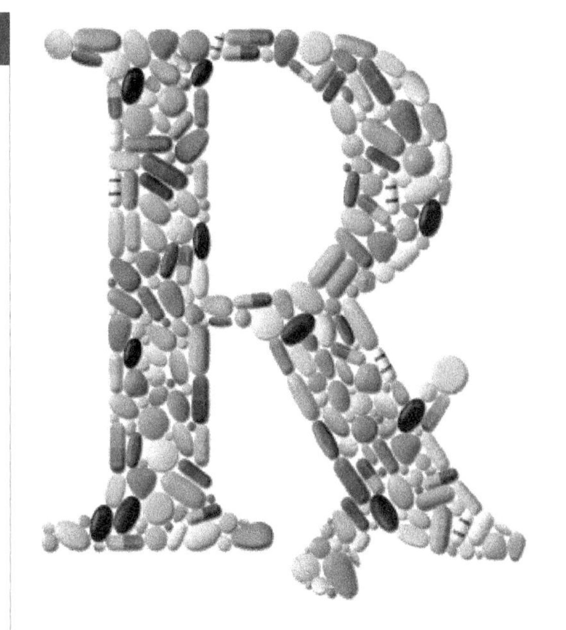

2 Always keep context in mind as '*cm*' can also mean centimeter.

Time

Abbreviation	*Meaning*	*Latin Root*
a.c.	before food, before meals	ante cibum
a.m.	morning	ante meridian
atc	around the clock	
b.i.d., bid	twice a day	bis in die
b.i.w., biw	twice a week	
h, °	hour	hora
h.s.	at bedtime	hora somni
p.c.	after meals	post cibum
p.m.	evening	post meridian
p.r.n., prn	as needed	pro re nata
q.i.d., qid	four times a day	quarter in die
q	each, every	quaque
q.d.	every day	quaque die
q_h, q_°	every__hour(s)	
qod	every other day	
stat	immediately	statim
t.i.d., tid	three times a day	ter in die
t.i.w., tiw	three times a week	

Measurement

Abbreviation	*Meaning*	*Latin Root*
i, ii, ...	one, two, etc.	
a.a., aa[3]	of each	ana
ad[4]	to, up to	ad
aq. ad	add water up to	
BSA	body surface area	
cc	cubic centimeter	
dil	dilute	dilutus
f, fl.	fluid	
fl. oz.	fluid ounce	
g, G, gm	gram	
gr.	grain	
gtt	drop(s)	guttae
l, L	liter	
mcg, µg	microgram	
mEq	milliequivalent	
mg	milligram	
ml, mL	milliliter	
q.s.	a sufficient quantity	quantum sufficiat
q.s. ad	add sufficient quantity to make	quantum sufficiat ad
ss[5]	one-half	
Tbs, T	tablespoon	
tsp, t	teaspoon	
U	unit	
>	greater than	
<	less than	

3 Always keep context in mind, as '*aa*' can also mean affected area when applying topical medications.
4 Always keep context in mind, as '*ad*' can also refer to the right ear.
5 Sometimes, it is easier to think of '*ss*' as meaning 0.5 instead of one-half.

Other

Abbreviation	*Meaning*	*Latin Root*
c	with	cum
disp.	dispense	
f, ft[6]	make, let it be made	fac, fiat, fiant
n/v	nausea and vomiting	
neb	nebulizer	
NR	no refill	
NS	normal saline	
s	without	sine
Sig	write, label	signatura
SOB	shortness of breath	
T.O.	telephone order	
ut dict, u.d.	as directed	ut dictum
V.O.	verbal order	

Practice Problems

Translate the following abbreviation statements to provide proper household directions.

1) i gtt ou bid x7d
2) i tab po q6h prn pain
3) i tab po qid pc
4) iss tsp po tid prn cough
5) iii gtt ad q4h x5d
6) i supp pr q4h prn n/v
7) i cap po tid ac + hs
8) i tab sl q5 minutes prn chest pain, may repeat up to 3 times.
9) ii tabs stat, then i tab po qid x10d

6 Be careful with this abbreviation, as '*f*' could also mean fluid and '*ft*' could also mean feet.

Worksheet 8-1

Name:

Date:

Your assignment for tonight is two fold. One, you must make a set of flash cards using the abbreviations presented to you on the preceding pages. Two, in the space provided below, you must make up ten abbreviation statements similar to the ones you just did, but this time **do not translate them**.

1)

2)

3)

4)

5)

6)

7)

8)

9)

10)

Learning the Parts of a Prescription and how to incorporate Medical Abbreviations

The word "prescription" stems from two Latin word parts, prae-, a prefix meaning before, and scribere, a word root meaning to write. Putting it all together, prescription means "to write before," which reflects the historical fact that a prescription traditionally had to be written before a drug could be mixed and administered to a patient.

Many ancient prescriptions were noted for their multiple ingredients and complexity of preparation. The importance of the prescription and the need for complete understanding and accuracy made it imperative that a universal and standard language be used. Thus, Latin was adopted, and its use was continued until approximately a generation ago.

Present day prescriptions are written in English, with doses usually being given in the metric system, but often you still find contracted Latin words and Roman numerals intertwined. The ancient "Rx" and the Latin "Signatura," abbreviated as Sig., the occasional Roman numeral, and a hand full of apothecary symbols are all that remain of the ancient art of the prescription.

Traditionally, a prescription is a written order for compounding, dispensing, and administering drugs to a specific client or patient and once it is signed by the physician it becomes a legal document! Prescriptions are required for all medications that require the supervision of a physician, those that must be controlled because they are addictive and carry the potential of being abused, and those that could cause health threats from side effects if taken incorrectly, for example, cardiac medications, controlled substances, and antibiotics.

The following is a list of the parts of a prescription, and in bold are the most significant portions:

- Patient Information
- **Superscription**
- **Inscription**
- **Subscription**
- **Signatura**
- Date
- Signature lines, signature, degree, generic substitution
- Prescriber information
- DEA# if required
- Refills
- Warnings

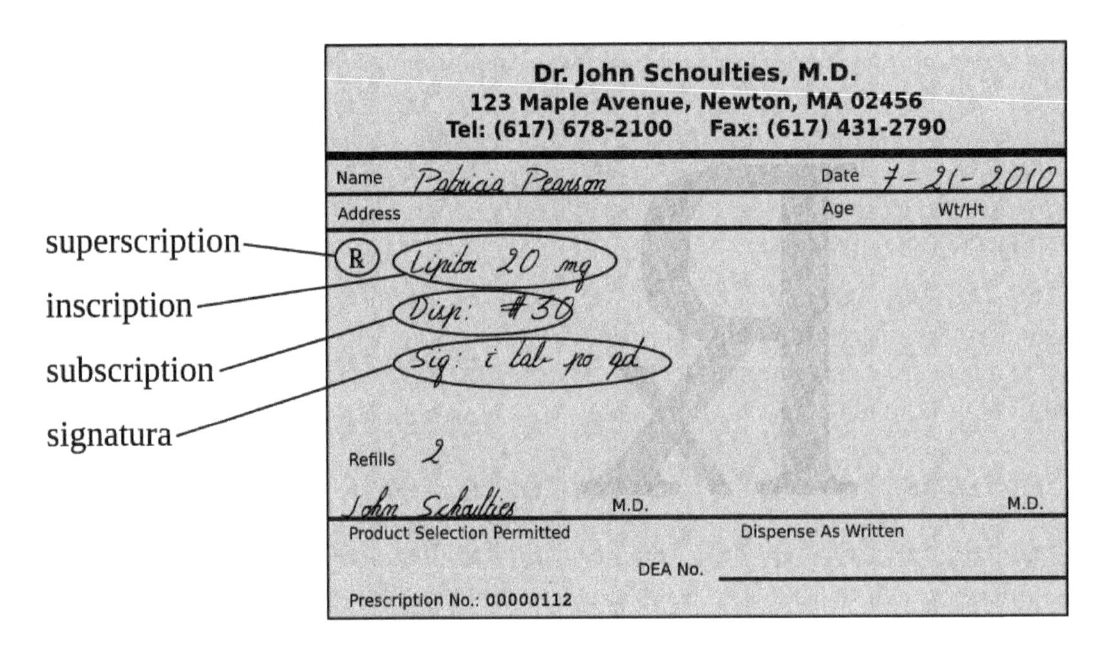

The **superscription** which consists of the heading where the symbol Rx (an abbreviation for recipe, the Latin for take thou) is found. The Rx symbol comes before the inscription.

The **inscription** is also called the body of the prescription, and provides the names and quantities of the chief ingredients of the prescription. Also in the inscription you find the dose and dosage form, such as tablet, suspension, capsule, syrup.

The **subscription**, which gives specific directions for the pharmacist on how to compound the medication. These directions to the pharmacist are usually expressed in contracted Latin or may consist of a short sentence such as: "make a solution," "mix and place into 10 capsules," or "dispense 10 tablets." However, that was in the old days. Today... doctors just name the pill!

The **signatura** (also called sig, or transcription), gives instructions to the patient on how, how much, when, and how long the drug is to be taken. These instructions are preceded by the symbol “S” or “Sig.” from the Latin, meaning "write" or "label." Whenever translating the signatura into instructions for a patient, begin it with an action verb such as take, inhale, spray, inject, place, swish, or whatever other verb seems appropriate for the medication.

Below the Sig line is room for special instructions, such as the number of times the prescription may be refilled, if any. You will also find the purpose of the prescription, special instructions, and warnings, followed by the signature of the prescriber.

You should also know and understand:

- The ***date*** and ***patient information***, which consists of the name of the party for whom it is designated and the address, usually occupies the upper part of the prescription. Sometimes age or weight is also added, though rarely.
- The instruction, "***take as directed***" is not satisfactory and should be avoided. The directions to the patient should include a reminder of the intended purpose of the medication by including such phrases as "for pain," "for relief of headache," or "to relieve itching"
- And if the patient is to receive a ***brand name medication***, rather then generic, the physician

enters NO SUBSTITUTIONS at the end of the prescription.

- If there are ***no refills*** to be dispensed, it is advisable not to enter the number 0, because it can be altered by adding numbers before the zero, thus making it a 10 to receive ten refills (or more!). Always write out the word *None*, or *No Refills*!!!
- The Drug Enforcement Administration (DEA) registration number system was implemented as a way to successfully ***track controlled substances*** from the time they are manufactured until the time they are dispensed to the patient.
- The ***DEA opposes use of the DEA number*** for other than its intended purpose, which is tracking controlled substances, and strongly opposes insurance company practice of requiring that a DEA number be placed on prescriptions for non-controlled substances.
- ***Not all medications require prescriptions***. There are certain medications on the market that can be purchased over the counter, thus their name, over-the-counter drugs (OTC.)

Now to put it all together, let's look at the previous example and translate the information on it:

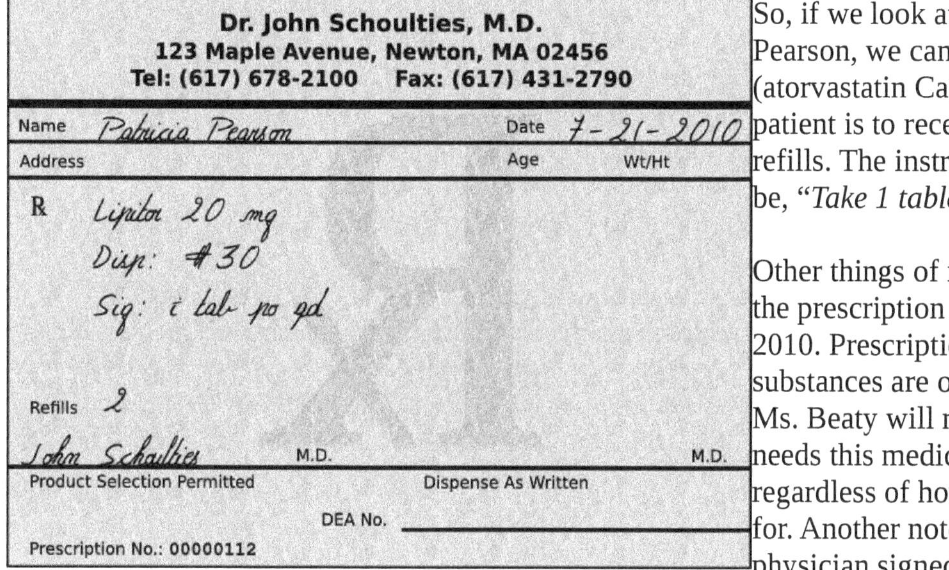

Dr. John Schoulties, M.D.
123 Maple Avenue, Newton, MA 02456
Tel: (617) 678-2100 Fax: (617) 431-2790

Name Patricia Pearson Date 7-21-2010
Address Age Wt/Ht

℞ Lipitor 20 mg
Disp: #30
Sig: i tab po qd

Refills 2

John Schoulties M.D. M.D.
Product Selection Permitted Dispense As Written

DEA No. ________________

Prescription No.: 00000112

So, if we look at this script for Patricia Pearson, we can see that it is for Lipitor (atorvastatin Ca) 20 mg tablets, and that the patient is to receive 30 of them with 2 refills. The instructions to the patient would be, "*Take 1 tablet by mouth daily*."

Other things of note include the date that the prescription is written for is July 21, 2010. Prescriptions for non-controlled substances are only good for one year, so Ms. Beaty will need a new script if she still needs this medication past July 21, 2011, regardless of how many refills were written for. Another noteworthy item is that the physician signed permitting product selection (*i.e.*, generic substitution). The last significant item on this label is that the physician did not include their DEA number. A DEA number should only be used for controlled substances.

This brings us to one last major concept in this chapter:

The Additional Prescription Requirements and Limitations when dealing with Controlled Substances

Besides over the counter medications (OTC) such as aspirin and ibuprofen, behind the counter medications (BTC) such as Allegra-D (fexofenadine with pseudoephedrine), and prescription medications (Rx legend) such as amoxicillin and digoxin, there is another group of medications to be concerned with called controlled substances. Controlled substances are medications with further restrictions due to abuse potential. There are 5 schedules of controlled substances with various prescribing guidelines based on abuse potential, as determined by the Drug Enforcement Administration and individual state legislative branches. Let's look at the table on the next page.

Schedule	Characteristics	Examples
CI	Unaccepted medical use Highest potential for abuse Not available by a prescription	Heroin and LSD
CII	High potential for abuse or misuse	Oxycodone, morphine, and amphetamines
CIII	Potential risk for abuse, misuse, and dependence	Tylenol with Codeine tablets and Vicodin
CIV	Low potential for abuse and limited risk of dependence	Phenobarbital, benzodiazepines, and other sedatives and hypnotics
CV	Low potential for abuse or misuse	Cough medicines that contain a limited amount of codeine, and antidiarrheal medications that contain a limited amount of an opiate such as Lomotil

- CI medications are not available via a prescription.
- CII medications may be written for a maximum 90 day supply excluding hospice patients. No refills are allowed on schedule II medications.
- CIII-IV medications may only be written for a 6 month supply.
- CV medications may be written for up to I year. Many states limit this to 6 months.

Many problems associated with drug abuse are the result of legitimately-manufactured controlled substances being diverted from their lawful purpose into the illicit drug traffic. Many of the narcotics, depressants and stimulants manufactured for legitimate medical use are subject to abuse, and have therefore been brought under legal control. The goal of controls is to ensure that these "controlled substances" are readily available for medical use, while preventing their distribution for illicit sale and abuse.

Under federal law, all businesses which manufacture or distribute controlled drugs, all health professionals entitled to dispense, administer or prescribe them, and all pharmacies entitled to fill prescriptions must register with the DEA. Authorized registrants receive a "DEA number". Registrants must comply with a series of regulatory requirements relating to drug security, records accountability, and adherence to standards. Any properly licensed medical professional that wishes to prescribe a controlled substance must include their DEA number on the prescription.

A physician's DEA number is a two letter seven digit number designed in such a way that a pharmacy can verify it via a mathematical algorithm. An example of a DEA number would be:

BP4567890

Let's go to the next several pages and practice thoroughly translating some prescriptions.

Worksheet 8-2

Name:

Date:

Translate the following prescriptions, and make note of anything that you find interesting. I will be breaking you into teams that are responsible for thorough translations on specific scripts. I've provided two examples of what I'm looking for.

Example 1:

Calvin J. Robins, M.D.
Contemporary Physician Group Practice
3459 5th Avenue, Pittsburgh, PA 15206
Tel: (412) 555-1234 Fax: (412) 555-2345

Name Margaret Adams Date 7-21-2010
Address Age Wt/Ht

℞ Nitrol 2% ung
Disp: i tube
Sig: apply 2" q8°

Refills 5

Calvin Robins M.D. M.D.
Product Selection Permitted Dispense As Written
DEA No.
Prescription No.: 00004001

This script for Margaret Adams is for one tube of Nitrol 2% ointment (nitroglycerin 2% ointment) and the patient is allowed 5 refills. The instructions to the patient would be, "*Apply 2 inches every 8 hours.*"

Things to note: This is interesting because NTG ung is usually measured in inches. A patient should know to rotate sites and apply to well cleaned areas that have minimal hair. Also, you should probably check with the physician to see if they want the patient to receive a nitrate free interval or not.

Example 2:

Donna Johns, M.D.
Contemporary Physician Group Practice
3459 5th Avenue, Pittsburgh, PA 15206
Tel: (412) 555-1234 Fax: (412) 555-2345

Name James Wilson Date 7-21-2010
Address Age Wt/Ht

℞ Compazine Supp 25 mg
#12
Sig: 1 pr q6h prn severe nausea

Refills NR

Donna Johns M.D. M.D.
Product Selection Permitted Dispense As Written
DEA No.
Prescription No.: 00005007

Mr. James Wilson's script is for twelve 25 mg Compazine (prochlorperazine) suppositories with no refills. The instructions to the patient would be "*Insert 1 suppository rectally every 6 hours as needed for severe nausea.*"

Things to note: Female patients may need to be informed to only use this suppository rectally as it will not have the correct systemic effects if given vaginally.

Prescription 1

David M. Ferguson, M.D.
Contemporary Physician Group Practice
3459 5th Avenue, Pittsburgh, PA 15206
Tel: (412) 555-1234 Fax: (412) 555-2345

Name *Okla Beaty* Date *7-21-2010*
Address Age Wt/Ht

℞ *Flonase Nasal Spray*
Disp: 1
Sig: i spray each nare qam

Refills *prn*

M.D. *David Ferguson* M.D.
Product Selection Permitted Dispense As Written
DEA No. ______
Prescription No.: 00000107

Prescription 2

David M. Ferguson, M.D.
Contemporary Physician Group Practice
3459 5th Avenue, Pittsburgh, PA 15206
Tel: (412) 555-1234 Fax: (412) 555-2345

Name *Okla Beaty* Date *7-21-2010*
Address Age Wt/Ht

℞ *Nitrostat 1/150 gr*
Disp: 25
Sig: i SL q5 min prn chest pain
may repeat X3

Refills *3*

David Ferguson M.D. M.D.
Product Selection Permitted Dispense As Written
DEA No. ______
Prescription No.: 00000108

Prescription 3

David M. Ferguson, M.D.
Contemporary Physician Group Practice
3459 5th Avenue, Pittsburgh, PA 15206
Tel: (412) 555-1234 Fax: (412) 555-2345

Name Okla Bealy Date 7-21-2010
Address Age Wt/Ht

℞ NitroDur 0.4 mg
Disp: #30
Sig: i patch on 8 a.m.,
off 10 p.m. qd

Refills 3

David Ferguson M.D. M.D.
Product Selection Permitted Dispense As Written

DEA No.

Prescription No.: 00000109

Prescription 4

David M. Ferguson, M.D.
Contemporary Physician Group Practice
3459 5th Avenue, Pittsburgh, PA 15206
Tel: (412) 555-1234 Fax: (412) 555-2345

Name Okla Bealy Date 7-21-2010
Address Age Wt/Ht

℞ Coumadin 5 mg
Disp: 1 month supply
Sig: ss tab on S-T-T-S,
i tab on M-W-F

Refills NR

David Ferguson M.D. M.D.
Product Selection Permitted Dispense As Written

DEA No.

Prescription No.: 00000110

Prescription 5

David M. Ferguson, M.D.
Contemporary Physician Group Practice
3459 5th Avenue, Pittsburgh, PA 15206
Tel: (412) 555-1234 Fax: (412) 555-2345

Name *Okla Beaty* Date *7-21-2010*

Address Age Wt/Ht

℞ *Spiriva*

Disp: #30

Sig: inhale i cap po qd

Refills *3*

David Ferguson M.D. M.D.

Product Selection Permitted Dispense As Written

DEA No. ______

Prescription No.: 00000111

Prescription 6

Dr. John Schoulties, M.D.
123 Maple Avenue, Newton, MA 02456
Tel: (617) 678-2100 Fax: (617) 431-2790

Name *Patricia Pearson* Date *7-21-2010*

Address Age Wt/Ht

℞ *Lipitor 10 mg #90*

i po qd

Refills *N/R*

John Schoulties M.D. M.D.

Product Selection Permitted Dispense As Written

DEA No. ______

Prescription No.: 00000212

Prescription 7

Dr. John Schoulties, M.D.
123 Maple Avenue, Newton, MA 02456
Tel: (617) 678-2100 Fax: (617) 431-2790

Name *Patricia Pearson* Date *7-21-2010*

Address Age Wt/Ht

℞ *Humulin R 10 ml*

Disp: 1 vial

8 u SC ā bkfst, 8 u ā lunch,

& 11 u ā supper

Refills *2*

John Schoulties M.D. M.D.

Product Selection Permitted Dispense As Written

DEA No. ____________

Prescription No.: 00000213

Prescription 8

Dr. John Schoulties, M.D.
123 Maple Avenue, Newton, MA 02456
Tel: (617) 678-2100 Fax: (617) 431-2790

Name *Patricia Pearson* Date *7-21-2010*

Address Age Wt/Ht

℞ *Novolin N 10 ml*

Disp: 1 vial

24 u SC qam

& 22 u SC qpm

Refills *5*

John Schoulties M.D. M.D.

Product Selection Permitted Dispense As Written

DEA No. ____________

Prescription No.: 00000214

Prescription 9

Dr. John Schoulties, M.D.
123 Maple Avenue, Newton, MA 02456
Tel: (617) 678-2100 Fax: (617) 431-2790

Name *Patricia Pearson* Date *7-21-2010*

Address Age Wt/Ht

℞ *Cardizem CD 240 mg #90*
i po qd

Refills *NR*

John Schoulties M.D. M.D.

Product Selection Permitted Dispense As Written

DEA No. ____________________

Prescription No.: 00000215

Prescription 10

Dr. John Schoulties, M.D.
123 Maple Avenue, Newton, MA 02456
Tel: (617) 678-2100 Fax: (617) 431-2790

Name *Patrick Pearson* Date *7-21-2010*

Address Age Wt/Ht

℞ *Hytrin 1 mg*
Disp: 1 month supply
Sig: 1 po qhs X3d
then, 2 qhs X5d
then, 4 qhs thereafter

Refills *2*

John Schoulties M.D. M.D.

Product Selection Permitted Dispense As Written

DEA No. ____________________

Prescription No.: 00000216

Prescription 11

Dr. John Smith, M.D.
739 Stockton Street, Waltham, MA 02454
Tel: (781) 333-2121 Fax: (781) 734-6340

Name *Richard Stallman* Date *7-21-2010*

Address Age Wt/Ht

℞ *Ambien 5 mg #30*
i po q hs prn sleep

Refills *one*

John Smith M.D. M.D.

Product Selection Permitted Dispense As Written

DEA No. *AS3456325*

Prescription No.: 00000317

Prescription 12

Dr. John Smith, M.D.
739 Stockton Street, Waltham, MA 02454
Tel: (781) 333-2121 Fax: (781) 734-6340

Name *Richard Stallman* Date *7-21-2010*

Address Age Wt/Ht

℞ *Adderall XR 25 mg*
Disp: #30
Sig: i po qd

Refills *NR*

John Smith M.D. M.D.

Product Selection Permitted Dispense As Written

DEA No. *AS3456325*

Prescription No.: 00000318

Prescription 13

Dr. John Smith, M.D.
739 Stockton Street, Waltham, MA 02454
Tel: (781) 333-2121 Fax: (781) 734-6340

Name *Richard Stallman* Date *7-21-2010*
Address Age Wt/Ht

℞ *Augmentin 400 mg/5 ml*
Disp: 100 cc
Sig: i tsp po q12° X10d

Refills *NR*

John Smith M.D. M.D.
Product Selection Permitted Dispense As Written

DEA No.

Prescription No.: 00000319

Prescription 14

Dr. John Smith, M.D.
739 Stockton Street, Waltham, MA 02454
Tel: (781) 333-2121 Fax: (781) 734-6340

Name *Richard Stallman* Date *1-3-2012*
Address Age Wt/Ht

℞ *Tobrex ophthalmic drops*
Sig: ii gtt OS q2h on day 1 and
ii gtt q4h on days 2 and 3
Call physician if eye infection persists

Refills *NR*

John Smith M.D. M.D.
Product Selection Permitted Dispense As Written

DEA No.

Prescription No.: 00000320

Prescription 15

Dr. John Smith, M.D.
739 Stockton Street, Waltham, MA 02454
Tel: (781) 333-2121 Fax: (781) 734-6340

Name *Richard Stallman* Date *7-21-2010*

Address Age Wt/Ht

℞ *Sinemet 25/100*

Disp: #180

Sig: ii PO TID

Refills *5*

John Smith M.D. M.D.

Product Selection Permitted Dispense As Written

DEA No. __________

Prescription No.: 00000322

Prescription 16

Dr. Andrew Yountz, M.D.
888 NW 27th Ave., Miami, FL 98885
Tel: (247) 555-6613 Fax: (247) 555-6340

Name *Barbara Erickson* Date *7-21-2010*

Address Age Wt/Ht

℞ *Imitrex 25 mg tab*

Disp: #9

Sig: i q6h prn migraine

Refills *6*

Andrew Yountz M.D. M.D.

Product Selection Permitted Dispense As Written

DEA No. __________

Prescription No.: 00006327

Prescription 17

Dr. Andrew Yountz, M.D.
888 NW 27th Ave., Miami, FL 98885
Tel: (247) 555-6613 Fax: (247) 555-6340

Name *Kurt Thomas* Date *7-21-2010*

Address Age Wt/Ht

℞ *Oph Sol: Trusopt 2%*

Sig: i gtt ou TID

Refills *6*

Andrew Yountz M.D. M.D.

Product Selection Permitted Dispense As Written

DEA No. ______________________

Prescription No.: **00006328**

Prescription 18

Dr. Andrew Yountz, M.D.
888 NW 27th Ave., Miami, FL 98885
Tel: (247) 555-6613 Fax: (247) 555-6340

Name *Tania Beltran* Date *4-19-2010*

Address Age Wt/Ht

℞ *Fosamax 70 mg*

Disp: #4

Sig: 1 tab weekly

Refills *PRN*

Andrew Yountz M.D. M.D.

Product Selection Permitted Dispense As Written

DEA No. ______________________

Prescription No.: **00006329**

ISMP's List of Error-Prone Abbreviations, Symbols, and Dose Designations

The abbreviations, symbols, and dose designations found on the following tables have been reported to the Institute for Safe Medication Practices (ISMP) through the ISMP Medication Error Reporting Program (MERP) as being frequently misinterpreted and involved in harmful medication errors. According to the ISMP, they should NEVER be used when communicating medical information. This includes internal communications, telephone/verbal prescriptions, computer-generated labels, labels for drug storage bins, medication administration records, as well as pharmacy and prescriber computer order entry screens. The truth is that all the items we are about to discuss ARE ACTUALLY USED and with that in mind we should look over these to help us not make errors in interpreting these abbreviations.

Abbreviations	*Intended Meaning*	*Misinterpretation*	*Correction*
µg	Microgram	Mistaken as "mg"	Use "mcg"
AD, AS, AU	Right ear, left ear, each ear	Mistaken as OD, OS, OU (right eye, left eye, each eye)	Use "right ear", "left ear", or "each ear"
OD, OS, OU	Right eye, left eye, each eye	Mistaken as AD, AS, AU (right ear, left ear, each ear)	Use "right eye", "left eye", or "each eye"
BT	Bedtime	Mistaken as "BID" (twice daily)	Use "bedtime"
cc	Cubic centimeter	Mistaken as "u" (units)	Use "mL"
D/C	Discharge or discontinue	Premature discontinuation of medications if D/C (intended to mean "discharge") has been misinterpreted as "discontinued" when followed by a list of discharge medications	Use "discharge" and "discontinue"
IJ	Injection	Mistaken as "IV" or "intrajugular"	Use "injection"
IN	Intranasal	Mistaken as "IM" or "IV"	Use "intranasal" or "NAS"
HS **hs**	Half-strength At bedtime, hours of sleep	Mistaken as bedtime Mistaken as half-strength	Use "half-strength" or "bedtime"
IU	International unit	Mistaken as IV (intravenous) or 10 (ten)	Use "units"
o.d. Or OD	Once daily	Mistaken as "right eye" (OD – oculus dexter), leading to oral medications administered in the eye	Use "daily"

Abbreviations	*Intended Meaning*	*Misinterpretation*	*Correction*
OJ	Orange juice	Mistaken as OD or OS (right or left eye); drugs meant to be diluted in orange juice may be given in the eye	Use "orange juice"
Per os	By mouth, orally	The "os" can be mistaken as "left eye" (OS – oculus sinister)	Use "PO," "by mouth," or "orally"
q.d. Or QD	Every day	Mistaken as q.i.d., especially if the period after the "q" or the tail of the "q" is misunderstood as an "i"	Use "daily"
qhs	Nightly at bedtime	Mistaken as "qhr" or every hour	Use "nightly"
qn	Nightly or at bedtime	Mistaken as "qh" (every hour)	Use "nightly" or "at bedtime"
q.o.d. or QOD	Every other day	Mistaken as "q.d." (daily) or "q.i.d." (four times daily) if the "o" is poorly written	Use "every other day"
q1d	Daily	Mistaken as q.i.d. (four times daily)	Use "daily"
q6PM, etc.	Every evening at 6 PM	Mistaken as every 6 hours	Use "daily at 6 PM" or "6 PM daily"
SC, SQ, sub q	Subcutaneous	SC mistaken as SL (sublingual); SQ mistaken as "5 every;" the "q" in "sub q" has been mistaken as "every"	Use "subcut" or "subcutaneously"
ss	Sliding scale (insulin) or ½ (apothecary)	Mistaken as "55"	Spell out "sliding scale;" use "one-half" or "½"
SSRI **SSI**	Sliding scale regular insulin Sliding scale insulin	Mistaken as selective-seritonin reuptake inhibitor Mistaken as Strong Solution of Iodine (Lugol's)	Spell out "sliding scale (insulin)"
i/d	One daily	Mistaken as "tid"	Use "1 daily"
TIW or tiw	3 times a week	Mistaken as "3 times a day" or "twice in a week"	Use "3 times weekly"

Abbreviations	*Intended Meaning*	*Misinterpretation*	*Correction*
U or u	Unit	Mistaken as the number 0 or 4, causing a 10-fold overdose or greater (e.g., 4U seen as "40" or 4u seen as "44"); mistaken as "cc" so dose given in volume instead of units (e.g., 4u seen as 4cc)	Use "unit"
UD	As directed ("ut dictum")	Mistaken as unit dose (e.g., diltiazem 125 mg IV infusion "UD" misinterpreted as meaning to give entire infusion as unit [bolus] dose)	Use "as directed"
Dose Designations and Other Information	***Intended Meaning***	***Misinterpretation***	***Correction***
Trailing zero after decimal point (e.g., 1.0 mg)	1 mg	Mistaken as 10 mg if the decimal point is not seen	Do not use trailing zero for doses expressed in whole numbers
No leading zero before a decimal point (e.g., .5 mg)	0.5 mg	Mistaken as 5 mg if the decimal point is not seen	Use zero before a decimal point when the dose is less than a whole unit
Drug name and dose run together (especially problematic for drug names that end in "l" such as Inderal40 mg; Tegretol300 mg)	Inderal 40 mg Tegretol 300 mg	Mistaken as Inderal 140 mg Mistaken as Tegretol 1300 mg	Place adequate space between the drug name, dose, and unit of measure

Dose Designations and Other Information	*Intended Meaning*	*Misinterpretation*	*Correction*
Numerical dose and unit of measure run together (e.g., 10mg, 100mL)	10 mg 100 mL	The "m" is sometimes mistaken as a zero or two zeros, risking a 10- to 100-fold overdose	Place adequate space between the drug name, dose, and unit of measure
Abbreviation such as mg. or mL. With a period following the abbreviation	mg mL	The period is unnecessary and could be mistaken as the number 1 if written poorly	Use mg, mL, etc. without a terminal period
Large doses without properly placed commas (e.g., 100000 units; 1000000 units)	100,000 units 1,000,000 units	100000 has been mistaken as 10,000 or 1,000,000; 1000000 has been mistaken as 100,000	Use commas for dosing units at or above 1,000, or use words such as 100 "thousand" or 1 "million" to improve readability
Drug Name Abbreviations	*Intended Meaning*	*Misinterpretation*	*Correction*
ARA A	vidarabine	Mistaken as cytarabine (ARA C)	Use complete drug name
AZT	zidovudine (Retrovir)	Mistaken as azathioprine or aztreonam	Use complete drug name
CPZ	Compazine (prochlorperazine)	Mistaken as chlorpromazine	Use complete drug name
DPT	Demerol-Phenergan-Thorazine	Mistaken as diptheria-pertusis-tetanus (vaccine)	Use complete drug name
DTO	Diluted tincture of opium, or deodorized tincture of opium (Paregoric)	Mistaken as tincture of opium	Use complete drug name
HCl	Hydrochloric acid or hydrochloride	Mistaken as potassium chloride (the "H" is misinterpreted as "K")	Use complete drug name unless expressed as salt of drug

Drug Name Abbreviations	Intended Meaning	Misinterpretation	Correction
HCT	hydrocortisone	Mistaken as hydrochlorothiazide	Use complete drug name
HCTZ	hydrochlorothiazide	Mistaken as hydrocortisone (seen as HCT250 mg)	Use complete drug name
$MgSO_4$	magnesium sulfate	Mistaken as morphine sulfate	Use complete drug name
MS, MSO_4	morphine sulfate	Mistaken as magnesium sulfate	Use complete drug name
MTX	methotrexate	Mistaken as mitoxantrone	Use complete drug name
PCA	procainamide	Mistaken as patient controlled analgesia	Use complete drug name
PTU	propylthiouracil	Mistaken as mercaptopurine	Use complete drug name
T3	Tylenol with codeine No. 3	Mistaken as liothyronine	Use complete drug name
TAC	triamcinolone	Mistaken as tetracaine, adrenalin, cocaine	Use complete drug name
TNK	TNKase	Mistaken as "TPA"	Use complete drug name
$ZnSO_4$	zinc sulfate	Mistaken as morphine sulfate	Use complete drug name
Stemmed Drug names	***Intended Meaning***	***Misinterpretation***	***Correction***
"Nitro" drip	Nitroglycerin infusion	Mistaken as sodium nitroprusside infusion	Use complete drug name
"Norflox"	norfloxacin	Mistaken as Norflex	Use complete drug name
"IV Vanc"	intravenous vancomycin	Mistaken as Invanz	Use complete drug name
Symbols	***Intended Meaning***	***Misinterpretation***	***Correction***
ʒ ♍	Dram Minim	Symbol for dram mistaken as "3" Symbol for minim mistaken as "mL"	Use metric system
x3d	For three days	Mistaken as "3 doses"	Use "for three days"

Symbols	*Intended Meaning*	*Misinterpretation*	*Correction*
> and <	Greater than and less than	Mistaken as opposite of intended of intended; mistakenly use incorrect symbol; “<10” mistaken as “40”	Use “greater than” or “less than”
/ (slash mark)	Separates two doses or indicates “per”	Mistaken as the number 1 (e.g., “25 units/10 units” misread as “25 units and 110 units”)	Use “per” rather than a slash mark to separate doses
@	At	Mistaken as “2”	Use “at”
&	And	Mistaken as “2, 3, 4, or 8”	Use “and”
+	Plus or and	Mistaken as “4”	Use “and”
°	Hour	Mistaken as a zero (e.g., q2° seen as q 20)	Use “hr,” “h,” or “hour”

Worksheet 8-3

Name:

Date:

Match the following abbreviations with their English translations.

Route

1) _____ a.d.	a. by mouth
2) _____ a.s.	b. by nebulizer
3) _____ a.u.	c. each ear
4) _____ IM	d. each eye
5) _____ IV	e. intramuscular
6) _____ IVP	f. intravenous
7) _____ IVPB	g. intravenous push
8) _____ KVO	h. intravenous piggyback
9) _____ n.g.t.	i. keep vein open
10) _____ n.p.o.	j. left ear
11) _____ nare	k. left eye
12) _____ o.d.	l. naso-gastric tube
13) _____ o.s.	m. nostril
14) _____ o.u.	n. nothing by mouth
15) _____ per neb	o. rectally
16) _____ p.o.	p. right ear
17) _____ p.r.	q. right eye
18) _____ p.v.	r. subcutaneously
19) _____ SC, SQ	s. sublingually
20) _____ S.L.	t. topically
21) _____ top.	u. vaginally

Form

22) _____ amp	a. ampule
23) _____ aq, aqua	b. capsule
24) _____ caps	c. cream
25) _____ cm.	d. elixir
26) _____ elix.	e. liquid
27) _____ liq.	f. ointment
28) _____ sol.	g. slow/extended release
29) _____ supp.	h. solution
30) _____ SR, XR, XL	i. suppository
31) _____ syr.	j. syrup
32) _____ tab.	k. tablet
33) _____ ung., oint	l. water

Measurement

34) _____ i, ii

35) _____ a.a. or aa

36) _____ ad

37) _____ aq. ad

38) _____ BSA

39) _____ cc

40) _____ dil.

41) _____ f, fl.

42) _____ fl. oz.

43) _____ g, G, gm

44) _____ gr.

45) _____ gtt

46) _____ l, L

47) _____ mcg, µg

48) _____ mEq.

49) _____ mg

50) _____ ml, mL

51) _____ q.s.

52) _____ q.s. ad

53) _____ ss

54) _____ tbsp., T

55) _____ tsp., t

56) _____ U

57) _____ >

58) _____ <

a. a sufficient quantity

b. add sufficient quantity to make

c. add water up to

d. body surface area

e. cubic centimeter

f. dilute

g. drops

h. fluid

i. fluid ounce

j. grain

k. gram

l. greater than

m. less than

n. liter

o. microgram

p. milliequivalent

q. milligram

r. milliliter

s. of each

t. one, two, etc

u. one-half

v. tablespoon

w. teaspoon

x. to, up to

y. unit

Time

59) _____ a.c.

60) _____ a.m.

61) _____ atc

62) _____ b.i.d., bid

63) _____ b.i.w., biw

64) _____ h

65) _____ h.s.

66) _____ p.c.

67) _____ p.m.

68) _____ p.r.n., prn

69) _____ q.i.d., qid

70) _____ q

71) _____ q.d.

72) _____ q_h

73) _____ qod

74) _____ stat

75) _____ t.i.d., tid

76) _____ t.i.w., tiw

a. after meals

b. around the clock

c. as needed

d. at bedtime

e. before food, before meals

f. each, every

g. evening

h. every __ hour(s)

i. every day

j. every other day

k. four times a day

l. hour

m. immediately

n. morning

o. three time a day

p. three times a week

q. twice a day

r. twice a week

Other

77) _____	c	a. as directed
78) _____	disp.	b. dispense
79) _____	f, ft.	c. make, let it be made
80) _____	neb	d. nausea and vomiting
81) _____	n/v	e. nebulizer
82) _____	NR	f. no refill
83) _____	NS	g. normal saline
84) _____	s	h. shortness of breath
85) _____	Sig.	i. telephone order
86) _____	SOB	j. verbal order
87) _____	T.O.	k. with
88) _____	ut dict., u.d.	l. without
89) _____	V.O.	m. write, label

Choose the best answer for the following multiple choice questions.

90) The directions for use of a medication are "gtt ii os bid." The route of administration is:

a) right eye
b) left eye
c) right ear
d) left ear

91) The directions for use of a medication are "Tylenol 80 mg pr q6h prn." What dosage form should be dispensed?

a) chew tab
b) syrup
c) suppository
d) enema

92) The directions for use are "Nitrostat 1/200 gr S.L. prn." How should this medication be administered?

a) in the left ear
b) very slowly
c) under the tongue
d) under the skin

93) Which of the following ways would be the best way for a physician to write a prescription for levothyroxine?

a) levothyroxine .1 mg qam
b) levothyroxine 0.100 mg qam
c) levothyroxine .100 mg qam
d) levothyroxine 0.1 mg qam

Answer the following questions.

94) Why is a physician supposed to avoid using the abbreviation "U" for units?

95) Why should physicians not use apothecary symbols when writing prescriptions?

96) Should you have a trailing zero after a decimal point? Why or why not?

97) Should you place a lead zero before a number that is less than one? Why or why not?

CHAPTER 9

BASIC MEDICATION CALCULATIONS

A student enters a pharmacy and asks, "Do you have a pill for math?" The pharmacist says "Wait just a moment", and goes back into the storeroom and brings back a whopper of a pill and plunks it on the counter. "I have to take that huge pill for math?" inquires the student. The pharmacist replies "Well, you know math always was a little hard to swallow."

--Unknown Author

This chapter is intended to provide an overview of some basic concepts involved in pharmacy math including:

- calculating dosages when giving medications in tablet or capsule form,
- calculating dosages when giving medications in liquid form, and
- percentage strength of solutions.

An important thing to keep in mind while doing this chapter is that many of the concepts taught in this section will still be applied in the coming weeks.

Calculating Dosages When Giving Medications in Tablet or Capsule Form

On a regular basis, you will need to figure out how many tablets or capsules to dispense. These problems are usually made easier with the “5 Step Method” and dimensional analysis. Key things to look for are: the quantity of medication per dose, the strength of the tablet, how often the doses are being given, and how long of a time frame we need to cover with these doses. You may need to include some conversion factors.

Without wanting to over explain the process, let's look at an example problem on the next page where we just want to find a number of tablets needed per dose.

Example

A 100 mg dose of medication is ordered, and the tablet size available is 25 mg. How many tablets will be needed per dose?

QUESTION

How many tablets will be needed per dose?

DATA

$$\frac{100\ mg}{dose} \qquad \frac{25\ mg}{tablet}$$

METHOD/FORMULA

dimensional analysis

DO THE MATH

$$\frac{100\ mg}{dose} \times \frac{tablet}{25\ mg} = \mathbf{4\ tablets/dose}$$

DOES THE ANSWER MAKE SENSE?

Yes

An important item to point out is that you need to be able to interpret the data in the word problem to pick out the necessary information. An example would be that when the problem says "A 100 mg dose of medication is ordered....." the ability to express it mathematically as 100 mg/dose. Another thing to point out is that when you are using dimensional analysis you need to line up your units the correct way. In the above problem, we need tablets on top and dose on the bottom, which was why we needed to 'flip' 25 mg/tablet so that tablet was on top.

Using what we've just discussed, let's attempt another practice problem.

Practice Problem

The ordered medication is gr xxx/dose. The available capsules are gr v. How many capsules are needed per dose?

QUESTION

DATA

MATHEMATICAL METHOD / FORMULA

DO THE MATH

DOES THE ANSWER MAKE SENSE?

6 capsules/dose

Provided that the practice problem on the previous page made sense, let's do a few more practice problems before we expand on this idea.

Practice Problems

Calculate the number of capsules or tablets needed per dose.

1) 300 mg ordered; 50 mg/capsule available

2) gr xvi ordered; gr iv/tab available

3) gr ii ordered; gr ss/tab available

4) 250 mg ordered; 125 mg/cap available

5) gr v ordered; 100 mg/cap available

1) 6 capsules/dose 2) 4 tablets/dose 3) 4 tablets/dose 4) 2 capsules/dose 5) 3 capsules/dose

Besides just figuring out how many tablets or capsules we will need for a particular dose, we will need to look at how many we need to cover a particular time frame in order for us to provide a sufficient quantity for the patient. Most institutional settings provide medications in 24 hour increments, although this may vary between various institutions.

Example

If gr v t.i.d. is ordered and 100 mg capsules are available, then what is the total number of capsules needed per day?

QUESTION

What is the total number of capsules needed per day?

DATA

$$\frac{5\ gr}{dose} \qquad \frac{3\ doses}{day} \qquad \frac{100\ mg}{cap}$$

And you will need a conversion factor for grains to milligrams, which is either:

$$\frac{60\ mg}{gr} \quad or \quad \frac{65\ mg}{gr}$$

MATHEMATICAL METHOD / FORMULA

dimensional analysis

DO THE MATH

$$\frac{5\ gr}{dose} \times \frac{3\ doses}{day} \times \frac{60\ mg}{gr} \times \frac{cap}{100\ mg} = \mathbf{9\ \textit{capsules/ day}}$$

DOES THE ANSWER MAKE SENSE

Yes

In the above scenario you needed to use 60 mg/gr. If you had used 65 mg/gr you would have ended up with 9.75 capsules. While tablets can be cut in half (and sometimes in quarters), capsules can only be given in whole numbers; therefore 60 mg/gr works out much better in this problem.

Now, using what we've just discussed, let's attempt a practice problem on the next page using the "5 Step Method".

Practice Problems

A physician orders 90 mg q.4h. of a medication; and you have gr ss/cap in stock on this medication. What is the total number of capsules needed per day?

QUESTION

DATA

MATHEMATICAL METHOD / FORMULA

DO THE MATH

DOES THE ANSWER MAKE SENSE?

18 capsules/day

Provided that the above practice problem made sense, let's do a few more practice problems.

Practice Problems

Calculate the number of capsules or tablets needed per day.

1) 0.4 g q.2h ordered; 200 mg/tab available

2) 60 mg b.i.d. ordered; gr ss/cap available

3) 150 mg q.h. ordered; 0.3 g/tab available

4) gr ss t.i.d. ordered; 65 mg/tab available

5) 600 mg q.8h. ordered; 50 mg/tab available

1) 24 tablets/day 2) 4 capsules/day 3) 12 tablets/day 4) 1.5 tablets/day 5) 36 tablets/day

Worksheet 9-1

Name:

Date:

Calculate the number of capsules or tablets needed for the following doses.

1) If 200 mg/dose of a medication is ordered and it is available as 50 mg/cap, how many caps are needed per dose?

2) If gr iv/dose of a medication is ordered and it is available as gr ss/tab, how many tabs are needed per dose?

3) If 150 mg/dose of a medication is ordered and it is available as 25 mg/tab, how many tabs are needed per dose?

4) If gr viii/dose of a medication is ordered and it is available as gr ii/cap, how many caps are needed per dose?

5) If 75 mg/dose of a medication is ordered and it is available as 150 mg/tab, how many tabs are needed per dose?

6) If 500 mg/dose of a medication is ordered and it is available as 125 mg/cap, how many caps are needed per dose?

7) If 400 mg/dose of a medication is ordered and it is available as 0.1 g/tab, how many tabs are needed per dose?

8) If 30 gr/dose of a medication is ordered and it is available as 6 gr/tab, how many tabs are needed per dose?

9) If 0.2 g/dose of a medication is ordered and it is available as 100 mg/tab, how many tabs are needed per dose?

10) If gr ii/dose of a medication is ordered and it is available as 60 mg/tab, how many tabs are needed per dose?

Calculate the number of capsules or tablets needed over 24 hours.

11) If 50 mg/dose of a medication is ordered t.i.d. and it is available as 25 mg/tab, how many tabs are needed per day?

12) If gr xx/dose of a medication is ordered q.4h. and it is available as gr v/tab, how many tabs are needed per day?

13) If 10 mg/dose of a medication is ordered q.4h. and it is available as 5 mg/tab, how many tabs are needed per day?

14) If 100 mg/dose of a medication is ordered b.i.d. and it is available as 25 mg/cap, how many caps are needed per day?

15) If gr xvi/dose of a medication is ordered q.3h. and it is available as gr viii/cap, how many caps are needed per day?

16) If 10 mg/dose of a medication is ordered q.h. and it is available as 5 mg/tab, how many tabs are needed per day?

17) If 0.2 g/dose of a medication is ordered q.4h. and it is available as 100 mg/tab, how many tabs are needed per day?

18) If gr ss/dose of a medication is ordered b.i.d. and it is available as 60 mg/tab, how many tabs are needed per day?

19) If 100 mg/dose of a medication is ordered b.i.d. and it is available as 0.1 g/tab, how many tabs are needed per day?

20) If gr ii/dose of a medication is ordered q.8h. and it is available as 60 mg/tab, how many tabs are needed per day?

Calculate the number of capsules or tablets needed to fill the following prescriptions.

21) Rx Ambien 5 mg/tab
Disp: 14 day supply
Sig: 1 tab po hs

22) Rx furosemide 20 mg/tab
Disp: 7 day supply
Sig: 1 tab po bid

23) Rx alprazolam 0.5 mg/tab
Disp: 7 day supply
Sig: 1 tab po tid

24) Rx Lopressor 25 mg/tab
Disp: 30 day supply
Sig: 12.5 mg po bid

25) Rx zafirlukast 20 mg/tab
Disp: 7 day supply
Sig: 1 tab bid

26) Rx tetracycline 250 mg/cap
Disp: 3 day supply
Sig: 1 cap qid

27) Rx repaglinide 0.5 mg/tab
Disp: 21 day supply
Sig: 1 tab po tid

28) Rx dipyridamole 50 mg
Disp: 21 day supply
Sig: 1 tab po qid

29) Rx zidovudine 100 mg/cap
Disp: 90 day supply
Sig: 300 mg po q12h

Calculating Dosages When Giving Medications in Liquid Form

Frequently you will need to transfer solutions from manufacturers' containers to patient specific containers, which will contain just enough solution for the patient's needs. These problems are usually made easier with the "5 Step Method" and dimensional analysis or ratio proportions. Key things to look for are: the quantity of medication per dose, the concentration of the liquid, how often the doses are being given, and how long of a time-frame we need to cover with these doses. Just like in the previous section, you may need to include some conversion factors.

Without wanting to over explain the process, let's look at an example problem where we just want to find out how many mL need to be withdrawn from a vial.

Example

An order for 50 mg of a drug is received. A 10 mL vial with 100 mg/mL is available. How many mL should be withdrawn from the vial?

QUESTION

How many mL should be withdrawn from the vial?

DATA

$$\frac{50\ mg}{dose} \quad \frac{10\ mL}{vial} \quad \frac{100\ mg}{mL}$$

MATHEMATICAL METHOD / FORMULA

Dimensional Analysis or Ratio Proportion

DO THE MATH

dimensional analysis

$$\frac{50\ mg}{dose} \times \frac{mL}{100\ mg} = \mathbf{0.5\ mL/\ dose}$$

ratio proportion

$$\frac{50\ mg}{N} = \frac{100\ mg}{mL}$$

$$N = \mathbf{0.5\ mL}$$

DOES THE ANSWER MAKE SENSE?

Yes

Two things worth noting are: the fact that you had more information than was needed to solve the problem (the 10 mL vial was not necessary information to do the math) and that you could have solved this with either dimensional analysis or ratio proportion. Using the above problem as a template, let's attempt the practice problem on the next page using the "5 Step Method".

Practice Problem

A physician orders azithromycin 500 mg IV. In the pharmacy you have 250 mg/5 mL in stock. Calculate the number of mL required to fill this order.

QUESTION

DATA

MATHEMATICAL METHOD / FORMULA

DO THE MATH

DOES THE ANSWER MAKE SENSE?

10 mL

Provided that the above practice problem made sense, let's do a few more practice problems before we expand on this idea.

Practice Problems

Assume that all of the solutions below are in 10 mL vials. Calculate the amount of solution to be withdrawn from the vial.

1) 20 mg is ordered; 50 mg/mL is available.

2) 60 mg is ordered; 40 mg/mL is available.

3) 80 mg is ordered; 50 mg/mL is available.

4) 40 mg is ordered; 50 mg/mL is available.

5) 75 mg of drug ordered; 100 mg/cc available

1) 0.4 mL 2) 1.5 mL 3) 1.6 mL 4) 0.8 mL 5) 0.75 cc

Sometimes drugs are ordered in units. A unit is a measurement for the amount of a substance, based on measured biological activity or effect. The unit is used for vitamins, hormones, some medications, vaccines, blood products, and similar biologically active substances. Calculating a dosage in units is no different than what we have already been doing with other labels (mg, gr, etc.). Let's take a moment and look at an example problem using units.

Example

8,000 units of a drug is ordered. A 10 cc vial containing 10,000 units/cc is available. How many cc should be withdrawn from the vial?

QUESTION

How many cc should be withdrawn from the vial?

DATA

$$\frac{8{,}000\ units}{dose} \quad \frac{10\ cc}{vial} \quad \frac{10{,}000\ units}{cc}$$

MATHEMATICAL METHOD / FORMULA

Dimensional Analysis or Ratio Proportion

DO THE MATH

dimensional analysis

$$\frac{8{,}000\ units}{dose} \times \frac{cc}{10{,}000\ units} = \mathbf{0.8\ cc/dose}$$

ratio proportion

$$\frac{8{,}000\ units}{N} = \frac{10{,}000\ units}{cc}$$

$$N = \mathbf{0.8\ cc}$$

DOES THE ANSWER MAKE SENSE?

Yes

Just like the previous problems, the above problem was able to be solved two different ways. Using the above problem as a template, let's attempt a few practice problems below.

Practice Problems

Assume that all of the solutions below are in 10 mL vials. Calculate the amount of solution to be withdrawn from the vial.

1) 6000 units ordered; 5000 units/cc is available.

2) 300,000 units ordered; 200,000 units/mL is available.

3) 500,000 units ordered; 1,000,000 units/mL is available.

1) 1.2 cc 2) 1.5 mL 3) 0.5 mL

Sometimes medications will come as a lyophylized powder to provide longer stability and/or easier shipping, and they need to be reconstituted prior to patient use. Let's look at an example problem to demonstrate this concept.

Example

300,000 units is to be administered from a vial containing 1,000,000 units reconstituted with 5 cc of diluent. The reconstituted vial contains 200,000 units/cc. How many cc should be withdrawn from the reconstituted vial?

QUESTION

How many cc should be withdrawn from the reconstituted vial?

DATA

$$\frac{300{,}000\ units}{dose} \quad \frac{200{,}000\ units}{cc} \quad \frac{1{,}000{,}000\ units}{vial} \quad reconstituted\ with\ 5\ cc$$

MATHEMATICAL METHOD / FORMULA

Dimensional Analysis or Ratio Proportion

DO THE MATH

dimensional analysis

$$\frac{300{,}000\ units}{dose} \times \frac{cc}{200{,}000\ units} = \mathbf{1.5\ cc/dose}$$

ratio proportion

$$\frac{300{,}000\ units}{N} = \frac{200{,}000\ units}{cc}$$

$$N = \mathbf{1.5\ cc}$$

DOES THE ANSWER MAKE SENSE?

Yes

Once again, you had unnecessary information and multiple ways to choose to solve these problems. Using the above problem as a template, let's attempt a couple of practice problems below.

Practice Problems

1) 200,000 units is to be administered from a vial containing 500,000 units reconstituted with 5 cc of diluent. The reconstituted vial now contains 100,000 units/cc. How many cc should be withdrawn from the reconstituted vial?

2) 750 mg of methylprednisolone injection is ordered. A vial containing 1000 mg is reconstituted with 7.4 mL of sterile water for injection and has 0.6 ml of powder volume. The final concentration in the vial is 125 mg/mL. How many mL should be withdrawn from the reconstituted vial?

1) 2 cc 2) 6 mL

Worksheet 9-2

Name:

Date:

Calculate the quantity of volume that needs to be withdrawn from the following vials to fill their orders.

1) 75 mg of drug ordered; a 10 cc vial with a concentration of 100 mg/cc is available.

2) 40 mg of drug ordered; a 10 cc vial with a concentration of 100 mg/cc is available.

3) 50 mg of drug ordered; a 10 cc vial with a concentration of 100 mg/cc is available.

4) 20 mg of drug ordered; a 10 cc vial with a concentration of 50 mg/cc is available.

5) 60 mg of drug ordered; a 10 cc vial with a concentration of 40 mg/cc is available.

6) 75 mg of drug ordered; a 10 cc vial with a concentration of 40 mg/cc is available.

7) 80 mg of drug ordered; a 10 cc vial with a concentration of 50 mg/cc is available.

8) 40 mg of drug ordered; a 10 cc vial with a concentration of 50 mg/cc is available.

9) A physician orders 37.5 mg of methotrexate, and the 2 mL stock vial in the pharmacy has a concentration of 25 mg/mL. How many mL will you need to dispense to provide the appropriate dose?

10) A physician orders 1050 mg of fluorouracil, and the stock vial in the pharmacy has a concentration of 50 mg/mL. How many mL will you need to dispense to provide the appropriate dose?

11) A physician orders 162.5 mg of methotrexate, and the 10 mL stock vials in the pharmacy have a concentration of 25 mg/mL. How many mL will you need to dispense to provide the appropriate dose?

12) 6,000 units of drug ordered; a 10 cc vial with a concentration of 5,000 units/cc is available.

13) 12,500 units of drug ordered; a 10 cc vial with a concentration of 5,000 units/cc is available.

14) 300,000 units of drug ordered; a 10 cc vial with a concentration of 200,000 units/cc is available.

15) 500,000 units of drug ordered; a 10 cc vial with a concentration of 1,000,000 units/cc is available.

16) A TPN requires the addition of 15 units of regular insulin. A 10 mL vial with a concentration of 100 units/mL is available. How many mL of insulin should be added to the TPN?

17) A diabetic patient with insulin resistance is to receive a very concentrated form of regular insulin. Its concentration is 500 units/mL and it comes in a 20 mL vial. How many mL will the patient receive for a 140 unit dose of concentrated regular insulin?

18) You receive an order for heparin 12,000 units to be added to an IV bag. If the concentration of the heparin available is 5,000 units/mL, how many mL of heparin should you use?

19) 200,000 units are ordered ; a vial containing 1,000,000 units is reconstituted with 10 cc of diluent. The reconstituted solution now contains 100,000 units/cc.

20) 300,000 units are ordered; a vial containing 2,000,000 units is reconstituted with 10 cc of diluent. The reconstituted solution now contains 200,000 units/cc.

21) 50,000 units are ordered; a vial containing 1,000,000 units is reconstituted with 10 cc of diluent. The reconstituted solution now contains 100,000 units/cc.

22) 750,000 units are ordered; a vial containing 1,000,000 units is reconstituted with 5 cc of diluent. The reconstituted solution now contains 200,000 units/cc.

23) 150,000 units are ordered; a vial containing 2,000,000 units is reconstituted with 10 cc of diluent. The reconstituted solution now contains 200,000 units/cc.

24) An order is written for 4,000,000 units of penicillin G potassium. You have a vial containing 20,000,000 units. The directions are to add 32 mL of sterile water for injection to reconstitute to a concentration of 500,000 units/mL (the vial contains 8 mL of powder volume). How many mL of the reconstituted solution will you need to dispense?

25) An oncologist orders 850 mg of gemcitabine for a patient. The pharmacy has 1 gram vials of gemcitabine in stock. The reconstitution instructions suggest using 25 mL of diluent to obtain a concentration of 38 mg/mL (the vial contains approximately 1.3 mL of powder volume). How many mL of the reconstituted solution will be required to prepare the patient's dose of gemcitabine?

Percentage Strength Solutions

There are many ways to express the concentration of a drug; one of the most common is percentage strength (*hint: remember that "percent" means "per 100"*) .

There are three kinds of percentage strength that you will frequently use when doing dosage calculations:

> **weight/weight** (w/w) – examples include ointments, creams, etc.
> **volume/volume** (v/v) – a common example is an alcohol preparation
> **weight/volume** (w/v) – this is the most common group and includes items such as solutions, suspensions, etc.

Weight/Weight (w/w)

A weight/weight percentage strength expresses the number of parts in 100 parts of a preparation. Typically it is expressed as number of grams per 100 grams, but it could be expressed in any unit of weight (grams, grains, ounces, pounds, etc.) as long as the units on the top and bottom match. Let's look at an example to help this make more sense.

Example

What would be the weight, expressed in grams, of zinc oxide (zinc oxide is the active ingredient) in 120 grams of a 10% zinc oxide ointment (the ointment is the total mixture)?

QUESTION

How many grams of zinc oxide are in the ointment?

DATA

$120\ g\ of\ ointment$ $\quad 10\% = \frac{10\ g\ zinc\ oxide}{100\ g\ ointment}$

MATHEMATICAL METHOD / FORMULA

Ratio Proportion (This can be done other ways, but this is the easiest method to explain thoroughly.)

DO THE MATH

$$\frac{N}{120\ g\ ointment} = \frac{10\ g\ zinc\ oxide}{100\ g\ ointment}$$

$$\boldsymbol{N = 12\ g\ zinc\ oxide}$$

DOES THE ANSWER MAKE SENSE?

Yes

Let's look at a couple of practice problems based on w/w percentage strength.

Practice Problem 1

What would be the percentage strength of a zinc oxide ointment if you prepared 90 grams of an ointment that contained 5 grams of zinc oxide? *(Hint: In this problem you have the weight of the active ingredient and the weight of the mixture. You need to find the percentage strength which would be out of 100 grams of mixture.)*

QUESTION

DATA

MATHEMATICAL METHOD / FORMULA

DO THE MATH

DOES THE ANSWER MAKE SENSE?

5.6% zinc oxide ointment

Now, let's look at another w/w practice problem.

Practice Problem 2

If you were to prepare 150 g of a coal tar ointment containing 12 g of coal tar, what is the percent strength of coal tar in the ointment?

QUESTION

DATA

MATHEMATICAL METHOD / FORMULA

DO THE MATH

DOES THE ANSWER MAKE SENSE?

8% coal tar ointment

Volume/Volume (v/v)

Volume/volume percentage strength problems are worked out in a similar manner to w/w percentage strength problems, except now the ingredients are liquids. Typically it is expressed as number of milliliters per 100 milliliters, but it could be expressed in any unit of volume (liters, pints, fluid ounces, etc.) as long as the units on the top and bottom match. Look at the example problem below, and then complete the practice problem.

Example

How mL of isopropyl alcohol are in a 480 ml bottle of 70% isopropyl alcohol?

QUESTION

How many mL of isopropyl alcohol are in the bottle?

DATA

$480\ mL\ of\ mixture$ $\quad 70\%\, isopropyl\ alcohol = \frac{70\ mL\ isopropyl\ alcohol}{100\ mL\ mixture}$

MATHEMATICAL METHOD / FORMULA

Ratio Proportion (This can be done other ways, but this is the easiest method to explain thoroughly.)

DO THE MATH

$$\frac{N}{480\ mL\ mixture} = \frac{70\ mL\ isopropyl\ alcohol}{100\ mL\ mixture}$$

$\boldsymbol{N = 336\ mL\ isopropyl\ alcohol}$

DOES THE ANSWER MAKE SENSE?

Yes

Practice Problem

How many mL of lysol would be required to make 4000 mL of a 2% lysol solution?

QUESTION

DATA

MATHEMATICAL METHOD / FORMULA

DO THE MATH

DOES THE ANSWER MAKE SENSE?

80 mL of lysol

Weight/Volume (w/v)

Weight/volume (w/v) percentage strengths are the most common percentages worked with in pharmacy. The units in this type of problem are **always** grams of drug dissolved in 100 milliliters of solution. Let's look at a practice problem.

Example

How many grams of sodium chloride are in a 500 mL bag of 0.9% sodium chloride?

QUESTION

How many g of sodium chloride are in the bag?

DATA

$500\ mL\ of\ mixture$ $\quad 0.9\%\,sodium\ chloride = \frac{0.9\ g\ sodium\ chloride}{100\ mL\ mixture}$

MATHEMATICAL METHOD / FORMULA

Ratio Proportion

DO THE MATH

$$\frac{N}{500\ mL\ mixture} = \frac{0.9\ g\ sodium\ chloride}{100\ mL\ mixture}$$

$$N = \mathbf{4.5\ g\ sodium\ chloride}$$

DOES THE ANSWER MAKE SENSE?

Yes

As w/v percentage strength problems are the ones we will be working with most frequently, (and what we are going to concentrate on in this chapter), it is worth pointing out that our ratio proportions for w/v percentage strength problems always have grams on top and milliliters on the bottom. Let's look at a practice problem.

Practice Problem

How many grams of neomycin are needed to prepare 500 mL of a 1% neomycin solution?

QUESTION

DATA

MATHEMATICAL METHOD / FORMULA

DO THE MATH

DOES THE ANSWER MAKE SENSE?

5 g of neomycin

Let's look at some additional practice problems.

Practice Problems

1) You dissolve 10 g of mannitol in 50 mL of water. What is the percentage of mannitol?

2) How much solution can you make from 25 g of drug if you want a 10% concentration?

3) How much drug is in 50 cc of a 1% solution?

4) How many cc of a 20% solution can be made with 5 g of drug?

1) 20% 2) 250 mL 3) 0.5 g 4) 25 cc

Worksheet 9-3

Name:

Date:

Make the calculations needed to prepare these solutions.

1) What weight of drug is in 10 cc of a 15% solution?

2) A 100 mL solution contains 10 g of a drug. What is the percent concentration of this solution?

3) How many cc of a 30% solution can be made with 15 g of drug?

4) What weight of drug is in 15 mL of a 10% solution?

5) A 200 cc solution contains 20 g of a drug. What is the percent concentration of this solution?

6) How many cc of a 10% solution can be made with 15 g of a drug?

7) What weight of drug is in 20 mL of an 8% solution?

8) What weight of drug is in 50 cc of a 1% solution?

9) A 200 cc solution contains 40 g of a drug. What is the percent concentration of this solution?

10) A 300 mL solution contains 3 g of a drug. What is the percent concentration of this solution?

11) How many cc of a 5% solution can be made with 10 g of a drug?

12) How many mL of a 2% solution can be made with 6 g of a drug?

13) How many cc of a 10% solution can be made with 5 g of a drug?

14) What weight of drug is in 30 mL of a 15% solution?

15) A solution of 80 cc contains 50 g of a drug. What is the percent concentration of this solution?

Worksheet 9-4

Name:

Date:

Solve the following practical percentage strength problems. *Hint: All the problems on this worksheet are w/v percentage strength problems.*

1) You have a patient with severe renal impairment and the physician wants him to receive a TPN with 10 g of amino acid. How many mL of a 5.2% amino acid (Aminosyn-RF) solution will you need to add to the TPN?

2) You have an order for 1 g of calcium gluconate IV. It is available in a 10 mL vial with a 10% concentration. How many mL will you need to draw up?

3) You have a 10 mL vial of 50% magnesium sulfate, and you receive an order for 4 g of magnesium sulfate IV. How many mL will you need to draw up?

4) You receive an order for 2500 mg of magnesium sulfate IV. How many mL would you withdraw from a 10 mL vial of 50% magnesium sulfate?

5) A CRNA intern needs to administer 100 mg of tetracaine. The vial he has on his block cart in the OR has a concentration of 1%. How many mL should you tell him to administer?

6) You added 15.4 mL of a 14.6% sodium chloride solution to a 1000 mL bag of sterile water for injection. How many grams of sodium chloride are in the final bag?

7) A patient on an ICU unit was suffering from calcium channel blocker toxicity so a physician had a nurse give 10 cc of 10% calcium chloride injection stat. The nurse is asking you how many grams of calcium chloride she should chart as having given the patient.

8) How many mL of a 0.025% ophthalmic solution could be made with 1 mg of active ingredient?

9) A nurse hung a 500 mL IV bag with 20% mannitol, but only wants to infuse 50 g. How many mL should you tell her to infuse?

10) A patient needs 750 mg of oral potassium chloride solution. If the oral potassium chloride solution has a concentration of 10%, how many tablespoons of solution should the patient receive for this dose?

Worksheet 9-5

Name:

Date:

Calculate the number of capsules or tablets needed per dose.

1) 150 mg of drug ordered; 75 mg/tab available

2) 0.3 g/dose ordered; 100 mg/tab available

3) gr viii/dose ordered; 240 mg/tab available

4) 150 mg of drug ordered; 100 mg/tab available

5) 0.2 g/dose ordered; 200 mg/cap available

Calculate the total number of capsules or tablets needed over 24 hours.

6) If 0.4 g/dose of a medication is ordered q.i.d. and it is available as 200 mg/tab, how many tabs are needed per day?

7) If 120 mg/dose of a medication is ordered b.i.d. and it is available as gr iv/tab, how many tabs are needed per day?

8) If 0.1 g/dose of a medication is ordered q.6h. and it is available as 50 mg/cap, how many caps are needed per day?

9) If gr iiss/dose of a medication is ordered t.i.d. and it is available as gr ss/tab, how many tabs are needed per day?

10) If 240 mg/dose of a medication is ordered h.s. and it is available as gr ii/cap, how many caps are needed per day?

Calculate the amount of solution to be withdrawn from a 10 cc vial.

11) 60 mg of drug ordered; 100 mg/cc available

12) 40 mg of drug ordered; 100 mg/cc available

13) 60 mg of drug ordered; 40 mg/cc available

14) 80,000 units are ordered. A vial containing 1,000,000 units is reconstituted with 10 cc of diluent. The reconstituted solution now contains 100,000 units/cc.

15) 50,000 units are ordered. A vial containing 200,000 units is reconstituted with 10 cc of diluent. The reconstituted solution now contains 20,000 units/cc.

Solve the following (*w/v*) percent strength problems.

16) What weight of drug is in 10 mL of a 8% solution?

17) A 500 cc solution contains 15 g of a drug. What is the percent concentration of this solution?

18) How many mL of a 20% solution can be made with 18 g of drug?

19) What weight of drug is in 5 mL of a 10% solution?

20) A 150 mL solution contains 15 g of a drug. What is the percent concentration of this solution?

Chapter 10

Basic Infusion Calculations

Who wouldn't want to solve an alligation, after all it's more like playing Tic-Tac-Toe than math.
--Sean Parsons

Now the math starts to get really exciting as we look at calculations involving infusions. We will look at five major concepts in this section:

- diluting stock solutions,
- infusion rates,
- dosages based on body weight,
- body surface area, and
- pediatric dosing.

Diluting Stock Solutions

A stock solution is a concentrated solution from which less-concentrated solutions can be made. The stock solution is diluted with a solvent (sometimes also referred to as a diluent), which may be water or some other liquid substance. Two questions must be answered to solve a dilution problem.

1) What volume of the stock solution must be diluted to make the ordered solution?
2) What volume of solvent must be added to perform the dilution?

These questions can be answered using ratio-proportions, the dilution formula, or the alligation method . We will review all three possible methods.

Solving dilutions using the ratio-proportion method.

We have previously learned how to solve problems using the ratio-proportion method, but in order to solve dilution problems you will actually need to perform two ratio-proportions. Let's look at an example problem on the next page to demonstrate how this works.

Example

If 500 mL of a 5% solution is ordered, how much of a 25% stock solution is needed to prepare the 5% solution?

QUESTION

How much stock solution is needed (we are looking for a volume)?

DATA

final volume is 500 mL — final concentration is 5% and $5\% = \frac{5\ g}{100\ mL}$

stock volume is ??? — stock concentration is 25% and $25\% = \frac{25\ g}{100\ mL}$

MATHEMATICAL METHOD / FORMULA

ratio-proportion

DO THE MATH

First, we need to figure out how much active ingredient is in our final solution:

$$\frac{5\ g}{100\ mL} = \frac{N}{500\ mL}$$

When you solve for *N* you will find:

$$N = 25\ g$$

Which means we will need to figure out how much volume of the stock solution is required to provide 25 g of drug.

$$\frac{25\ g}{100\ mL} = \frac{25\ g}{N}$$

When you solve for *N* you will find:

$N = \mathbf{100\ mL}$ ***of stock solution is required.***

DOES THE ANSWER MAKE SENSE?

Yes

Even though there is an extra problem involved, the ratio-proportion method is convenient because it is building on skills you have previously learned. Looking again at the problem we just solved, you will often also need to know how much solvent or diluent is required. In this scenario, since we know what our final volume is, and what volume of stock solution is required we can logically ascertain that our final volume minus our stock volume will equal how much diluent is required:

500 mL – 100 mL = 400 mL of diluent

Try the practice problem on the next page to reinforce these concepts.

Practice Problem

A 500 mL bag of 20% mannitol is ordered. 50% mannitol is the available stock solution. How much of the stock solution is needed to make this bag?

QUESTION

DATA

MATHEMATICAL METHOD / FORMULA

DO THE MATH

DOES THE ANSWER MAKE SENSE?

200 mL of stock solution

Now that you've solved that, also figure out how much diluent would be required to prepare the above IV bag.

300 mL of diluent

You will have more opportunities to practice this problem solving method shortly, but first I would like to introduce two new problem solving methods: the dilution formula and the alligation method. We will look at the dilution formula first.

Solving dilutions using the dilution formula.

The dilution formula is exactly what it sounds like, a formula to make it quicker to solve dilutions. Without any further ado, let's concentrate our attention on the formula below:

The Dilution Formula

$$C_1Q_1 = C_2Q_2$$

C_1 = concentration of available stock solution
Q_1 = quantity of available stock solution needed
C_2 = concentration of ordered solution
Q_2 = quantity of ordered solution needed

Let's look at our previous example and solve it using the dilution formula.

Example

If 500 mL of a 5% solution is ordered, how much of a 25% stock solution is needed to prepare the 5% solution?

QUESTION

How much stock solution is needed (we are looking for a quantity)?

DATA

C_1 = stock concentration is 25%
Q_1 = stock quantity is ???
C_2 = final concentration is 5%
Q_2 = final quantity is 500 mL

MATHEMATICAL METHOD / FORMULA

dilution formula

DO THE MATH

$C_1Q_1 = C_2Q_2$

$(25\%)(Q_1)=(5\%)(500\ mL)$

To solve for Q_1 we will need to get it by itself (isolate it) by dividing both side by 25%.

$$\frac{(\cancel{25\%})(Q_1)}{\cancel{25\%}}=\frac{(5\%)(500\ mL)}{25\%}$$

$Q_1=\mathbf{100\ mL}$ ***of stock solution***

DOES THE ANSWER MAKE SENSE?

Yes, especially since it is the same answer we derived using the ratio-proportion method.

And if we needed to solve for the quantity of diluent required, we would use the same logic as last time, our final volume minus our stock volume will equal how much diluent is required:

500 mL – 100 mL = 400 mL of diluent

Try the practice problem on the following page to reinforce this concept.

Practice Problem

You are instructed to make 1000 cc of a 0.8% solution. You have in stock a 95% solution. How much of the stock solution will you use?

QUESTION

DATA

MATHEMATICAL METHOD / FORMULA

DO THE MATH

DOES THE ANSWER MAKE SENSE?

8.4 cc of stock solution

Now that you've solved that, also figure out how much diluent would be required to make the above solution.

991.6 cc of diluent

Solving dilutions using the alligation method.

The dilution formula is a very useful tool as it is quick and easy, but it will not work in certain circumstances, such as when no diluent is being used but instead multiple stock solutions of varying concentration are being mixed together. One tool that might be useful then is the alligation method. Also, depending on what information is present, the alligation method can also be useful for solving many dilution problems.

Below, we have a diagram explaining the alligation method. Based on the appearance of an alligation, you can see why it is sometimes referred to as the Tic-Tac-Toe method.

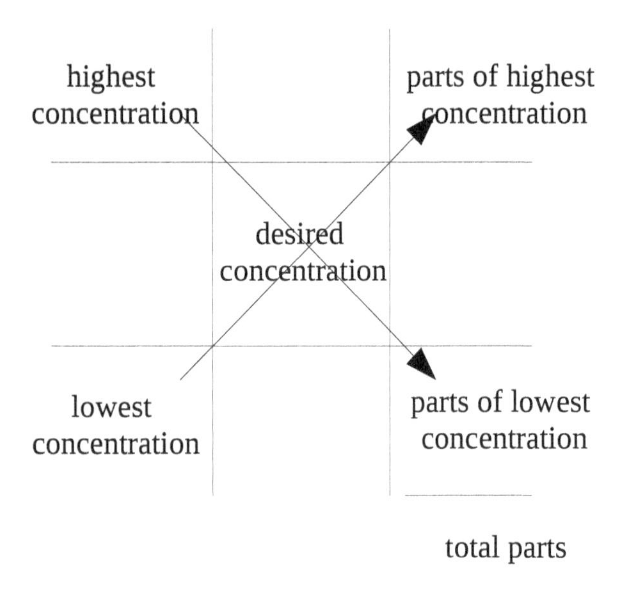

- Place the highest concentration in the upper left-hand corner
- Place the lowest concentration in the lower left-hand corner
- Place the desired concentration in the center
- Find the difference between the highest concentration and the desired concentration to find the parts of lowest concentration
- Find the difference between the lowest concentration and the desired concentration to find the parts of highest concentration
- Add the parts of highest concentration and the parts of lowest concentration to find the total parts
- This provides you with a ratio that you can use to finish solving the problem.

The alligation method looks very abstract at first, but becomes much easier when we start using real numbers. Let's attempt the same example problem we've previously used with both the ratio-proportion method and the dilution formula on the next page.

Example

If 500 mL of a 5% solution is ordered, how much of a 25% stock solution is needed to prepare the 5% solution? How much diluent is needed?

QUESTION

How much stock solution is needed (we are looking for a quantity)?
How much diluent is needed (we are looking for a quantity)?

DATA

high concentration = 25%
desired concentration = 5%
low concentration = 0% (the diluent has no drug in it, therefore it has a 0% concentration)
final quantity = 500 mL

MATHEMATICAL METHOD / FORMULA

alligation method

DO THE MATH

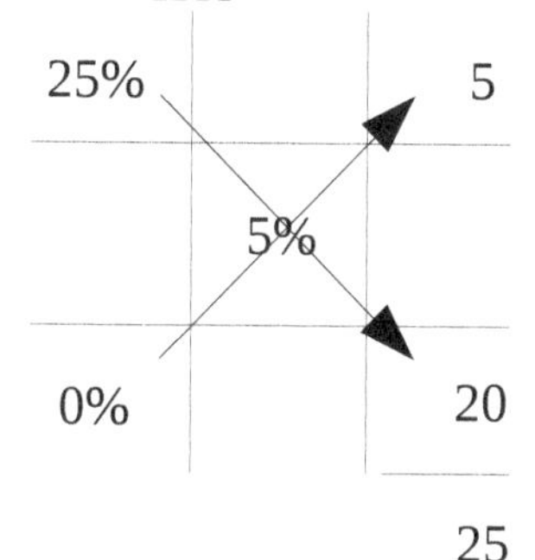

$\frac{5}{25} \times 500\ mL = \mathbf{100\ mL\ of\ 25\%\ solution.}$

$\frac{20}{25} \times 500\ mL = \mathbf{400\ mL\ of\ diluent.}$

DOES THE ANSWER MAKE SENSE?

Yes, although we may need to look at some more practice problems to help this method make sense.

Let's look to the next page and attempt another problem using this same method.

Practice Problem

200 cc of a 15% solution is ordered. A 30% solution is available. How much stock solution is needed, and how much solvent must be added to prepare the 15% solution?

QUESTION

DATA

MATHEMATICAL METHOD / FORMULA

DO THE MATH

DOES THE ANSWER MAKE SENSE?

100 cc of a 30% solution; 100 cc of solvent

Now, let's try a few more practice problems, but use whichever method(s) you wish to perform the necessary calculations. The answers are on the next page.

1) 1000 cc of a 2% solution is needed. A 40% stock solution is available. How much stock solution is needed, and how much solvent must be added to prepare the 2% solution?

2) Respiratory needs you to make 3 mL of 3% sodium chloride. Your stock solution has a concentration of 14.6% sodium chloride. How many mL of stock solution will you need and how much sterile water will you need to add?

3) You are instructed to make 240 mL of a 0.45% solution. You have a 100% stock solution. How much of the stock solution will you use, and how much diluent will be needed?

4) Prepare 3 mL of a 1% phenobarbital solution from a 6.5% stock solution. How much stock solution and how much diluent are needed to make the 1% solution?

1) 50 cc of stock solution; 950 cc of solvent 2) 0.62 mL of stock solution; 2.38 mL of sterile water 3) 1.08 mL of stock solution; 238.92 mL of diluent 4) 0.46 mL of stock solution; 2.54 mL of diluent

There will be times when only one of the three methods may be appropriate for solving a particular dilution or alligation, but in this section most of the problems will be written in such a way that any of the three methods (ratio-proportion, dilution formula, or alligation method) will be viable options for solving the problems.

Worksheet 10-1

Name:

Date:

Solve the following dilution problems.

1) 40 cc of a 10% solution is ordered. A 50% stock solution is available. How much stock solution is needed, and how much solvent must be added to prepare the 10% solution?

2) 1000 cc of a 10% solution is ordered. How much of a 25% stock solution is needed to prepare the 10% solution? How much solvent must be added to prepare the 10% solution?

3) 500 mL of a 20% solution is ordered. How much of a 50% stock solution is needed to prepare the 20% solution? How much solvent must be added to prepare the 20% solution?

4) 500 cc of a 2% solution is needed. A 40% stock solution is available. How much stock solution is needed and how much solvent must be added to prepare the 2% solution?

5) 200 cc of an 8% solution is ordered. A 20% stock solution is available. How much stock solution is needed, and how much solvent must be added to prepare the 8% solution?

6) 20 cc of a 10% solution is ordered. A 40% stock solution is available. How much stock solution is needed, and how much solvent must be added to prepare the 10% solution?

7) A physician makes a special request for a 500 mL bag of 3% sodium chloride (hypertonic saline solution). How much 14.6% sodium chloride and how much sterile water for injection will you need to fulfill his request?

8) You receive an order for a 250 mL bag with 20% mannitol. You have 50% stock bottles of mannitol. How many mL of mannitol and how many mL of diluent will you need to make this order?

9) A physician orders 8 fl. oz. of a 1% povidone-iodine wash. You have a 10% povidone-iodine wash in stock. How many mL of stock solution and how many mL of diluent will you need to prepare the physician's order?

10) A physician orders a liter of 0.25% sodium hypochlorite solution (often referred to as half-strength Dakin's Solution). On hand in the pharmacy is a 5.95% stock solution of sodium hypochlorite. How many mL of stock solution and how many mL of diluent will you need to prepare the physician's order?

Problem 11 can only be solved by doing three ratio-proportions.

11) If you mix 100 mL of a 1% solution with 350 mL of a 0.5% solution, what is the percentage strength of the final solution?

Problem 12 can only be solved using the alligation method.

12) You have on hand 70% dextrose stock solution and 40% dextrose stock solution. You are to prepare 1000 mL of 45% dextrose solution. How many mL of each stock solution will you need?

Infusion Rates

Infusion rates can be requested in many different ways:

- Infuse at 125 mL/hr
- Infuse 1000 mL over 8 hours
- Infuse 10 mg per minute
- Infuse at a drip rate of 32 gtt/min

So when discussing parenterals, we can define an infusion rate as a quantity of drug per a quantity of time:

$$\frac{\textit{Quantity of Drug}}{\textit{Time}} = \textit{Infusion Rate}$$

When looking at infusion rates, let's start with drip rates. In many modern settings, pumps will be used to infuse IV solutions, but sometimes when there are equipment failures, or times of excessively high census, IVs may need to be timed with an old fashioned drip chamber (also called a venoclysis set). You can use factor label to solve these kinds of problems to find your rate of flow. A drip rate is a specific kind of infusion rate; it is the quantity of drops being infused every minute.

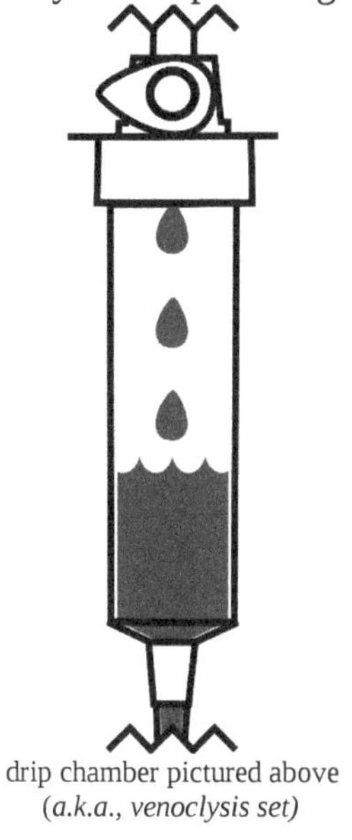

drip chamber pictured above
(a.k.a., venoclysis set)

$$\textit{drip rate} = \frac{\textit{quantity of drops}}{\textit{minute}}$$

Something to keep in mind is that whenever measuring a drip rate, you must use a whole number of drops as you can not cut a drop in half while it is falling. Therefore, you should use general rounding rules. If something calculated out to be 15.5 drops/minute, you would set the drip rate to be 16 drops/minute, and if another drip rate worked out to be 33.3 drops/minute, you would set the drip rate at 33 drops/minute. Another point to make is that the drip rate should seem reasonable, such as you could have a drip rate of 20 drops/minute, but not 2,000 drops per minute as you can't accurately count drops that fast.

In order to solve these problems, we also need to introduce a new term, drop factor. A drop factor, simply put, is the number of drops that add up to 1 cc; and it is important to note that various administration sets will produce different sizes of drops. So if a particular administration set had a drop factor of 15, it would mean that you would have to count 15 drops in the drip chamber to equal 1 cc.

$$\boldsymbol{drop\ factor\ of\ 15 = \frac{15\ drops}{1\ cc}}$$

Common drop factors for standard tubing include 10,15, and 20 drops/mL. Microdrip tubing sets have a drop factor of 60 drops/mL. So, let's look at an example problem where we are solving for the drip rate. You will find that dimensional analysis tends to be a useful method for solving these problems.

Example

A 1000 cc bag of NS is set to run for 5 hours. The infusion set has a drop factor of 20. What is the drip rate?

QUESTION

What is the drip rate?

DATA

1000 cc   5 hours   drop factor of 20 = 20 gtt/cc

a potentially useful conversion:

1 hour/60 minute

MATHEMATICAL METHOD / FORMULA

dimensional analysis

DO THE MATH

$$\frac{1000\,cc}{5\,hours} \times \frac{20\,gtt}{cc} \times \frac{1\,hour}{60\,minutes} = 66.67\,gtt/minute = \mathbf{67\ gtt/minute}$$

DOES THE ANSWER MAKE SENSE?

Yes

Notice how we can conveniently divide our amount of solution by the quantity of time it is being infused over, and then just start canceling everything out until we arrive at the units we are looking for. Let's use the above example as a template to attempt a practice problem on the next page.

Practice Problem

An I.V. bottle containing 500 cc is set to run for 3 hours with a drop factor of 15 gtt/cc. What is the drip rate?

QUESTION

DATA

MATHEMATICAL METHOD / FORMULA

DO THE MATH

DOES THE ANSWER MAKE SENSE?

Now, lets do a few more practice problems.

42 gtt/minute

Practice Problems

1) An IV bottle containing 1000 cc of solution is set to run 5 hours with a drop factor of 10 gtt/cc. What is the drip rate?

2) Calculate the drip rate for an I.V. solution containing 2000 cc of solution with a drop factor of 15 set to run 15 hours.

3) An I.V. solution of 1200 cc is set to run 8 hours with a drop factor of 20. What is the drip rate?

4) An I.V. solution of 500 cc is using a microdrip tubing set to run over 10 hours with a drop factor of 60. What is the drip rate?

1) 33 gtt/minute 2) 33 gtt/minute 3) 50 gtt/minute 4) 50 gtt/minute

Continuing with the concept of infusion rates, let's look at some other infusion rate examples:

1) If a 1 liter bag of D5W is run through an IV over eight hours, what is the rate of infusion in mL/hr?

$$\frac{1\,L}{8\,hours} \times \frac{1000\,mL}{1\,L} = \mathbf{125\,mL/hr}$$

2) An order is for Heparin IV to infuse at 1000 units/hr. What will be the flow rate in mL/hr for a 500 mL bag of D5W with 25,000 units of heparin?

$$\frac{1000\,units}{1\,hour} \times \frac{500\,mL}{25{,}000\,units} = \mathbf{20\,mL/hr}$$

3) If a 1 liter bag of D5W is started on a patient at 1400 hours on Tuesday, when will the next bag be needed if it is running at a rate of 50 mL/hr?

$$\frac{1\,hour}{50\,mL} \times \frac{1000\,mL}{1} = 20\,hours$$

This means the next bag won't be needed till **1000 on Wednesday**.

Now, lets do a few more practice problems.

Practice Problems

1) If a 500 mL bag of 0.9% NaCl is run over 8 hours, what is the rate of infusion in mL/hr?

2) A patient is on a heparin drip, 12,500 units in 250 mL of half-normal saline. He is to receive 1500 units per hour. At what rate (mL/hr) should the drug be infused?

3) If a 1 liter bag of NSS is started on a patient at 1200 hours on Tuesday, when will the next bag be needed if it is running at a rate of 100 mL/hr?

1) 62.5 mL/hr 2) 30 mL/hr 3) 2200 on Tuesday

Worksheet 10-2

Name:

Date:

Solve the following problems.

1) An I.V. solution of 1000 cc is set to run over 8 hours with a drop factor of 15, what is the drip rate?

2) The physician orders 3000 mL to be infused over 24 hours and the IV set has a drop factor of 20 gtt/mL, what will the drip rate need to be?

3) A medication order calls for 1 L of D5W to be administered over an 8 hour period. Using an IV administration set that delivers 10 gtt/mL, how many drops/minute should be delivered to the patient?

4) Ten mL of 10% calcium gluconate injection and 10 mL of multivitamin infusion are mixed with 500 mL of D5W. The IV solution is to be infused over 5 hours. If the administration set has a drop factor of 15 what should the drip rate be? (*Hint: don't forget to calculate the total volume that is being infused.*)

5) A 50 mL IVPB bag of ampicillin 500 mg in Normal Saline is to be run in over 20 minutes. What is the infusion rate in mL/hr?

6) What is the flow rate in mL/hr for a TPN containing 500 mL of D10W, 500 mL of 7% amino acids, and 36 mL of micronutrients if it is run in over 16 hours?

7) A 50 mL IVPB contains 500,000 units of penicillin G potassium. The rate of infusion ordered by the physician is 120 mL/hr. How long will the IVPB take to infuse?

8) If a 1 liter bag of D5W is started on a patient at 2200 hours on Tuesday, when will the next bag be needed if it is running at a rate of 200 mL/hr?

9) A 100 mL bag containing 0.5 mg octreotide is to be infused at 50 mcg/hr. How many mL/hr will you need to set the pump to?

10) A patient received a 500 mL whole blood transfusion starting at 2113 at a rate of 180 mL/hr. At what time was the transfusion completed?

11) A physician orders a 1200 mL hyperalimentation to infuse over 24 hours. The pharmacy has in stock standard tubing with a drop factor of 10 and microdrip tubing with a drop factor of 60. Which tubing should the pharmacy send with the hyperalimentation and why?

Dosages Based on Body Weight

Many drugs need to be calculated based on body weight. Some of the drugs where you will see this most often include:

- chemotherapy,
- steroids,
- antibiotics,
- heparinoids, and
- drugs for pediatric and geriatric patients

If a medication dose is to be based on a body weight, it will usually be requested in mg/kg.

If a medication is ordered as 5 mg/kg, and the patient weighs 100 kg, you will need 500 mg of drug.

$$\frac{100\,kg}{1} \times \frac{5\,mg}{kg} = \mathbf{500\,mg}$$

Example

A physician orders cyclophosphamide to be given 5 mg/kg qid in 50 mL D5W. The patient weighs 132 pounds. If the concentration of the drug available is 500 mg/10 mL, how many mL should be added to each bag?

QUESTION

How many mL should be added to each bag?

DATA

5 mg/kg qid 50 mL of D5W 132 lbs 500 mg/10 mL

a potentially useful conversion:

1 kg/2.2 lbs

MATHEMATICAL METHOD / FORMULA

dimensional analysis

DO THE MATH

$$\frac{132\,lbs}{1} \times \frac{1\,kg}{2.2\,lbs} \times \frac{5\,mg}{kg} \times \frac{10\,mL}{500\,mg} = \mathbf{6\,mL}$$

DOES THE ANSWER MAKE SENSE?

Yes

Notice that there was a lot of unnecessary information in the problem, and that you needed to use a conversion to solve the problem.

Sometimes you will need to incorporate additional units (such as time) into weight based dosage calculations. Let's look at an example of this.

Example

You are asked to set the infusion rate for a patient's Remodulin, which is being infused via a CAD pump. The physician ordered 104 ng/kg/min. You pull up the patient's profile and verify the troprostenil cassette with 20 mg of troprostenil and 98 mL of NS has a total volume of 100 mL, and you confirm the patient's weight of 58 kg that you have on file is correct. How many mL/day will you need to set this pump to?

QUESTION

How many mL/day will you need to set this pump to?

DATA

104 ng/kg/min   20 mg   98 mL NS   100 mL total volume   58 kg

a potentially useful conversion:

1 mcg/1000 ng   1 mg/1000 mcg   60 min/hr   24 hr/day

MATHEMATICAL METHOD / FORMULA

dimensional analysis

DO THE MATH

$$\frac{104\ ng}{kg\ min}\times\frac{58\ kg}{1}\times\frac{60\ min}{hr}\times\frac{24\ hr}{day}\times\frac{100\ mL}{20\ mg}\times\frac{1\ mg}{1000\ mcg}\times\frac{1\ mcg}{1000\ ng}=\mathbf{43.4\ mL/\ day}$$

DOES THE ANSWER MAKE SENSE?

Yes

Notice that minutes from *104 ng/kg/min* were kept with the kg for mathematical purposes. Often, when given a value with multiple '/' the unit after the *kg* will be kept with the *unit of time*. Besides *minutes*, you will also see *hours, days*, and *doses* kept with the *kg*.

Let's do a few practice problems.

Practice Problems

To do these next four problems, let's create a scenario where the 165 pound author of this textbook is in a terrible car accident and requires the spinal cord protocol for two methylprednisolone infusions.

1) First, you need to make a bolus dose of 30 mg/kg in 50 mL of NS. How many mg of methylprednisolone will you need to make the bolus dose? (The bolus dose will be infused over 1 hour.)

2) If methylprednisolone comes in a double chamber vial with a concentration of 1 g/8 mL, how many mL will you need to add to the 50 mL bag of NS?

3) After the bolus dose is infused, your instructor will need a continuous infusion of methylprednisolone at a rate of 5.4 mg/kg/hr for 23 hours. How many mg of methylprednisolone will you need to make the continuous infusion?

4) The continuous infusion is to be administered with normal saline as the diluent. If the final bag has to have an exact volume of 1000 mL, how many mL methylprednisolone will be needed since it comes in a double chamber vial with a concentration of 1 g/8 mL, and how many mL of NS will be needed?

1) 2,250 mg 2) 18 mL 3) 9,315 mg 4) 74.5 mL of methylprednisolone and 925.5 mL of normal saline

Worksheet 10-3

Name:

Date:

Solve the following problems.

1) Ceftazidime is ordered for a 55 pound pediatric patient at a dosage of 150 mg/kg/day to be infused over 24 hours. The drug when reconstituted has a concentration of 100 mg/mL. How many mL of the reconstituted solution will be required?

2) Succinylcholine is available in a concentration of 20 mg/ml in a 10 mL vial. The order reads 40 mcg/kg IV push every 5 to 10 minutes as a maintenance dose for a patient that weighs 198 lbs. How many ml will the CRNA give IV push for each dose?

3) A patient, weighing 165 lbs, is ordered phenobarbital 5 mg/kg at bedtime. The phenobarbital is available in a 1 mL vial with a concentration of 65 mg/mL. How many mL will the patient need for this dose?

4) Gentamycin is ordered 5 mg/kg for a patient who weighs 220 pounds. Gentamycin is available in a 20 mL MDV with a concentration of 40 mg/mL. How many mL of gentamycin should the patient receive?

5) A physician orders a dose of vincristine, based on the patient's body weight, as 25 mcg/kg. The maximum dose of vincristine is capped at 2 mg, so if your calculated dose exceeds that it is automatically reduced to 2 mg. The drug is available as 1 mg/mL vials. The patient weighs 143 pounds. How many mL are used for a dose?

6) A 209 pound patient with an acute spinal cord injury is ordered a methylprednisolone bolus and drip.

   a) The bolus dose is 30 mg/kg of body weight in 50 ml of normal saline. How many milligrams of methylprednisolone will the patient receive in their bolus?

   b) The continuous infusion is 5.4 mg/kg/hour for 23 hours in normal saline with a total volume of 1000 ml. How many milligrams of methylprednisolone will the patient receive in their continuous infusion?

7) A physician writes for a 154 pound patient to receive a cyclophosphamide induction dose of 40 mg/kg equally divided into two doses. After that the patient is to receive a maintenance dose of 10 mg/kg every seven days.

   a) How many mg of cyclophosphamide will the patient receive for each of their induction doses?

   b) How many mg of cyclophosphamide will the patient receive for each of their maintenance doses?

8) A two day old 5 pound 6 ounce neonate with a severe infection is ordered a continuous infusion of penicillin g potassium at a rate of 4,000 units/kg/hr. The nursing unit wants to hang a new bag every 12 hours, how many units of penicillin g potassium should be in each bag?

Dosage Calculations Based on Body Surface Area

The body surface area (BSA) is the measured or calculated surface of a human body, and it is measured in square meters (m^2). For many clinical purposes, BSA is a better indicator of metabolic mass than body weight because it is less affected by abnormal adipose mass.

Examples of uses for BSA include:

- renal clearance is usually divided by the BSA to gain an appreciation of the true required glomelular filtration rate (GFR),
- chemotherapy is often dosed according to the patient's BSA, and
- glucocorticoid dosing can also be expressed in terms of BSA for calculating maintenance doses or to compare high dose use with maintenance requirement.

There are number of ways to calculate BSA:

such as various formulas:

- Dubois & Dubois; BSA (m^2) = $0.007184 \times \text{weight in kg}^{0.425} \times \text{height in cm}^{0.725}$
- Mosteller; BSA (m^2) = $\sqrt{\dfrac{\text{weight in kg} \times \text{height in cm}}{3600}}$
- Haycock; BSA (m^2) = $0.024265 \times \text{weight in kg}^{0.5378} \times \text{height in cm}^{0.3964}$

using a nomogram (a chart based method pictured on the next page),
or some more peculiar ways:

- geometry,
- thoroughly detailed 3D scans,
- even wrapping patients in aluminum foil.

A nomogram is a common method and is the way we will concentrate on. A nomogram pictured on the next page has three columns:

- height based in both centimeters and inches,
- body surface area in m^2, and
- weight based in both kilograms and pounds.

How does someone use a nomogram?

The height and the weight of the patient are found on the nomogram, and then a straight line is drawn connecting the two values. The BSA for that patient is found where the line intersects the BSA column. As an example, find 5' 3" (63 inches) and 110 pounds. If you draw a line connecting the two values, you should get a BSA of 1.5 m^2.

Nomogram for Determination of Body Surface Area from Height and Weight

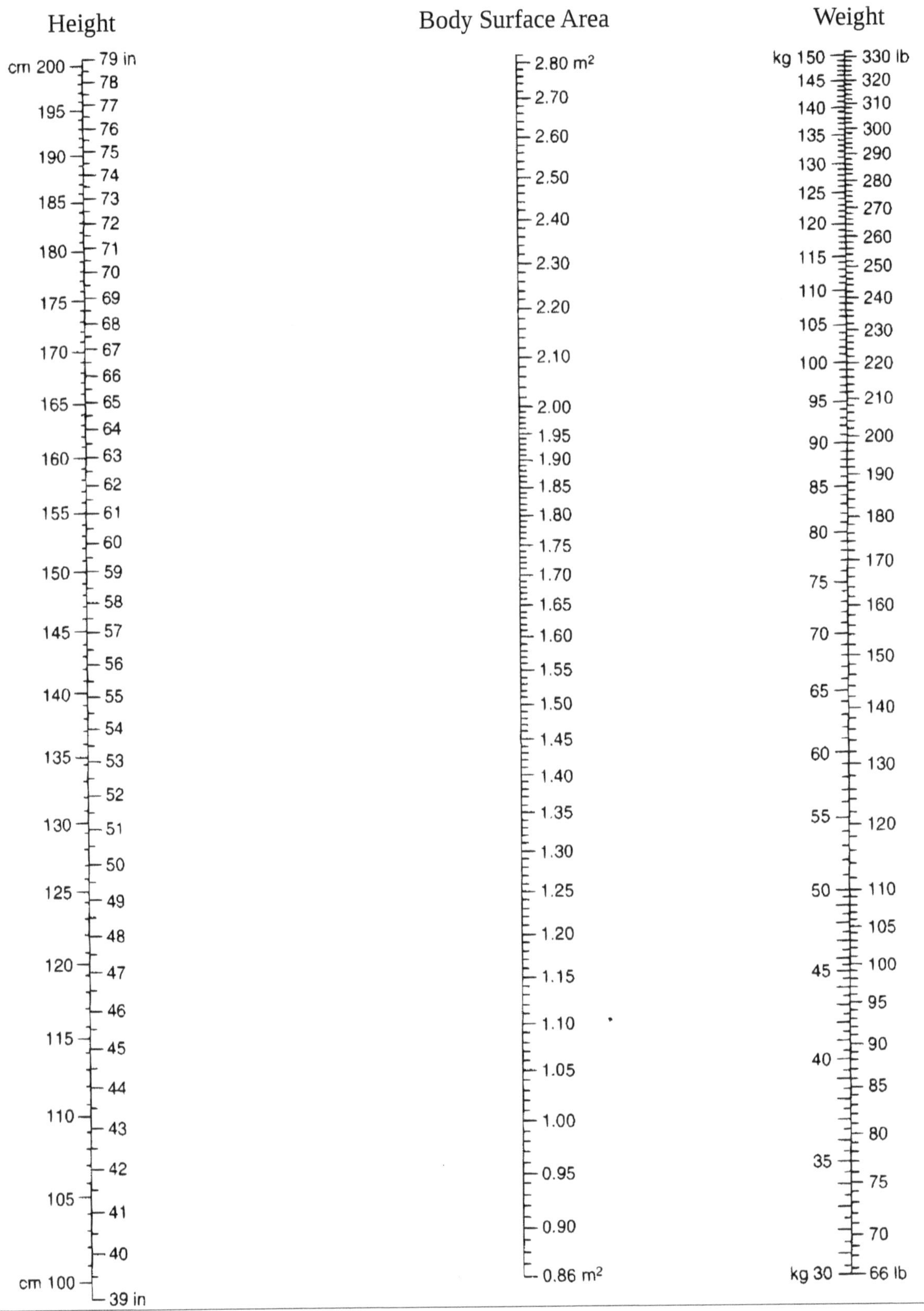

Nomogram is based on the Dubois & Dubois formula.

BSA (m^2) = 0.007184 x weight in $kg^{0.425}$ x height in $cm^{0.725}$

Let's look at an example problem.

Example

QUESTION

A physician orders a bolus dose of doxorubicin for a 6'1" patient that weighs 165 lbs. The drug dose is 75 mg/m^2. What is the bolus dose in mg?

DATA

6'1"

165 lbs

75 mg/m^2

FORMULA/METHOD

Use a nomogram to find BSA, then use factor label to do the calculation.

MATH

BSA = 1.98 m^2

$$\frac{1.98\,m^2}{1} \times \frac{75\,mg}{m^2} = \mathbf{148.5\,mg}$$

DOES THE ANSWER MAKE SENSE?

Yes

Now you should attempt a practice problem.

Practice Problem

If the dose of Taxol (paclitaxel) in the treatment of metastatic ovarian cancer is 135 mg/m^2, what would be the dose for a patient 155 cm tall and weighing 53 kg?

202.5 mg

Worksheet 10-4

Name:

Date:

Solve the following problems.

1) Leukine (a drug used to increase neutrophils in patients receiving chemotherapy for Leukemia) is to be administered by IV at 250 mcg/m^2/day for 21 days. How many mcg will a patient receive each day if they are 5'10" and weigh 156 lbs?

2) A patient weighs 176 lbs and is 71 inches tall. Their physician orders doxorubicin (an antineoplastic agent) 25 mg/m^2 in 50 ml of NS. The doxorubicin is supplied as 50 mg vials with a concentration of 4 mg/ml.

   a) How many mg will the patient need?

   b) How many mL of doxorubicin will you add to the NS bag?

   c) How many vials of doxorubicin will you need to gather to make this IV?

3) Methotrexate when reconstituted has a concentration of 25 mg/mL. A patient being treated for osteosarcoma has a BSA of 1.60 m^2. The patient is ordered methotrexate 12 g/m^2 by IV infusion every week. How many mL of the reconstituted solution will the patient receive?

4) A physician orders bleomycin in a dose of 20 units/m^2 twice weekly. The reconstituted bleomycin has a concentration of 30 units/mL. The patient is 5'3" tall and weighs 110 lbs. How many mL will the patient need for a single dose?

5) An order is received in the pharmacy for 5-FU IV for a 5'6" patient weighing 125 lbs. The 5-FU

solution is available in a 50 mg/mL concentration. The dosage schedule is as follows:

Initial dose: 800 mg/m^2/day for 5 days IV push
How many grams of 5-FU has the patient received after the first 5 days?

Pediatric Dosing

We will be reviewing a multitude of ways to calculate a pediatric patient's medication dose.

Some pediatric doses may simply be calculated based on **body weight** or **body surface area** (BSA). Those kinds of problems are fairly straight forward in problems like:

A neonate weighing 4000 g is ordered tobramycin q8h at 2.5 mg/kg/dose, how many mg of tobramycin should the neonate receive?

$$\frac{4000\,g}{1} \times \frac{1\,kg}{1000\,g} \times \frac{2.5\,mg}{kg/dose} = 10\,mg/dose$$

A medication is dosed as 100 mg/m^2, what would be the dose if a pediatric patient had a BSA of 0.87 m^2?

$$\frac{0.87\,m^2}{1} \times \frac{100\,mg}{m^2} = 87\,mg$$

But, both of the aforementioned ways treat pediatric dosing as if children were simply miniature adults. We will take some time and look at other ways to calculate pediatric doses.

Clark's Rule
(based on weight for children ages 2-17)

$$\frac{Weight\,(in\ pounds) \times Adult\ dose}{150\,(average\ weight\ of\ adult\ in\ pounds)} = Dose\ for\ child$$

Young's Rule
(based on age for all children)

$$\frac{Age}{Age+12} \times Adult\ dose = Dose\ for\ child$$

Fried's rule
(based on age up to 24 months)

$$\frac{Age\,(in\ months) \times Adult\ dose}{150\ months} = Dose\ for\ infant$$

Cowling's rule

$$\frac{Age\ at\ next\ birthday\,(in\ years) \times Adult\ dose}{24\ years} = Dose\ for\ child$$

Pediatric nomogram, when adult dose is known
(based on BSA for all children while treating the average adult BSA as 1.73 m^2)

$$\frac{Child's\ BSA}{1.73\,m^2} \times Adult\ dose = Dose\ for\ child$$

You will need this nomogram to help you solve the problems on the next page.

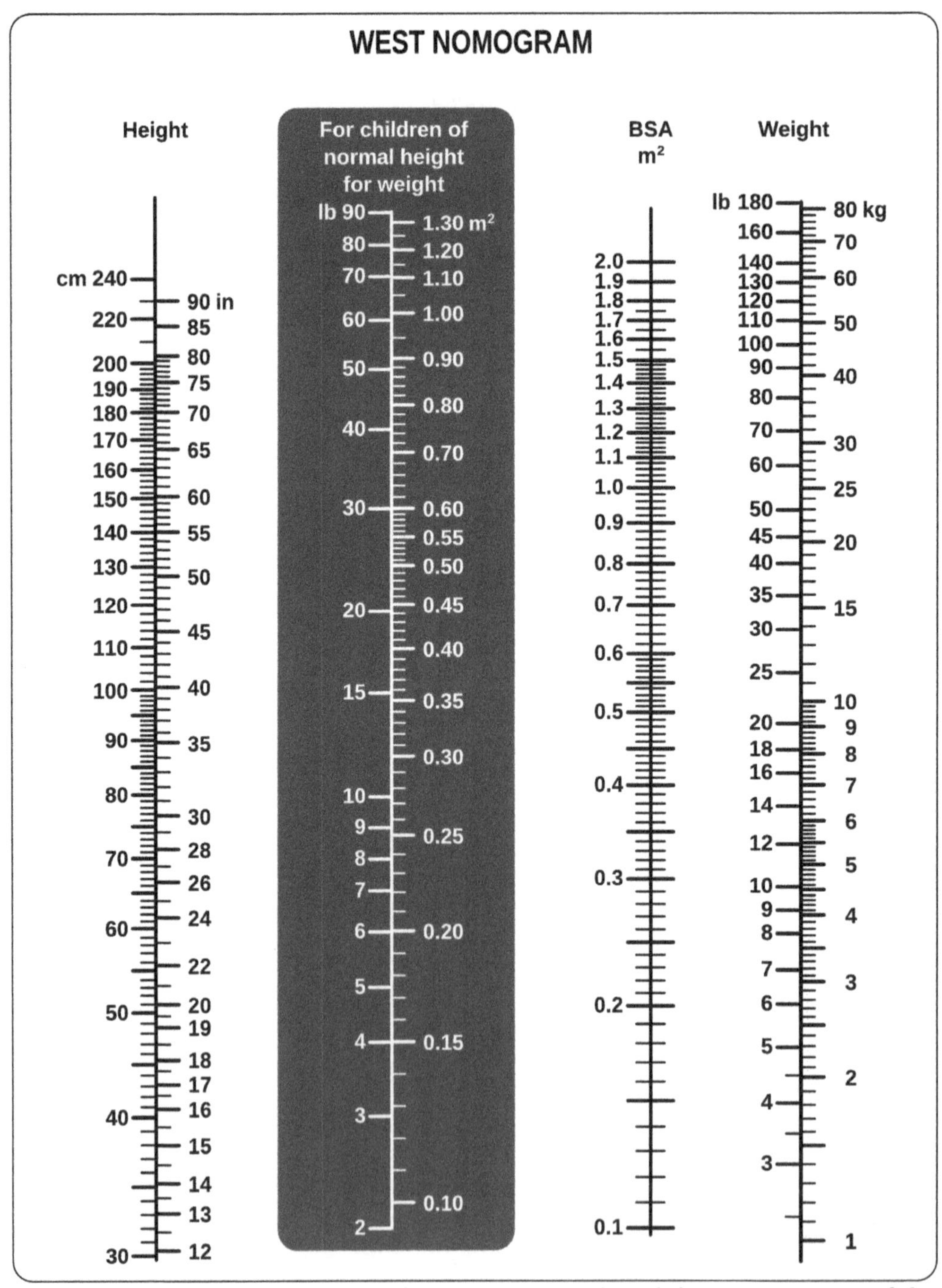

Notice that you can use the West Nomogram the same way as the adult nomogram, or if the child is a normal height for their weight you may use the darkened column instead as a shortcut.

Worksheet 10-5

Name:

Date:

Solve the following questions.

1) A newborn (we'll count them as 1 month for calculation purposes) weighs 5.6 lbs, has a height of 17 inches, and is ordered gentamicin to treat a meningitis infection they've contracted.

   a) If based strictly on body weight, premature and full-term neonates should receive gentamicin 2.5 mg/kg every 12 hours. Based only on body weight, how much gentamicin should the neonate receive each day?

   b) An average adult dose of gentamicin for treating meningitis is 350 mg/day given in three equally divided doses. Based on this information, what should this neonate's daily gentamicin be based on Cowling's rule?

   c) Using the above information, what would be the average daily quantity of gentamicin be based on Fried's rule?

   d) Using the above information, what would be the average daily quantity of gentamicin be based on a pediatric nomogram, when the adult dose is known?

2) Calculate the dose for a child, 4 years of age, 39 in. in height, weighing 32 lbs, for a drug with an adult dose of 100 mg, using the following:

   a) Young's rule

   b) Cowling's rule

   c) Clark's rule

d) Pediatric nomogram when the adult dose is known

3) The daily dose of diphenhydramine HCl for a child may be determined on the basis of 5 mg/kg of body weight or on the basis of 150 mg/m^2. Calculate the dose on each basis for a 4 year old child weighing 55 lbs and measuring 3' 4" in height.

a) Based on weight?

b) Based on BSA?

An average adult dose of injectable diphenhydramine is 400 mg per day (often given in 4 divided doses). Based on this, how much diphenhydramine should they receive according to Clark's rule and Young's rule?

c) Clark's rule

d) Young's rule

Worksheet 10-6

Name:

Date:

Solve the following problems.

1) How many milliliters of a 14.6% stock solution of sodium chloride would be required to make a 250 mL bag of 3% sodium chloride? How many mL of sterile water for injection (SWFI) will be required to make this bag?

2) How many mL of a 2% stock solution should be used to prepare 1 L of a 0.025% solution?

3) How many mL of 95% ethyl alcohol and how many mL of NS will be needed to prepare 2.5 L of 50% ethyl alcohol? (*NS will be used as the diluent in this problem*)

4) How many mL of a 10% cyclosporine solution would be required to make 1 fl. oz. of a 2% cyclosporine solution?

5) How many mL of 5% potassium permanganate stock solution and how many mL of diluent will be required to make 180 mL of 0.5% potassium permanganate solution?

6) A medication order calls for a 1 liter TPN to be infused over 6 hours. If the infusion set has a drop factor of 20, what drip rate should the drop chamber be set to (*in gtt/min*)?

7) A medication order calls for a 500 mL bag with 20 mEq of potassium chloride to be infused over 4 hours. What will be the infusion rate in mL per hour?

8) If the IV bag in the previous problem needed to be infused using a venoclysis set calibrated to 15 gtt/mL, what will be the drip rate?

9) If a 50 mL IVPB containing 1 gram of cefazolin is being infused at 200 mL/hr, how long will it take to infuse?

10) If a 1,500 mL IV solution is being infused over 24 hours, at what drip rate will you set the drip chamber if it has a drop factor of 20?

11) The daily dose of a drug is 12 mcg/kg of body weight. How many mg should be administered each day to a woman who weighs 55 kg?

12) If the dose of a drug is 0.25 mg/kg, how many milligrams should be administered to a man weighing 175 lbs?

13) The usual dose of lucanthone hydrochloride is 5 mg/kg/dose given t.i.d. for one week. How many g would be required to cover an entire week for a youth weighing 120 lbs?

14) Ceftazidime is ordered for a 66 pound pediatric patient at a dosage of 150 mg/kg/day to be infused over 24 hours. The drug when reconstituted has a concentration of 100 mg/mL. How many mL of the reconstituted solution will be required?

15) Gentamycin is ordered 5 mg/kg for a patient who weighs 165 pounds. Gentamycin is available in a 20 mL MDV with a concentration of 40 mg/mL. How many mL of Gentamycin should the patient receive?

16) A physician orders a bolus dose of doxorubicin for a patient with a BSA of 1.56 m^2. The drug dose is 75 mg/m^2. What is the bolus dose in mg?

17) A patient weighs 117 lbs and is 5'2" tall. Using the nomogram on page 260 to determine their BSA, find the patient's dose of vincristine if the physician ordered it as 1.4 mg/m^2. The maximum dose of vincristine is capped at 2 mg, so if your calculated dose exceeds that it is automatically reduced to 2 mg.

18) 5FU 750 mg/m^2 as IVP is ordered for a patient (73" and 165 lbs). 5FU is available as 50 mg/mL in 10 mL vials.

   a) What is the patient's dose in mg?

   b) How many mL will the patient receive?

   c) How many vials will be needed to prepare this dose?

19) Methotrexate, when reconstituted, has a concentration of 25 mg/mL. A patient with a BSA of 1.80 m^2 is ordered 12 g/m^2 by IV infusion every week. How many mL of the reconstituted solution will the patient receive?

20) An injectable medication is available in a 10 cc vial and has a concentration of 100 mg/mL. If a 98 lb, 4'10" tall patient receives an order for 500 mg/m^2, how many milligrams should the patient receive?

21) Is Young's rule or Clark's rule based on age?

Use the following information to solve problems 22-26:

A 9 year old child that is 4'2" and weighs 50 lbs is ordered diphenhydramine. The daily dose of diphenhydramine HCl for a child may be determined on the basis of 5 mg/kg/day of body weight or on the basis of 150 mg/m^2/day. Also, an average adult dose of injectable diphenhydramine is 400 mg per day (often given in 4 divided doses).

22) Based strictly on body weight, what should she receive daily?

23) Based strictly on BSA, what should she receive daily?

24) Using Clark's rule, what should she receive daily?

25) Using Young's rule, what should she receive daily?

26) Using Cowling's rule, what should she receive daily?

UNIT 3

COMMUNITY PHARMACY MATH

What does community pharmacy math consist of?

Despite popular belief, community pharmacy requires pharmacy technicians to do significantly more than just count by fives. Pharmacy technicians need to translate the physician's prescribed quantity into the quantity that an insurance provider will cover. Extemporaneous compounding requires a significant quantity of mathematical manipulations. Technicians often need to maintain price databases, work with insurance pricing formulas, maintain inventories, and streamline cash flow. A pharmacy technician can carry a lot of responsibility in today's community pharmacy and the goal of this unit is to provide a solid foundation for those skills.

What are the specific learning objectives in this unit?

- Days' supply,
- Adjusting refills and short-fills,
- Extemporaneous compounding,
- Billing for extemporaneous compounds,
- Calculating usual and customary prices,
- Common insurance reimbursement formulas,
- Gross profits and net profits,
- Medication inventory control,
- Medication storage,
- Daily cash reports,
- Calculating professional/dispensing fees, and
- Depreciating capitol expenditures.

CHAPTER 11

DAYS' SUPPLY

"They should really pay someone to do these calculations."
"True enough, but the person they're supposed to pay to do these calculations will be you."

--Classroom discussion between an instructor and one of his students.

In this chapter you'll be learning how to do the following calculations:

- Days' Supply
- Adjusting Refills, and
- Short-fills

Days' Supply

Physicians often write their prescriptions with various time frames in mind. Common prescribing time frames include 5 days, 1 week, 10 days, 2 weeks, 1 month, 50 days, 90 days or even 100 days. Physicians often just include a quantity of medication to dispense and directions on how frequently to use it. They usually don't include the actual intended time frame. As a pharmacy technician, you will need to translate that into *Days' Supply*.

In this chapter, *Days' Supply* is referring to how long a prescription order will last. Often it is not as simple as giving a tablet once a day for 30 days; you will frequently need to do calculations for oral liquid medications, injectables, nasal sprays, and inhalers, and make estimations for PRN's, ointments and creams, lotions, eye and ear drops, and ophthalmic ointments.

Days' Supply for tablets, capsules, and liquid medications

The first things to look at are tablets, capsules, and liquid medications because they're the most common and the most straight-forward to perform calculations with. Without wanting to over explain this process, let's look at some example problems.

Examples

1) A prescription is written for amoxicillin 250 mg capsules #30 i cap t.i.d. What is the days' supply?

$$\frac{30\,caps}{1} \times \frac{1\,dose}{1\,cap} \times \frac{day}{3\,doses} = \mathbf{10\ days}$$

2) A prescription is written for amoxicillin 250 mg/5 mL 150 mL i tsp tid. What is the days' supply?

$$\frac{150\,mL}{1} \times \frac{tsp}{5\,mL} \times \frac{dose}{1\,tsp} \times \frac{day}{3\,doses} = \mathbf{10\,days}$$

Now that you've seen a couple of examples, solve the following practice problems.

Practice Problems

1) A prescription is written for cephalexin 250 mg #28 i cap q6h. What is the days' supply?

2) A prescription is written for cephalexin 250 mg/5 mL 100 mL i tsp q6h. What is the days' supply?

1) 7 days 2) 5 days

The item to be careful about when it comes to tablets, capsules, and liquid medications are PRN medications, especially ones with variable doses and variable frequencies. In general, you should perform the calculations using the shortest interval with the highest dose. This will provide the shortest span of time in which they could use all the medication dispensed. Let's look at an example problem.

Example

A prescription is written for Ultram 50 mg #60 i-ii tabs po q4-6h prn pain. What is the days' supply?

$$\frac{60\,tabs}{1} \times \frac{1\,dose}{2\,tabs} \times \frac{4\,hours}{1\,dose} \times \frac{1\,day}{24\,hours} = \mathbf{5\,days}$$

Conveniently, this example came out to an even number of days. Sometimes your calculations will come out to a decimal number of days and you may need to use some professional judgment to determine whether to drop the decimal or round up. If you are not sure, it is usually better to drop the decimal. Attempt the following practice problem working with a PRN medication.

Practice Problem

1) A prescription is written for hydromorphone 2 mg #30 i-ii tabs po q3-4h prn pain. What is the days' supply?

1) Even though the answer comes out to 1.875 days, I would expect the medication to last 2 days based on my own professional judgment

Days' Supply for insulins

Most insulins are called U-100 insulins meaning that each mL contains 100 units. Also most insulin vials are either 10 mL vials or boxes of 5 syringes containing 3 mL in each syringe for a total of 15 mL in a box. A 10 mL vial of U-100 strength insulin would contain 1000 units and a 15 mL box of syringes with U-100 insulin would contain 1500 units. This is good information to help you make quick work of the vast majority of insulin calculations. With insulin problems, whenever you come out with a decimal number of days you should always just drop the decimal as you never want a diabetic patient to run out of their insulin. The last thing to keep in mind with respect to an insulin vial is that it should not be kept for longer than 30 days after it has been opened. Determining how long a box will last is different since each syringe is only good for 30 days after it is started, but there are five syringes. Let's look at an example problem with insulin.

Example

A prescription is written for Humulin N U-100 insulin 10 mL 35 units SC qd. What is the days' supply?

$\frac{10\,mL}{1} \times \frac{100\,units}{mL} \times \frac{1\,day}{35\,units} = 28.57\,days$ which means **28 days** because you should drop the decimal.

Let's look at some practice problems.

Practice Problems

When solving the next two problems, treat the first problem as a 10 mL vial with a U-100 concentration and the second problem as 15 mL (5 syringes with 3 mL each) with a U-100 concentration.

1) A prescription is written for Humulin R 1 vial 8 units SQ before breakfast, 8 units before lunch, and 11 units before supper. What is the days' supply?

2) A prescription is written for Novolin N prefilled syringes #1 box Sig: 22 units SC q am and 24 units q pm. What is the days' supply?

1) 30 days 2) 32 days

Days' Supply for inhalers and sprays

Whenever you see instructions on a product for a patient to receive a particular number of sprays or puffs of a given drug, you should stop and actually look at the packaging to discover how many metered inhalations or how many metered sprays are actually in the container. Lets look at an example problem accompanied by a rendering of what the front of the container actually looks like.

Example

A prescription is written for ProAir HFA 8.5 g inhaler 2 puffs q.i.d. What is the days' supply? (*Hint: In order to solve this, you will need to look at the box pictured on the right.*)

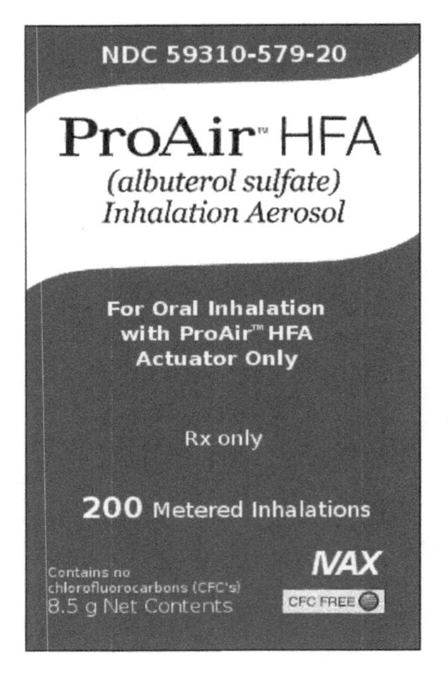

After looking at the box, you should realize that this particular package contains 200 *puffs* (metered doses), and you now have enough information to solve this problem.

$$\frac{200\,puffs}{1}\times\frac{1\,dose}{2\,puffs}\times\frac{day}{4\,doses}=\mathbf{25\,days}$$

Now that we've looked at an example, attempt the following practice problem.

Practice Problem

A prescription is written for Flonase 16 gm 2 sprays per nare qd for a total daily dose of 200 µg. What is the days' supply?

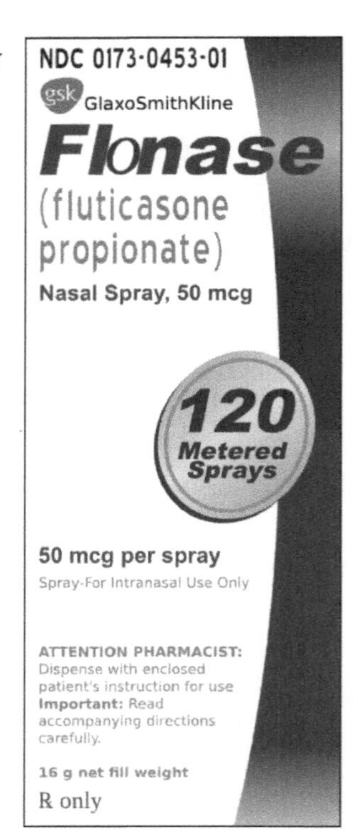

1) 30 days

Days' Supply for ointments and creams

Calculations for creams and ointments are a little more tricky because you usually don't know exactly how much will be used in a dose. The amount will depend on how large of an area is affected and how many areas it needs to be applied to. The amount applied usually does not exceed 500 mg to 1 g, so unless you know otherwise, use 1 gram for the dose for each affected area.

Example

A prescription is written for Mycolog II cream 15 g apply sparingly bid. What is the days' supply?

$$\frac{15\,g}{1}\times\frac{1000\,mg}{1\,g}\times\frac{dose}{500\,mg}\times\frac{1\,day}{2\,doses}=\mathbf{15\,days}$$

Practice Problem

1) A prescription is written for Nizoral cream 15 gm apply to the affected and surrounding areas once daily. What is the days' supply?

2) A prescription is written for Bactroban ung 15 gm apply a small amount to the affected area tid. What is the days' supply?

1) 7 days   2) 10 days (since the problem said *small amount*, each dose can be treated as 500 mg)

Days' Supply for ophthalmic and otic preparations

To solve this type of problem, you need to know a conversion factor from milliliters to drops. Unfortunately (or fortunately since most students despise the apothecary system), you *cannot* use the conversion from the apothecary system. The USP[1] in chapter 1101 has written regulations on standardizations of medicine droppers. Unless your specific medication notes something different, a dropper should be calibrated by the manufacturer to deliver between 18 and 22 drops per milliliter. Most people just split the difference and estimate **20 gtt/mL**. With that in mind, we can get a good *estimate* on how long the medication should last.

Another odd thing to keep track of when dealing with eye preparations are ophthalmic ointments. An ophthalmic ointment is typically applied as a very thin strip. Treat each dose of an ophthalmic ointment as 100 mg. Let's look at some example problems with respect to ophthalmic and otic preparations.

---

1 This important standard is contained in a combined publication that is recognized as the official compendium, the United States Pharmacopeia (USP) and the National Formulary (NF)

Examples

1) A prescription is written for timolol 0.25% Opth. Sol. 5 mL i gtt ou q.d. What is the days' supply?

$$\frac{5\,mL}{1} \times \frac{20\,gtt}{1\,mL} \times \frac{1\,dose}{2\,gtt} \times \frac{1\,day}{1\,dose} = \mathbf{50\ days}$$

2) A prescription is written for Neosporin Opth. ung 3.5 g apply a thin strip ou q3-4h. What is the days' supply?

$$\frac{3.5\,g}{1} \times \frac{1000\,mg}{1\,g} \times \frac{1\,dose}{200\,mg} \times \frac{3\,hours}{1\,dose} \times \frac{1\,day}{24\,hours} = \mathbf{2\ days}\ after\ you\ drop\ the\ decimal$$

Notice that both example problems were written for each eye, requiring the quantity of medication to be doubled in both cases.

Now, you should try a couple of practice problems.

Practice Problems

1) A prescription is written for Tobradex Opth. Susp. 5 ml 1-2 gtt os q4-6h. What is the days' supply?

2) A prescription is written for tobramycin Opth. Oint. 3.5 g apply od bid. What is the days' supply?

1) 8 days 2) 17 days

Days' Supply for various packs

Many medications (such as birth control, steroids, and antibiotics) may come in packs with very explicit instructions for use. These instructions often explain exactly how many days they will last and require no additional calculations. You must simply read the instructions on the package to know how long it will last. Let's look at an example problem on the next page.

Example

1) A prescription is written for methylprednisolone 4 mg tabs taper pack use as directed. What is the days' supply?

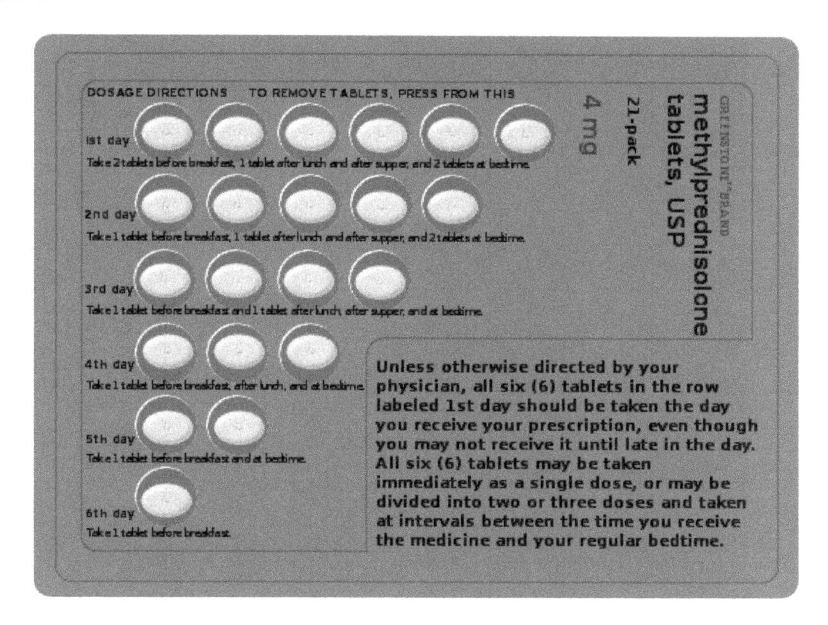

It requires nothing more than observation to realize it will last ***6 days***.

Days' Supply for other miscellaneous medications

Unfortunately, there are many medications that have their own specific rules, such as estrogens given for hormone replacement therapy that are cycled on and off. The cycle is days 1-25 on, followed by five days off or it may be cycled for three weeks on, followed by one week off. Items like a vial of nitroglycerin sublingual tablets or spray are expected to last a patient 30 days. With items, like vaginal preparations, you'll need to know how much is delivered via the applicator. Lotions can be a challenge depending on their viscosity. A good rule of thumb for a lotion is to expect 2 mL to be used on each affected area per application. As you can tell, there are many individual little rules to try and work with when estimating how long a particular medication will last.

Worksheet 11-1

Name:

Date:

Solve how many days each of the following prescriptions will last.

1) Advair 250/50 #1 1 puff q12h (*Hint: use the image on the right hand side of the page.*)

2) Atrovent inhalation solution 0.02% 2.5 ml Disp 25 vials Sig: 1 vial per neb q.i.d.

3) Rx Augmentin 400 mg-57 mg/5 mL Disp. 100 mL Sig: i tsp po q12h till all of medicine is gone

4) Axid 150 mg #60 i cap po bid before meals

5) Rx Betoptic 0.5% Disp 10 mL Sig: 2 gtt OS b.i.d.

6) Cortisporin Otic 10 mL iv gtt ad qid

7) Crestor 10 mg #90 1 tab po qd

8) DexPak 13 day TaperPak take u.d.

9) Rx Duragesic 50 mcg/hr Disp: 10 patches Sig 1 patch q72h

10) E.E.S. 400 mg #42 i tab po tid c meals

11) Fentora 100 mcg #28 i tablet in buccal cavity q4-6h prn breakthrough pain

12) Fosamax 70 mg #4 i tab po q week

13) Rx Glucotrol 10 mg Disp: 120 tabs Sig: 20 mg po bid

14) Rx Humulin R 10 mL Disp: 1vial Sig: 11 units SQ before breakfast, 11 units before lunch, and 14 units before supper.

15) Ibuprofen 400 mg #60 i tab po q6-8h prn pain

16) Januvia 100 mg #30 1 tab po qd

17) ketorolac 10 mg #20 1 tab po q4-6h prn pain. Not to exceed 40 mg/day.

18) Lac Hydrin 12% Lotion 150 mL aa bid rub in thoroughly. (*Hint: Use 1-2 mL for amount used per application unless you know a larger area is being treated; let's treat the patient as having 2 affected areas.*)

19) Lantus Insulin 10 mL #1 vial 40 units SQ q a.m.

20) Levothyroxine 100 mcg #100 i tab po qd

21) Rx Lexapro 20 mg Disp: 60 Sig: i cap po qd

22) Lipitor 20 mg #30 i tab po qd

23) methylprednisolone 4 mg tabs taper pack use ut dict (*Hint: Use the image below to determine how long it will last.*)

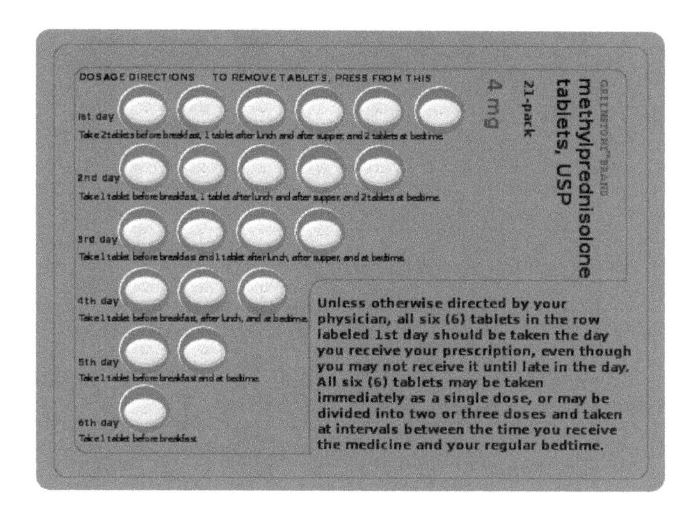

24) metoprolol tartrate 50 mg #180 i tab po b.i.d.

25) Rx MetroGel-Vaginal 0.75% Disp: 70 g tube Sig: 1 applicatorful pv bid ×5 days (*Hint: 1 applicator delivers 5 g of gel containing 37.5 mg of metronidazole.*)

26) Neosporin ophthalmic ointment 3.5 g apply thin strip od t.i.d ×10 days

27) Nexium 20 mg #28 i cap po bid

28) NitroDur 0.4 mg #30 i patch on 8 a.m., off 10 p.m. qd

29) Nitrostat 0.4 mg #100 i tab SL q5min prn chest pain. May repeat up to 3×.

30) ofloxacin otic solution 10 mL 10 gtt as bid ×10 days

31) Rx Patanase Disp: 1 bottle Sig: ii sprays each nare bid (*Hint: a bottle contains 240 metered sprays.*)

32) Plavix 75 mg #60 1 tab po qd

33) ProAir HFA i-ii puffs q4-6h prn relief of asthma symptoms.

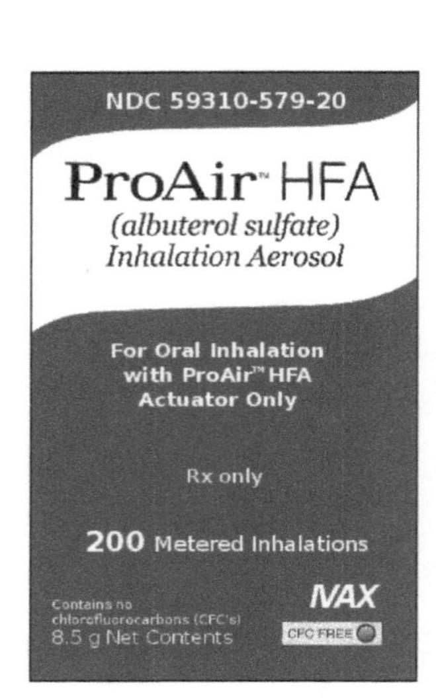

34) ProTopic 0.1% 100 g apply minimum amount to aa bid

35) Premarin 0.625 mg #63 i tab po qd cyclically (3 weeks on, 1 week off)

36) quinine 324 mg #42 2 caps po q8h

37) Requip 1 mg #270 1 tab po tid

38) simvastatin 20 mg #100 i tab po qd

39) Singulair 10 mg #30 i tab po q p.m.

40) Spiriva #30 inhale contents of 1 cap qd using HandiHaler

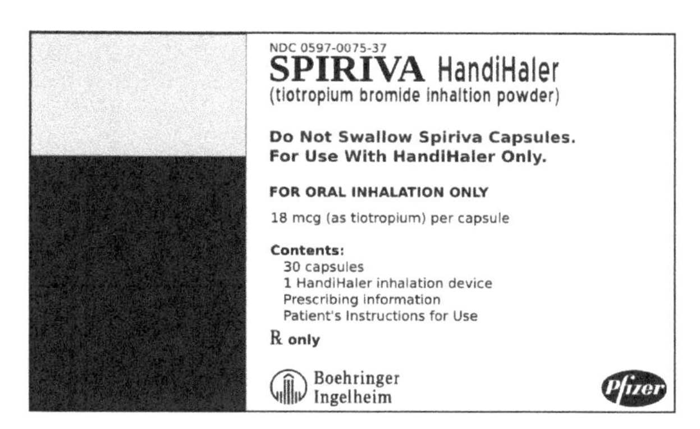

41) Terazol-7 cream 45 gm Insert 1 applicatorful (5 gm) hs × 7d

42) Rx theophylline elixir 80 mg/Tbs Disp 500 mL Sig: iv tsp po q6h atc

43) TobraDex opth. Susp 5 ml Instill 2 drops into the conjunctival sac os q6h

44) Valium 5 mg #180 i tab PO BID PRN anxiety

45) Vicodin #100 1-2 tabs q6h prn pain

46) wellbutrin 100 mg #90 i tab po tid

47) Xanax 0.25 mg #30 one t.i.d. prn

48) Rx Yaz Disp: 1 pack Sig: i tab po qd starting on first day of menstrual cycle (*Hint: look at pack below.*)

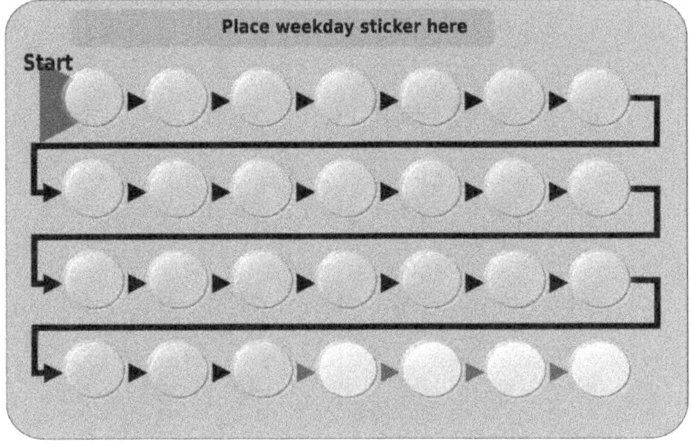

49) Z-Pak 500 mg PO on first day of therapy, then 250 mg PO once daily for 4 days. Total cumulative dose 1.5g

50) Zantac 150 mg #60 one b.i.d.

Adjusting Refills and Short-Fills

The amount of medication a pharmacy can dispense to a patient is restricted first, by the prescriber's guidelines and second by the insurer's guidelines. As we've previously stated, physicians tend to prescribe for the following time frames: 5 days, 1 week, 10 days, 2 weeks, 1 month, 50 days, 90 days or even 100 days. Meanwhile, insurances and other third party reimbursement plans tend to bill based on different time frames such as: 14 days, 21 days, 30 days, 32 days, 34 days, 90 days, etc. Calculations are often needed to first adjust the quantity dispensed to comply with the insurer's guidelines, and then the number of refills allowed. Pharmacy technicians often need to accurately estimate how long a medication will last with inadequate guidelines. Estimating days' supply is especially tricky when the dosage form is a lotion, cream, ointment, or inhalant.

When dispensing medications that come in handy convenience packages, such as a methylprednisolone dose-pack, estimate the days' supply when you have the package in your hand. You can visually examine how it is packaged and read from the labeling or package insert how long the contents of the package should last.

Let's look at a couple of example problems.

Examples

1) A prescription is written for Dyazide #50 i cap p.o. q.d. + 3 refills. The insurance plan has a 30 day supply limitation. How many capsules can be dispensed using the insurance plan guidelines and how many refills are allowed with the adjusted quantity?

   First calculate the total number of capsules allowed by the prescriber:

   $$50\,capsules \times 4\ total\ fills(original\ fill + 3\,refills) = 200\,capsules$$

   Then, using dimensional analysis we can figure out how many capsules will be needed for each fill.

   $$\frac{1\,cap}{dose} \times \frac{1\,dose}{day} \times \frac{30\,days}{fill} = \frac{30\,caps}{fill}$$

   Next, using dimensional analysis we can figure out how many fills from the insurer will be required to dispense the quantity written for by the physician.

   $$\frac{200\,caps}{1} \times \frac{fill}{30\,caps} = 6.666\ldots fills$$

   Therefore, there will be 5 refills after the initial fill is dispensed, but there is still a partial fill left.

   Once again using dimensional analysis, we will figure out how many capsules to dispense for the partial fill.

$$200\, capsules - \left(\frac{6\, fills}{1} \times \frac{30\, caps}{fill} \right) = 20\, capsules$$

Now, let's restate everything in short:

We are dispensing **30 caps for our initial fill**.
The patient can have **5 refills** of 30 caps,
and a **partial fill of 20 caps**.

2) A prescription is written for Rx Rondec-DM syrup Disp: 1 pint Sig: tsp i h.s. p.o. + 1 refill. The insurance plan has a 30 day supply limit. How many mL can be dispensed using the insurance plan guidelines and how many refills are allowed with the adjusted quantity?

First, calculate the total volume allowed by the prescriber in milliliters.

$$\frac{1\, pint}{prescriber\ fill} \times \frac{2\, prescriber\ fills}{1} \times \frac{480\, mL}{pint} = 960\, mL$$

Then, using dimensional analysis we can figure out how many milliliters will be needed for each fill.

$$\frac{1\, tsp}{dose} \times \frac{5\, mL}{tsp} \times \frac{1\, dose}{day} \times \frac{30\, days}{fill} = \frac{150\, mL}{fill}$$

Next, using dimensional analysis we can figure out how many fills from the insurer will be required to dispense the quantity written for by the physician.

$$\frac{960\, mL}{1} \times \frac{fill}{150\, mL} = 6.4\, fills$$

Therefore, there will be 5 refills after the initial fill is dispensed, but there is still a partial fill left.

Once again using dimensional analysis, we will figure out how many milliliters to dispense for the partial fill.

$$960\, mL - \left(\frac{6\, fills}{1} \times \frac{150\, mL}{fill} \right) = 60\, mL$$

Now, let's restate everything in short:

We are dispensing **150 milliliters for our initial fill**.
The patient can have **5 refills** of 150 milliliters,
and a **partial fill of 60 milliliters**.

Now let's stop and do a couple of practice problems.

Practice Problems

1) A prescription is written for Crestor 10 mg #100 i tab p.o. q.d. + 1 refill. The insurance plan has a 30 day dispensing limit. How many capsules can be dispensed using the insurance plan guidelines and how many refills are allowed with the adjusted quantity? Also, if there is a partial fill, how many tabs will be dispensed in this partial fill?

2) A prescription is written for Novolin N 10 mL #1 vial 22 units SC q am and 24 units q pm + 3 refills. The insurance plan has a 30 day supply limitation. How many vials can be dispensed using the insurance plan guidelines and how many refills are allowed with the adjusted quantity? Also, if there is a partial fill, how many vials will be dispensed in this partial fill?

1) We are dispensing 30 tabs for our initial fill. The patient can have 5 refills of 30 tabs, and a partial fill of 20 tabs.
2) We are dispensing 2 vials for our initial fill. The patient can have 1 refill of 2 vials, and there are no partial fills.

Worksheet 11-2

Name:

Date:

Appropriately adjust the following prescriptions to comply with the physician's prescribed quantities and the insurance guidelines in each scenario. In the following scenarios, treat each prescription as having a 30 day dispensing limit. Calculate the initially dispensed quantity, the number of refills, and any partial fills.

1) Rx Advair 250/50
   Disp: 1
   Sig: 1 puff q12h

   Refills: 3

2) Rx Atrovent inhalation sol. 0.02%
   Disp: 10 boxes (25 vials/box)
   Sig: 1 vial per neb q.i.d.

   Refills: 2

3) Rx Axid 150 mg
   Disp: 60
   Sig: i cap po bid before meals

   Refills: 5

4) Rx Crestor 10 mg
   #90
   Sig: i tab po qd

   Refills: 3

5) Rx Flonase 16 g
   #1
   Sig: 2 sprays per nare qd for a total daily dose of 200 mcg

   Refills: 2

NDC 0173-0453-01
GlaxoSmithKline
Flonase
(fluticasone propionate)
Nasal Spray, 50 mcg
120 Metered Sprays
50 mcg per spray
Spray-For Intranasal Use Only
ATTENTION PHARMACIST: Dispense with enclosed patient's instruction for use
Important: Read accompanying directions carefully.
16 g net fill weight
℞ only

6) Rx Fosamax 70 mg
   Disp: 4
   Sig: i tab po q week

   Refills: 2
   (*Some medications, such as once a week medications and birth control, are typically filled on 4 week cycles instead of 30 day cycles.*)

7) Rx Glucotrol 10 mg tabs
   Disp: 120
   Sig: 20 mg po bid

   Refills: 5

8) Rx Humulin R 10 mL
   Disp: 1 vial
   Sig: 11 units SC before breakfast, 9 units before lunch, and 13 units before supper

   Refills: 2

9) Rx: Januvia 100 mg
   #100
   Sig: 1 tab po qd

   Refills: 2

10) Rx Lantus 10 mL
Disp: 3 vials
Sig: 40 units SQ q a.m.

Refills: 1

11) Rx levothyroxine 100 mcg
#50
Sig: 1 tab po qd

Refills: 3

12) Rx Lexapro 20 mg
Disp: 60
Sig: i cap po qd

Refills: NR

13) Rx Lipitor 40 mg
#90
Sig: 1 tab po qd

Refills: 1

14) Rx metoprolol tartrate 25 mg tabs
#100
Sig: ss tab po bid

Refills: 2

15) Rx Nexium 20 mg
Disp: 112
Sig: i cap po bid

Refills: 1

16) Rx NitroDur 0.4 mg
#30 patches
Sig: i patch on 8 a.m., off 10 p.m. qd

Refills: 2

17) Rx Nitrolingual pumpspray
#3 bottles
Sig: i-ii spray SL q5min prn chest pain. May repeat up to 3 doses/15 min.

Refills: 3

18) Rx Patanase
Disp: 1 bottle
Sig: ii sprays each nare bid

Refills: 2

FOR INTRANASAL USE ONLY
Rx Only
NDC 0065-0332-30

Patanase®
(olopatadine hydrochloride)
Nasal Spray, 665mcg

240 Metered Sprays
Net Fill Weight 30.5g

19) Rx Plavix 75 mg
#60
Sig: 1 tab po qd

Refills: 4

20) Rx Requip 1 mg
#270
Sig: 1 tab po tid

Refills: 1

21) Rx simvastatin 20 mg
#100
Sig: i tab po qd

Refills: 2

22) Singulair 10 mg
#30
Sig: i tab po q p.m.

Refills: 11

23) Rx theophylline elixir 80 mg/Tbs 500 mL bottle
Disp 6 bottles
Sig: iv tsp po q6h atc (*Hint: dispense a whole number of bottles for this script.*)

Refills 2

24) Rx wellbutrin 100 mg
#90
Sig: i tab po tid

Refills: 5

25) Rx Zantac 150 mg
#50
Sig: one b.i.d.

Refills: 1

CHAPTER 12

COMPOUNDING MATH

Oh the things you can fill
For the folks who are ill.
With your bright shiny spatula
Oh, what a thrill.
-- unknown author

The art of pharmaceutical compounding has ancient roots dating back to early hunter gatherer societies. These ancient civilizations utilized pharmaceutical compounding for religion, grooming, keeping the healthy well, treating the ill and preparing the dead. These ancient compounders produced the first oils from plants and animals. They discovered poisons and the antidotes. They made ointments for wounded patients as well as perfumes for customers.

Today compounding is still a necessary skill for many pharmacists and pharmacy technicians. Extemporaneous compounding can be defined as the preparation, mixing, assembling, packaging, and labeling of a drug product based on a prescription from a licensed practitioner for the individual patient in a form that the drug is not readily available in (extemporaneous = impromptu, compounding = the act of combining things). Extemporaneous compounding is required for prescription orders that are not commercially available in the requested strength or dosage form.

In this chapter you will learn about:

- Reconstituting powders for oral suspensions,
- Mixing liquid preparations,
- Compounding ointments, gels, and creams,
- Medication sticks, and
- Advanced compounding calculations.

While these problems will be more complex than what you've previously done, you will find that you already know all the mathematical principles required to solve these. You will find the following skills helpful:

- Dimensional Analysis (Factor Label)
- Ratios / Proportions / Parts
- Percentage Strength
- The '5 Step Method'

Reconstituting powders for oral suspensions

Reconstituting oral suspensions is a good skill to understand as all pharmacy technicians will need to do it at some point and it is good to have forewarning of a common pitfall. It is often very important to reconstitute oral suspensions properly or you will end up with an overfilled bottle with a great deal of medication stuck in the bottom of the bottle. Let's look at an example briefly.

Example

To reconstitute a 150 mL bottle of amoxicillin for oral suspension 250 mg/5 mL, the manufacturer recommends 88 mL of distilled water is added in two divided portions. First loosen the powder in the bottle, then add approximately 1/3 of the total volume of water and shake the suspension. After the powder is wet, add the remaining water. How much water should you add each time?

QUESTION

How much water should you add each time?

DATA

final volume = 150 mL   concentration = 250 mg/5 mL   total water = 88 mL
add 1/3 of water initially   then add the rest of the water

MATHEMATICAL METHOD / FORMULA

Basic Math

DO THE MATH

$$\frac{88\,mL}{1} \times \frac{1}{3} = \textbf{\textit{add 29 mL of water initially}}$$

$$88\,mL - 29\,mL = \textbf{\textit{then add another 59 mL of water}}$$

DOES THE ANSWER MAKE SENSE?

Yes

Now, you should attempt the following practice problem.

Practice Problem

1) To reconstitute a 100 mL bottle of Augmentin for oral suspension 400 mg/5 mL, the manufacturer recommends 90 mL of distilled water is added in two divided portions. First loosen the powder in the bottle, then add approximately 1/3 of the total volume of water and shake the suspension. After the powder is wet, add the remaining water. How much water should you add each time?

1) Initially add 30 mL of water, then add another 60 mL.

Mixing liquid preparations

Sometimes you will need to determine how much to use of various products to fulfill a recipe written by a physician, sometimes you may take a recipe for a liquid medication and modify it for a different final volume, and other times you may need to either open up capsules or crush tablets and dissolve or suspend them in a liquid vehicle. Let's look at an example of each scenario.

Examples

1) How much clindamycin phosphate (the stock vial concentration is 150 mg/mL) and how much Cetaphil Lotion are needed to make the following compound?

    Rx clindamycin phosphate 1200 mg in Cetaphil Lotion
    Disp: 120 mL
    Sig: aa hs ud

    First, we should figure out how many mL of the clindamycin phosphate stock solution we should use:

$$\frac{1200\,mg}{1}\times\frac{mL}{150\,mg}=\mathbf{8\,mL\,of\,clindamycin\,phosphate}$$

    Then, we should figure out how much Cetaphil Lotion we'll need to qs this to 120 mL:

$$120\,mL-8\,mL=\mathbf{112\,mL\,of\,Cetaphil\,Lotion}$$

2) A prescription is written for a mouthwash containing 170 mL diphenhydramine elixir, 50 mL lidocaine viscous, 200 mL nystatin suspension, 52 mL erythromycin ethyl succinate suspension, and 28 mL of cherry syrup to make 500 mL of mouthwash. How much of each ingredient would be needed if you only wanted to prepare 240 mL of the mouthwash?

    First, I would make a ratio comparing each ingredient and specifying the total volume:

    170:50:200:52:28 to make 500 mL mouthwash

    Then, I would solve for how much of each ingredient is needed to make 240 mL of mouthwash:

$$\frac{170\,mL\,diphenhydramine\,elixir}{500\,mL\,mouthwash}=\frac{N}{240\,mL\,mouthwash}$$
$$N=\mathbf{81.6\,mL\,diphenhydramine\,elixir}$$

$$\frac{50\,mL\,lidocaine\,viscous}{500\,mL\,mouthwash}=\frac{N}{240\,mL\,mouthwash}$$
$$N=\mathbf{24\,mL\,lidocaine\,viscous}$$

$$\frac{200\,mL\,nystatin\,suspension}{500\,mL\,mouthwash}=\frac{N}{240\,mL\,mouthwash}$$
$$N=\mathbf{96\,mL\,nystatin\,suspension}$$

$$\frac{52\,mL\,erythromycin\,ethyl\,succinate\,suspension}{500\,mL\,mouthwash}=\frac{N}{240\,mL\,mouthwash}$$
$$N=\mathbf{25\,mL\,erythromycin\,ethyl\,succinate\,suspension}$$

$$\frac{28\,mL\,cherry\,syrup}{500\,mL\,mouthwash} = \frac{N}{240\,mL\,mouthwash}$$

$$N = \mathbf{13.4\,mL\,cherry\,syrup}$$

3) How many 25 mg tablets of metoprolol tartrate and how many milliliters of Ora-Plus and Ora Sweet are needed to compound the following prescription?

   Rx metoprolol tartrate 6.25 mg/tsp in a 50:50 mixture of Ora-Plus and Ora-Sweet
   Disp: 300 mL
   Sig: i tsp po bid

   First let's determine how many metoprolol tartrate tablets are needed:

$$\frac{tablet}{25\,mg} \times \frac{6.25\,mg}{tsp} \times \frac{tsp}{5\,mL} \times \frac{300\,mL}{1} = \mathbf{15\,tablets}$$

   You will often expect the powder volume from crushed tablets and opened capsules to be negligible, but since we don't know exactly, we will simply do the math for both liquids as if all the volume was from our two suspending agents. Since they are a 50:50 mixture it means that we only need half the volume for each suspension.

$$\frac{50}{100} = \frac{N}{300\,mL} \qquad N = \mathbf{150\,mL\,Ora-Plus}$$

$$\frac{50}{100} = \frac{N}{300\,mL} \qquad N = \mathbf{150\,mL\,Ora-Sweet}$$

Now you should try some practice problems.

Practice Problems

1) How much tobramycin (the stock vial concentration is 40 mg/mL) and how much Cetaphil Lotion are needed to prepare the following compound?

   Rx tobramycin 800 mg in Cetaphil Lotion
   Disp: 60 mL
   Sig: aa hs ud

2) A prescription is written for a G.I. Cocktail containing 120 mL Donnatal elixir, 120 mL of lidocaine viscous solution, and 480 mL of Mylanta to make a total of 720 mL of G.I. Cocktail. How much of each ingredient would be needed if you only need to prepare 120 mL of G.I. Cocktail?

3) A prescription is written for allopurinol liquid 20 mg/mL in Ora-Plus:Ora-Sweet 1:1 (label with a shelf life of 60 days). How many tablets of allopurinol 100 mg are needed to prepare 200 mL, and approximately how much Ora-Plus and Ora-Sweet are needed as well?

1) 20 mL tobramycin; 40 mL Cetaphil Lotion
2) 20 mL Donnatal elixir; 20 mL lidocaine viscous; 80 mL Mylanta
3) 40 allopurinol tablets; 100 mL Ora-Plus; 100 mL Ora-Sweet

Compounding ointments, gels, and creams

Sometimes compounding a semi-solid mixture (ointment, gel, or cream) can be as straight forward as mixing two semi-solids together, and other times it may require incorporating a medication into a semi-solid base. Let's look at an example of each.

Examples

1) A prescription is written for equal parts triamcinolone 0.1% cream and Lamisil cream, dispense 30 grams. How many grams of triamcinolone 0.1% cream are needed to fill the prescription? How many grams of Lamisil cream are needed to fill the prescription? What is the final percentage strength of triamcinolone in the compound?

   To solve this we need to first recognize that the ratio between the ingredients is 1:1 for a total of 2 parts. With that in mind, we know that half the total weight is how many grams of each ingredient we'll need.

   $\frac{1}{2}=\frac{N}{30\,g}$ Therefore, we will need **15 g of triamcinolone 0.1% cm** and **15 g of Lamisil cm**

   $N=15\,g$

   Next, we need to evaluate the final percentage strength of triamcinolone in the compound. There are 2 ways to do it, one is to calculate just how much triamcinolone is in the mixture and then figure out its percentage strength, the other is to also divide by 2 like we did the total weight. Both ways will be demonstrated, but recognize that you only have to do it one way to achieve the correct answer.

$$\frac{0.1\,g}{100\,g} = \frac{N}{15\,g}$$
$$N = 0.015\,g$$

$$\frac{0.015\,g}{30\,g} = \frac{N}{100\,g}$$
$$N = \mathbf{0.05\,\%}\ \textbf{\textit{triamcinolone}}$$

or 0.1% ÷ 2 = **0.05% *triamcinolone***

Obviously the second way was easier, but it is good to know that you will get the same answer either way.

It is also noteworthy that the methodology used in this example will apply to any compounding problem where you are mixing ingredients in equal parts.

2) If 50 g of salicylic acid ointment contains 10 grams of salicylic acid, what is the percentage strength of salicylic acid in the ointment?

   This problem is just a simple w/w percentage strength problem:

$$\frac{10\,g\,salicylic\,acid}{50\,g\,ointment} = \frac{N}{100\,g\,ointment}$$
$$N = \mathbf{20\,\%}\ \textbf{\textit{salicylic acid}}$$

Now we should once again look at some practice problems.

Practice Problems

1) How much hydrocortisone powder and how much Eucerin cream must be weighed out to prepare the following compound?

   Rx hydrocortisone 2.5% in Eucerin cream
   Disp: 60 g
   Sig: apply sparingly b.i.d. prn

2) A prescription is written for equal parts hydrocortisone 2.5% cream and Lamisil cream, dispense 60 grams. How many grams of hydrocortisone 2.5% cream are needed to fill the prescription? How many grams of Lamisil cream are needed to fill the prescription? What is the final percentage strength of hydrocortisone in this compound?

1) 1.5 g hydrocortisone powder; 58.5 g Eucerin cream
2) 30 g of hydrocortisone cream; 30 g of Lamisil cream; final mixture has 1.25% concentration of hydrocortisone

Medication Sticks

Medication sticks are a solid dosage form used in topical application of local anesthetics, sunscreens, antivirals, antibiotics, and of course cosmetics. Although cosmetic sticks are viewed as tools to improve appearance, they also may contain pharmaceutical active ingredients that serve to heal or protect. For example, a lip balm, which moisturizes the lips, may contain both an antiviral and a sunscreen for use in the treatment and prevention of herpes simplex outbreak. Sticks offer patients, physicians, and pharmacies a unique dosage form that is convenient , relatively stable, and fairly easy to prepare. The convenience comes from the fact that there are several formulas in which all you need to do is add your active ingredients. Let's look at an example problem.

Example

1) You receive the following prescription:

   Rx Acyclovir 1200 mg
   silica gel micronized 0.12 g
   PEG 4500 MW 6.5 g
   PEG 300 MW 15 mL
   Disp: tube i
   Sig: Apply to lips tid prn cold sores

   How many 200 mg acyclovir caps are needed to prepare this compound?

   If you look at this formula, you'll realize all the calculations are done for you other than figuring out how many acyclovir caps you will need to use. This calculation is fairly straight forward:

$$\frac{1200\,mg}{1} \times \frac{capsule}{200\,mg} = \mathbf{6\ acyclovir\ capsules}$$

Now let's look at a practice problem on the next page.

Practice Problem

1) A prescription is written for: valacyclovir 1000 mg, Silica gel micronized 0.12 gm, Polyethylene glycol 4500 MW 6.5 gm, Polyethylene glycol 300 MW 15 mL. How many 500 mg tablets of valacyclovir are needed to prepare this compound?

1) 2 valacyclovir tablets

With these basic compounding calculations, you are well prepared for the majority of things you will likely come in contact with in a compounding pharmacy, but some of the things that always make extemporaneous compounding exciting are the constant new and unique challenges.

Worksheet 12-1

Name:

Date:

Solve the following problems.

1) A prescription written for a toddler to receive:

    Rx cephalexin 125 mg/5 mL susp.
    Disp: 100 mL
    Sig: tsp ss po qid x10d

    When you retrieve the bottle from the shelf, you find the following reconstitution instructions: To reconstitute cephalexin 125 mg/5 mL (100 mL after reconstitution) Add 68 mL of water in two equally divided portions to the dry mixture in the bottle. Shake well after each addition. How many mL of water will you add each time?

2) How much clindamycin phosphate (the stock vial concentration is 150 mg/mL) and how much Cetaphil Lotion are needed to prepare the following compound?

    Rx clindamycin phosphate 600 mg in Cetaphil Lotion
    Disp: 60 mL
    Sig: Apply aa hs ud

3) A prescription is written for a mouthwash containing 170 mL diphenhydramine elixir, 50 mL lidocaine viscous, 200 mL nystatin suspension, 52 mL erythromycin ethyl succinate suspension, and 28 mL of cherry syrup to make 500 mL of mouthwash. How much of each ingredient would be needed if you only wanted to prepare 4 fluid ounces of this mouthwash?

4) How many 300 mg rifampin capsules are needed to compound the following solution?

    Rx Rifampin 600 mg/60 mL in Simple Syrup
    Dispense 240 mL
    Sig: 600 mg qd x 4 days

5) A SMOG enema is equal parts sorbitol solution, magnesium hydroxide suspension, mineral oil, and glycerin solution. How may mL of each would you need if you received an order for a 1 liter SMOG enema?

6) A prescription is written for ibuprofen 7.5% cream. How much ibuprofen powder is needed to prepare 60 grams of this compound?

7) A prescription is written for: acyclovir 1000 mg, Silica gel micronized 0.12 gm, Polyethylene glycol 4500 MW 6.5 gm, Polyethylene glycol 300 MW 15 mL. How many 200 mg capsules of acyclovir are needed to prepare this compound?

8) How much of each ingredient must be weighed out to prepare the following ointment?

   Rx testosterone 2% and
   menthol 4.33% in hydrophilic petrolatum
   Disp: 120 g
   Sig: apply q.i.d.

9) After you complete your calculations for the previous problem, you realize you are out of hydrophilic petrolatum. You find the following recipe to make 1000 g of hydrophilic petrolatum: cholesterol 30 g, stearyl alcohol 30 g, white wax 80 g, white petrolatum 860 g. How much of each ingredient will you need if the pharmacist asks you to make only 4 ounces?

10) You need to prepare 200 mL of metformin 100 mg/mL suspension in Ora-Plus:Ora-Sweet 1:1. How many metformin 1000 mg tablets will you need and approximately how much Ora-Plus and Ora-Sweet are needed as well?

Worksheet 12-2

Name:

Date:

Solve the following problems.

1) You are given the following recipe for compounding diclofenac gel: diclofenac sodium USP 4.8 g, ethanol 200 proof 4.8 mL, lipoil 28.8 mL, Polox 20% gel qs ad 120 g. What is the percentage strength of diclofenac sodium in this gel?

2) You are asked to compound 8 fl. oz. of glycopyrrolate 1% topical solution. Your recipe is as follows: glycopyrrolate 1 g, benzyl alcohol 0.96 mL, purified water qs 100 mL. How much of each ingredient will you need to compound 8 fluid ounces?

3) A prescription is written for phenytoin 10% in zinc oxide qs 60 gm. How many phenytoin 50 mg tablets are needed to prepare this compound?

4) You need to prepare 100 mL of potassium bromide 250 mg / mL. How much potassium bromide should you weigh?

5) You are preparing hydrocortisone 1.6 g in 160 mL Lubriderm lotion. What is the percent strength of the hydrocortisone?

6) A prescription for 240 mL of a syrup with a concentration of 10 mg of promethazine and 6.25mg of codeine per teaspoonful is ordered. Promethazine is available in a 50 mg/mL stock solution and codeine is available in a 12 mg/5 mL stock solution. You will q.s. the syrup with cherry syrup. How many mL of codeine stock solution will you need? How much promethazine will you need? How much cherry syrup will you need?

7) You need to prepare 60 mL of baclofen 10 mg/mL. How many tablets of baclofen 10 mg/tablet will you need?

8) You need to prepare 160 mL of amiodarone 5 mg/mL suspension. How many tablets of amiodarone 200 mg/tablet will you need?

9) You need to prepare 60 mL of celecoxib 100 mg/5 mL. How many capsules of 200 mg celecoxib/capsule will you need?

10) You receive the following script:

    Rx Mudd Mixture
    Disp: 184 mL
    Sig: swish and swallow 23 mL q6h x 2 days

    For every 23 mL of Mudd mixture, you have 20 mL of nystatin (100,000 units/mL), 2 mL of gentamicin (40 mg/mL), and 1 mL of colistimethate (20 mg/mL). How many milliliters of each ingredient are you going to need to fill this script?

11) You receive a prescription requesting 120 mL of an acetazolamide suspension 25 mg/mL in a 50:50 mixture of Ora-Plus and Ora-Sweet. How many 250 mg acetazolamide tablets and approximately how many mL each of Ora-Plus and Ora-Sweet are needed to compound this prescription?

12) You receive a prescription requesting 4 fl oz of a 1 mg/mL amlodipine suspension in a 50:50 mixture of Ora-Plus and Ora-Sweet. How many 5 mg amlodipine tablets and approximately how many mL each of Ora-Plus and Ora-Sweet are needed to compound this prescription?

13) You receive the following prescription for a pediatric patient:

> Rx atenolol suspension 2 mg/mL
> in Oral Diluent
> Disp: 150 mL
> Sig; i tsp po qd

How many atenolol 25 mg tablets will you need to compound this prescription?

14) You need to prepare 120 mL of azathioprine 50 mg/mL suspension by crushing 50 mg tablets of azathioprine and then qs with cherry syrup. How many azathioprine tablets do you need to prepare this suspension?

15) You are asked to make 60 mL of a 5 mg/mL baclofen suspension with 20 mg baclofen tablets, a small amount of glycerin to function as a levigating agent and then qs with simple syrup. How many baclofen tablets are needed?

Name:

Date:

Worksheet 12-3

Solve the following problems.

1) Using the directions on the following label, how many mL of water will you add each time?

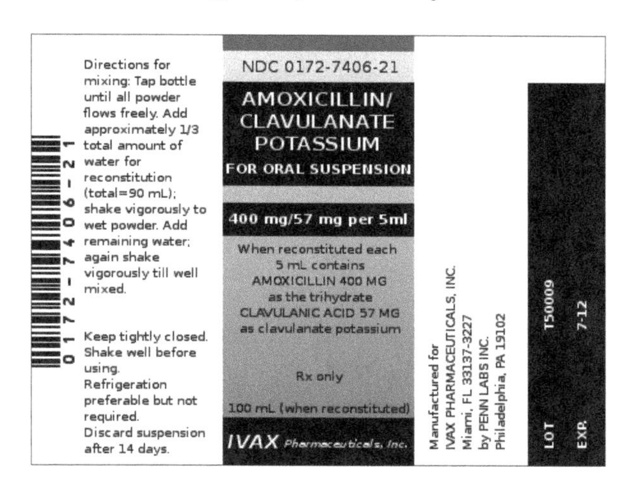

2) A prescription is written for a G.I. Cocktail containing 120 mL Donnatal elixir, 120 mL of lidocaine viscous solution, and 480 mL of Mylanta to make a total of 720 mL of G.I. Cocktail. How much lidocaine viscous would be needed if you only need to prepare 120 mL of the G.I. Cocktail?

3) A prescription is written for ichthammol ointment 2 oz. How much of each ingredient is needed to prepare 2 oz. if you are using the following formula: 100 g of ichthammol, 100 g of lanolin, and 800 g of white petrolatum to make 1000 g of ichthammol ointment.

4) You are asked to compound a pint of Schamberg's lotion. You are told to base your calculations off of the following formula: zinc oxide 8 g, menthol 0.25 g, phenol 0.5 g, calcium hydroxide solution 46 mL, olive oil qs ad 100 mL. How much of each ingredient will you need to prepare a pint of this compound?

5) You receive the following prescription:

Rx diclofenac sodium 8% in Pentravan cream
Disp: 60 grams
Sig: aa bid ut dict

How many grams of each ingredient will be needed to make this compound?

6) You receive the following prescription:

Rx tetracycline HCl susp. 125 mg/tsp in
50:50 mixture of Ora-Plus and Ora-Sweet
Disp: 300 mL
Sig: tsp i po bid

How many 250 mg tetracycline capsules and approximately how many mL of Ora-Plus and Ora-Sweet will you need to make this compound?

7) You receive the following prescription for hand rolled lozenges:

Rx benzocaine HCl	100 mg
powdered sugar	10 g
acacia	700 mg
water	qs
food coloring and flavoring	gtt v aa

Disp: M et Ft 10 lozenges c 10 mg benzocaine each
Sig: dissolve in mouth prn mouth sores

The pharmacists suggests you do your calculations for 12 lozenges, and when you're done you can simply discard the heaviest and the lightest lozenge. How much of each ingredient should you measure?

8) A prescription is written for: vitamin E 1,000 IU, zinc oxide 100 mg, Silica gel micronized 0.12 gm, Polyethylene glycol 4500 MW 6.5 gm, Polyethylene glycol 300 MW 15 mL. How many mL of vitamin E are needed to prepare this lip balm if your stock vitamin E has a concentration of 100 g/100 mL (1 mg = 1.1 IU)?

9) An erythromycin ophthalmic ointment requires you to mix 1 part erythromycin 2% (sterile) concentrate with 3 parts ophthalmic base (sterile) ointment. How many grams of each would you need to dispense 50 g of erythromycin ophthalmic ointment and what would be the percent concentration of erythromycin in the final product?

10) The pharmacy receives a prescription requesting 120 mL of a 0.1 mg/mL clonazepam suspension in a 50:50 mixture of Ora-Plus and Ora-Sweet. How many 1 mg clonazepam tablets and approximately how many mL each of Ora-Plus and Ora-Sweet are needed to compound this prescription?

11) You receive a request for 120 mL of 1 mg/mL bethanechol solution and you are to qs it with sterile water for irrigation. How many 10 mg bethanechol tablets are needed to compound this solution?

12) The pharmacy receives a prescription requesting 2 fl oz of a 5 mg/mL bethanechol suspension in a 50:50 mixture of Ora-Plus and Ora-Sweet. How many 10 mg bethanechol tablets are needed to compound this prescription?

13) The pharmacy receives the following prescription to prepare a 27 kg pediatric patient for a bone marrow transplant:

Rx busulfan suspension 2 mg/mL in Simple Syrup
Disp: 16 doses
Sig: 30 mg po q6h

a) How many milliliters of busulfan suspension will the patient receive for each dose?

b) How many milliliters of busulfan suspension will be needed to cover all 16 doses?

c) How many 2 mg busulfan tablets are required to make this suspension?

14) The pharmacy needs to prepare 300 mL of diltiazem suspension 12 mg/mL using 90 mg diltiazem tablets and a 50:50 mixture of Ora-Plus and Ora-Sweet. How many diltiazem tablets and approximately how many milliliters each of Ora-Plus and Ora-Sweet are needed to make this compound?

15) You receive the following prescription for a 9 lb 14 oz feline with lower back pain:

> Rx Gabapentin suspension chicken flavor 10 mg/mL
> Disp: 180 mL
> Sig: Day 1 ~ 20 mg po, Day 2 ~ 20 mg po bid, then 20 mg po tid thereafter

Your recipe for 100 mL of this suspension is as follows: gabapentin 1g, xanthum gum 0.4 g, stevia 0.75 g, acesulfame 0.75 g, sodium saccharin 0.1 g, magnasweet solution 0.2 mL, 1% citric acid approximately 0.1 mL (use to obtain pH of 5.5 to 6.5), sodium chloride 0.5 g, bitter stopping agent flavor 1 mL, glycerin 2 mL, chicken flavor 3 mL, bacteriostatic water q.s. 100 mL. How much of each ingredient will you need to compound the requested 180 mL?

Advanced Compounding Calculations

There are several things to consider with respect to advanced compounding calculations including:

- calculations involving specific gravity,
- accounting for excipients when compounding,
- suppositories and density factors, and
- determining shell sizes for extemporaneously compounded capsules.

Calculations Involving Specific Gravity

Sometimes when compounding mixtures, the recipes will provide you with weights of all the various ingredients including liquids. Obviously, it is easier to measure a liquid by volume than mass. Water has the unique advantage that 1 gram of water has a volume of 1 milliliter, but other liquids do not share this convenient conversion. When dealing with liquids other than water, you will need to know their specific gravity.

Specific gravity is commonly defined as the mass of 1 milliliter of a particular substance. Therefore, if I stated that a particular liquid had a specific gravity of 0.8, it means that every milliliter of it weighs 0.8 grams. If a particular compound requested 5 grams of a liquid known to have a specific gravity of 0.8, I could determine the volume as follows:

$$\frac{5\,g}{1} \times \frac{1\,mL}{0.8\,g} = \mathbf{6.25\,mL}$$

Some common specific gravities that you may end up working with include:

Substance	*Specific Gravity (g/mL)*
glycerin	1.249
honey	1.40-1.45
mineral oil	0.845-0.905
olive oil	0.910-0.915
stearyl alcohol	0.805-0.815
water	1

You will notice that with the list above there is often a slight range for specific gravity. When dealing with this range, it is generally acceptable to use the average value (add the low value of the range to the high value and divide by 2). As an example, if a prescription required 5 grams of honey, I would first determine the average specific gravity as follows:

$$\frac{1.40 + 1.45}{2} = 1.425$$

Then I would find the volume as follows:

$$\frac{5g}{1} \times \frac{mL}{1.425g} = \mathbf{3.5\ mL}$$

Now let's attempt a couple of practice problems.

Practice Problems

1) A prescription is written for 60 g of zinc oxide ointment, USP. If the formula for zinc oxide ointment, USP is as follows: 200 g of zinc oxide powder, 650 g of white ointment, and 150 gram of mineral oil to make 1000 grams of zinc oxide ointment, USP; how much of each ingredient will you need to fill this order? Using the information from the previous page (mineral oil has a specific gravity of 0.845 to 0.905) determine your quantity of mineral oil required for this prescription in milliliters. (*Note, since this preparation is commercially available you would not ordinarily compound it. This problem is intended for educational purposes.*)

2) Use the prescription below and information from the specific gravity chart on the previous page to determine how many grams of cocoa butter, how many grams of white petrolatum, and how many milliliters of olive oil you'll need to compound this prescription.

David M. Ferguson, M.D.
Contemporary Physician Group Practice
3459 5th Avenue, Pittsburgh, PA 15206
Tel: (412) 555-1234 Fax: (412) 555-2345

Name *Arid D Erma* Date *8-19-2010*
Address Age Wt/Ht

℞ *anhydrous emollient dry skin lotion*
1 part olive oil : 2 parts cocoa butter :
2 parts white petrolatum (1X/ : 1X/ : 1X/)
Disp: 2 ℥
Sig Apply to dry skin qhs prn

Refills *prn*

David Ferguson M.D. M.D.
Product Selection Permitted Dispense As Written

DEA No.

Prescription No.: 00000103

1) 12 g of zinc oxide powder, 39 g of white ointment, and 10.3 mL of mineral oil 2) 22.72 g of cocoa butter, 22.72 g of white petrolatum, 12.4 mL of olive oil

Accounting for Excipients when Compounding

Before we start looking at how to do calculations where we need to account for excipients, we should define what excipients are. Excipients are “inert” or inactive ingredients other than the active drug which are included in the manufacturing process or are contained in a finished pharmaceutical dosage form. They allow the proper delivery of the active compounds contained in nearly all over the counter and prescription medications.

At the risk of sounding like *'Mary Poppins'*, almost every drug needs a few inactive ingredients to help the medicine go down. Depending on the route of administration, and form of medication, various excipients may be used. According to the USP-NF, all excipients fit into one or more of 40 categories. The ingredients are classified by the functions they perform in a pharmaceutical dosage form. Some examples are antiadherents, binders, coatings, colors, disintegrants, fillers, flavors, glidants, lubricants, preservatives, sweeteners, and printing inks.

Example

1) A compound prescription order calls for 2200 mg of naproxen. You have available in your pharmacy 500 mg naproxen tablets. Each 500 mg naproxen tablet weighs 630 mg due to all the excipients required to manufacture it. To achieve this, we need to first figure out how many naproxen tablets you will need to pull out of stock to make this (you will need to automatically round up any decimal values). Then, you can triturate the naproxen tablets in a mortar and pestle and weigh out the required quantity of crushed naproxen tablets to provide the desired quantity of medication. Below is the math required to perform this task.

   $$\frac{2200\,mg}{1} \times \frac{tablet}{500\,mg} = 4.4\,tablets = \mathbf{5}\ \boldsymbol{tablets}\ will\ need\ removed\ from\ stock.$$

   Now that we know we will need 5 tablets from stock, you can crush them up and use a ratio proportion to determine how much you will need to weigh out.

   $$\frac{500\,mg\ naproxen}{630\,mg\ naproxen\ plus\ excipients} = \frac{2200\,mg\ naproxen}{N}$$

   $$N = \mathbf{2772}\ \boldsymbol{mg\ naproxen\ plus\ excipients}$$

   Therefore, after you triturate 5 naproxen tablets in your mortar and pestle, you will need to weigh out 2,772 mg of the crushed powder to acquire 2,200 mg of naproxen.

Let's proceed to the next page where you can do a practice problem involving calculations where you need to account for excipients.

Practice Problem

1) You need to prepare 120 mL of metronidazole suspension with a concentration of 59 mg/5 mL. How many 250 mg tablets will you need, and if each tablet weighs 430 mg, how will you determine the appropriate weight of the tablets to use after you triturate them for the requested suspension?

You will need to use 6 tablets, and after they are crushed, you will weigh out 2,436 mg.

Suppositories and Density Factors

Suppositories are defined by the USP-NF as follows:

> *Suppositories are solid bodies of various weights and shapes, adapted for introduction into the rectal, vaginal, or urethral orifice of the human body. They usually melt, soften, or dissolve at body temperature. A suppository may act as a protectant or palliative to the local tissue at the point of introduction or as a carrier of therapeutic agents for systemic or local action.*

Suppositories may be used when a local effect is needed in the rectum, vagina, or urethra.

Rectal suppositories (and to a lesser extent, vaginal suppositories) may also be used as carriers of systemic drugs. Rectal suppositories offer an alternative for the systemic delivery of drugs in patients who can not take drugs orally. Examples include patients who are unconscious, those who are vomiting or having seizures, and those who have obstructions in the upper gastrointestinal tract.

Some drugs that are ineffective orally may be successfully administered rectally or vaginally. Examples include drugs that are extensively metabolized by first-pass effect and drugs that are destroyed in the stomach or intestine. An example of a drug that is usually administered either rectally or vaginally for those reasons include progesterone.

Compounding suppositories is usually done as a last resort. This is because suppositories are, in general, more difficult to prepare than other dosage forms.

Suppositories are usually made of one of two bases:

- *polyethylene glycol (PEG)* which dissolves, or
- *cocoa butter* which melts at body temperature.

There are also two common methods for preparing suppositories:

- *hand-rolling suppositories*, which does not require any special equipment, but lacks pharmaceutical elegance, or
- *fusion suppositories*, provide much greater pharmaceutical elegance, but require either aluminum or disposable molds, and also require more calculations.

In this chapter, we are going to look at the more common (although more complex to calculate) cocoa butter based fusion suppositories. When doing these calculations, you will need to know how many grams of your base can be held by the suppository molds, and then figure out how much of your base to use when mixed with the active ingredients added to a suppository. Unfortunately, I can not just subtract the weight of the active ingredient from the weight of our aforementioned total weight to figure out how much cocoa butter is needed because the active ingredient does not occupy the same volume as that mass of cocoa butter (they have different densities). We can use the table on this page to find the active ingredient's density factor (DF) when compared to cocoa butter.

Density Factors for Cocoa Butter Suppositories[1] compared to the amount of weight (g) required to fill the same volume as 1 gram of cocoa butter.

Medication	Factor	Medication	Factor
Aloin	1.3	Iodoform	4.0
Alum	1.7	Menthol	0.7
Aminophylline	1.1	Morphine hydrochloride	1.6
Aminopyrine	1.3	Opium	1.4
Aspirin	1.1	Paraffin	1.0
Barbital	1.2	Pentobarbital	1.2
Belladonna extract	1.3	Perovian balsam	1.1
Benzoic acid	1.5	Phenobarbital	1.2
Bismuth carbonate	4.5	Phenol	0.9
Bismuth salicylate	4.5	Potassium bromide	2.2
Bismuth subgallate	2.7	Potassium iodide	4.5
Bismuth subnitrate	6.0	Procaine	1.2
Boric acid	1.5	Quinine hydrochloride	1.2
Castor oil	1.0	Resorcinol	1.4
Chloral hydrate	1.3	Salicylic acid	1.3
Cocaine hydrochloride	1.3	Secobarbital sodium	1.2
Codeine phosphate	1.1	Sodium bromide	2.3
Digitalis leaf	1.6	Spermaceti	1.0
Dimenhydrinate	1.3	Sulfathiazole	1.6
Diphenhydramine hydrochloride	1.3	Tannic acid	1.6
Gallic acid	2.0	White wax	1.0
Glycerin	1.6	Witch hazel fluid extract	1.1
Hydrocortisone acetate	1.5	Zinc oxide	4.0
Ichthammol	1.1	Zinc sulfate	2.8

Example

1) You receive a prescription requiring you to compound 6 cocoa butter based suppositories with

1 If a medicament of unknown density compared to cocoa butter is used, you may multiply the medicament's desired weight by 0.7 to obtain an estimate of how much cocoa butter is being displaced.

100 mg of aspirin in each. You know from previous work with these particular suppository molds that they can each hold 2 grams of cocoa butter. How much cocoa butter and how much aspirin powder will you need to compound this order?

First, we should determine how much cocoa butter these six suppositories could hold.

$$\frac{6\,supp}{1} \times \frac{2\,g}{1\,supp} = 12\,g\,cocoa\,butter$$

Now we should look at how much aspirin powder is required.

$$\frac{6\,supp}{1} \times \frac{100\,mg}{1\,supp} = 600\,mg\,aspirin\,powder$$

Since cocoa butter and aspirin have different densities, we will need to look at the chart on the previous page to determine how much mass of cocoa butter is displaced by our 600 mg of aspirin powder.

$$\frac{600\,mg\,aspirin}{1} \times \frac{1\,g\,cocoa\,butter}{1.1\,g\,aspirin} = 545\,mg\,cocoa\,butter$$

Now we can subtract the 545 mg from our previous number of 12 g to determine how much cocoa butter we actually need.

$$12\,g - 0.545\,g = 11.455\,g\,cocoa\,butter\,required$$

So now we know how much of each ingredient we need.

11.455 g cocoa butter
600 mg aspirin

Now, you should attempt a practice problem.

Practice Problem

1) You receive a prescription requiring you to compound 12 cocoa butter based suppositories with 30 mg of phenobarbital in each. You know from previous work with these particular suppository molds that they can each hold 2.08 grams of cocoa butter. How much cocoa butter and how much phenobarbital powder will you need to compound this order?

You will need to use 24.66 g of cocoa butter and 360 mg phenobarbital.

Determining Shell Sizes for Extemporaneously Compounded Capsules

The gelatin shells used in capsules are made of two parts. The base is the longer end and fits into the shorter end which is referred to as the cap. The cap is designed to fit over the base and then snap or lock into place with added pressure.

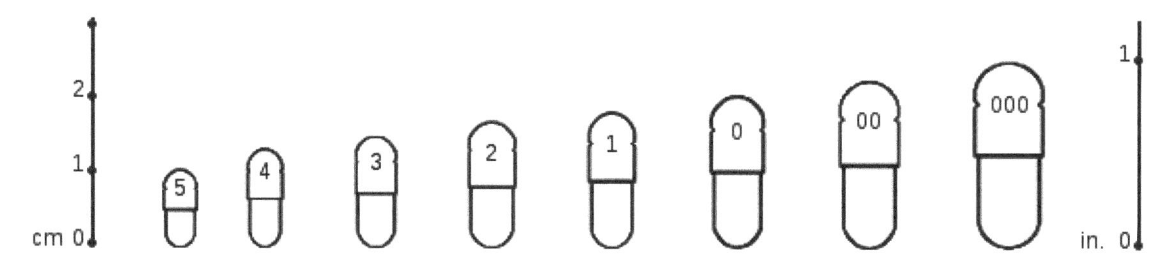

Capsules are oval in shape and available in eight different sizes for human use. These sizes are #5, #4, #3, #2, #1, #0, #00, and #000, with the smallest being a #5 and the largest being #000. The numbers used to designate size have no bearing on the volume that may be contained within. The capacity of a capsule is dependent on the density and physical characteristics of the powders used in the formula. (Larger capsules are available for veterinary use.)

The approximate capacity of various capsules can be found on the following chart:

	Capsule Size							
	5	4	3	2	1	0	00	000
Drug Substance	Capacity in grams of drug powder							
Acetaminophen	0.13	0.18	0.24	0.31	0.42	0.54	0.75	1.10
Aluminum hydroxide	0.18	0.27	0.36	0.47	0.64	0.82	1.14	1.71
Ascorbic acid	0.13	0.22	0.31	0.40	0.53	0.70	0.98	1.42
Aspirin	0.10	0.15	0.20	0.25	0.33	0.55	0.65	1.10
Bismuth subnitrate	0.12	0.25	0.40	0.55	0.65	0.80	1.20	1.75
Calcium carbonate	0.12	0.20	0.28	0.35	0.46	0.60	0.79	1.14
Calcium lactate	0.11	0.16	0.21	0.26	0.33	0.46	0.57	0.80
Corn starch	0.13	0.20	0.27	0.34	0.44	0.58	0.80	1.15
Lactose	0.14	0.21	0.28	0.35	0.46	0.60	0.85	1.25
Quinine sulfate	0.07	0.10	0.12	0.20	0.23	0.33	0.40	0.65
Sodium bicarbonate	0.13	0.26	0.32	0.39	0.52	0.70	0.97	1.43

When looking at the shell capacities for various drugs on the chart above, note that the number listed is the maximum quantity that you can pack in a shell that size. For example, if you wanted to place 300 mg (0.300 g) of acetaminophen in a capsule, you could fit it in a size #2 capsule, but if you wanted to put 325 mg (0.325 g) of acetaminophen, you would need to use a size #1 capsule.

When determining capsule size, it becomes more complex when looking at capsules with multiple drug additives. If one additive is going to take the majority of the volume, you may simply want to add the weights of all your drugs and then base your capsule size on the ingredient requiring the majority of the room. Let's look at an example problem to demonstrate this.

Example

1) You receive an order for a hospice patient requesting compounded capsules with 15 mg of hydrocodone bitartrate and 325 mg of acetaminophen. What size capsule shell will you need?

   As acetaminophen is the ingredient that is going to make up the bulk of the capsule, we will base our capsule size off that. First we need to determine our total weight.

   $325\,mg + 15\,mg = 340\,mg$

   Based on the chart from the previous page, you will need to use a size #1 capsule shell.

Practice Problems

1) What size capsule shell will you need if a physician requests 2 grains of aspirin to be dispensed in a capsule for a patient?

2) You receive an order for a hospice patient requesting 30 compounded capsules with 20 mg of oxycodone, 120 mg aspirin, and 300 mg acetaminophen. What size capsule shell will you need?

1) Size #4 capsules 2) Size #0 capsules

Worksheet 12-4

Name:

Date:

Solve the following problems.

1) You receive a request to make 'Pittsburgh Paste', which gets its name from its golden yellow color. It consists of Aquaphor, cholestyramine powder, and mineral oil (which acts as a levigating agent). You find the following compounding recipe for it:

   PITTSBURGH PASTE

Aquaphor	80 g
Cholestyramine Powder	5.9 g
Mineral Oil	20 mL

   Auxiliary Labeling: TOPICAL USE ONLY

   Expiration: 28 DAYS

   Mineral oil has a specific gravity of 0.845-0.905, so what is the percent concentration of cholestyramine powder in this compound?

2) A prescription is written for ibuprofen 7.5% cream. You are out of ibuprofen powder and will need to crush 200 mg ibuprofen tablets (each tablet weighs 0.33 g) to prepare this compound. How many tablets are you going to triturate and how much weight of these crushed tablets are needed to prepare 30 grams of this compound?

3) A pharmacy receives a prescription for 12 cocoa butter based suppositories with 200 mg of procaine each. The suppository mold in the pharmacy can hold 2.27 g of cocoa butter per suppository. How much cocoa butter and how much procaine will be needed to make these suppositories?

4) Based on the prescription below and the capsule chart earlier in this chapter, what size capsule shell should be used when preparing these capsules?

David M. Ferguson, M.D.
Contemporary Physician Group Practice
3459 5th Avenue, Pittsburgh, PA 15206
Tel: (412) 555-1234 Fax: (412) 555-2345

Name *Anna L. Gesia* Date *9-2-2010*
Address Age Wt/Ht

℞ *Codeine Phosphate* *gr 1/4*
Lactose anhydrous *200 mg*
M et Ft Capsules #10
Sig: prn pain x5D

Refills *NR*

David Ferguson M.D. M.D.
Product Selection Permitted Dispense As Written

DEA No. *BF6428521*

Prescription No.: **00000105**

5) While rarely seen today, Brompton Cocktails are still occasionally ordered for terminally ill patients, to provide them with comfort and promote sociability at end of life. Brompton Cocktail's received their name from where it was originally created during the early 19th century at the Royal Brompton Hospital in London, England. The recipe for the cocktail tends to vary between institutions. You are working at a hospital with a terminally ill cancer patient and the physician wants to give the patient a Brompton Cocktail. After a brief discussion, the pharmacist and the physician settle on the following recipe:

BROMPTON'S COCKTAIL

Morphine Solution 10 mg/mL	6 mL
Cocaine HCl powder	67.2 mg
Simple Syrup	15 mL
90% Ethanol	39 mL
TOTAL VOLUME	60 mL

Auxiliary Labeling: Keep Refrigerated, Oral Use Only

Expiration: Seven Days

The pharmacist is very busy and needs you to calculate several things for labeling the bottle.

a) What is the new percentage strength of the ethanol?

b) How many mg of morphine are in a dose of 1 teaspoon?

c) You are told that 1.12 g of cocaine HCl equals 1 g of cocaine, so how many mg of cocaine are in 1 teaspoon?

6) A prescription is written for a patient that has a hard time swallowing tablets and suffers from hypothyroidism:

Rx: Levothyroxine Na Suspension
Disp: 30 day supply
Sig: 100 mcg po qd

You find the following recipe in your compounding log:

LEVOTHYROXINE NA 25 MCG / mL SUSPENSION

Levothyroxine Na 0.1 mg tablets	25 tabs
Glycerin	40 mL
Distilled Water	q.s. 100 mL

Instructions: Crush levothyroxine Na tablets and triturate with glycerin, which acts as a levigating agent and rinse for mortar and pestle. Q.S. with distilled water to obtain a total volume of 100 mL.

Auxiliary Labeling: Shake Well, Oral Use Only, Protect From Light, Keep Refrigerated

How would you rewrite the recipe to provide the requested 30 day supply?

7) The pharmacist requests that you prepare 500 g of bentonite magma (used as a suspending medium for other drugs). The recipe for bentonite magma NF is as follows:

BENTONITE MAGMA

Bentonite	50 g
Purified Water	q.s. ad 1,000 g

Rewrite this recipe for 500 g, and how many mL of water will be needed since water has a specific gravity of 1?

8) A compound prescription order calls for 90 g of a 7% naproxen ointment. You have available in your pharmacy 500 mg naproxen tablets. Each 500 mg naproxen tablet weighs 630 mg due to all the excipients required to manufacture it. To achieve this, we need to first figure out how many naproxen tablets you will need to pull out of stock to make this. Then, you can triturate the naproxen tablets in a mortar and pestle and weigh out the required quantity of crushed naproxen tablets to provide the desired quantity of medication to compound this prescription. How many tablets will need triturated and what weight of crushed up tablets will then need to be weighed out?

9) A pharmacy receives a prescription for 12 cocoa butter based Rectal Rocket® suppositories with 200 mg of procaine, 100 mg of hydrocortisone acetate, and 60 mg of witch hazel fluid extract (witch hazel has a specific gravity of 0.979-0.983) each. These suppository molds can hold 4 g of cocoa butter per suppository.

a) How many grams of procaine will be needed to make these suppositories?

b) How many grams of hydrocortisone acetate will be needed to make these suppositories?

c) How many grams and how many milliliters of witch hazel fluid extract are needed to make these suppositories?

d) How many grams of cocoa butter will be needed to make these suppositories?

10) A physician has a patient with legitimate pain issues but also has a history of prescription drug abuse. After a brief discussion between the pharmacist and the patient's physician, they decide on dispensing 30 capsules with 5 mg of hydrocodone bitartrate, 325 mg of acetaminophen, and 5 mg of capsaicin each. What size capsule shells will you need to compound this prescription?

Worksheet 12-5

Name:

Date:

Solve the following problems.

1) A prescription written for a pediatric patient to receive:

    Rx cefadroxil 250 mg/5 mL susp.
    Disp: 100 mL
    Sig: i tsp po bid x10d

    When you retrieve the bottle from the shelf you find the following reconstitution instructions: To reconstitute cefadroxil 250 mg/5 mL (100 mL after reconstitution) tap bottle lightly to loosen powder. Add 61 mL of water in two equally divided portions to the dry mixture in the bottle. Shake well after each addition. How many mL of water will you add each time?

2) A prescription written for a pediatric patient to receive:

    Rx cefixime 100 mg/5 mL susp.
    Disp: 75 mL
    Sig: iss tsp po qd x10d

    When you retrieve the bottle from the shelf you find the following reconstitution instructions: To reconstitute cefixime 100 mg/5 mL (75 mL after reconstitution) tap bottle lightly to loosen powder. Add 52 mL of water in two equally divided portions to the dry mixture in the bottle. Shake well after each addition. How many mL of water will you add each time?

3) How much lidocaine HCl (the stock vial concentration is 40 mg/mL) and how much Cetaphil Lotion are needed to prepare the following compound?

    Rx lidocaine HCl 1200 mg in Cetaphil Lotion
    Disp: 120 mL
    Sig: apply lightly to aa q4h prn itching and burning

4) From the following formula, calculate the number of grams of each ingredient required to prepare 60 grams of this ointment:

Precipitated sulfur	10 g
Salicylic acid	2 g
Hydrophilic ointment	88 g

5) Using the same formula as in the previous problem, determine how many grams of each ingredient would be required to prepare a pound of this ointment.

6) A patient comes in suffering from mouth sores and presents the following prescription:

Rx Magic Swizzle
Disp: 480 mL
Sig: i Tbs swish and spit q6h prn mouth pain

Magic Swizzle is a 1:1:1 ratio of mixing viscous lidocaine, diphenhydramine elixir, and magnesium aluminum hydroxide suspension. How many mL of each are needed to prepare this solution?

7) You receive the following script:

Rx Mudd Mixture
Disp: 4 day supply
Sig: swish and swallow 23 mL q6h

For every 23 mL of Mudd Mixture, you have 20 mL of nystatin (100,000 units/mL), 2 mL of gentamicin (40 mg/mL), and 1 mL of colistimethate (20 mg/mL).

a) How many milliliters of Mudd Mixture do you need to dispense in total?

b) How many milliliters of each ingredient are you going to need to fill this script?

c) Nystatin is available in 1 pint bottles, gentamicin is available in 20 mL vials, and colistimethate comes as a lyophilized powder with 150 mg/vial. How many bottles of nystatin, vials of gentamicin, and vials of colistimethate will you need to compound this order?

8) Due to a product shortage, the pharmacist asks you to compound 60 mL of a 15 mg/mL oseltamivir suspension using 75 mg capsules and cherry syrup. How many 75 mg oseltamivir capsules will be needed for compounding this suspension?

9) The following is a recipe intended for veterinary use:

METRONIDAZOLE AND SILVER SULFADIAZINE CREAM

Metronidazole	1 g
Silver sulfadiazine	1 g
Glycerin	5 g
Hydrophyllic ointment	q.s. 100 g

a) How many grams of each ingredient would be needed to prepare 2 oz of this cream?

b) Considering that glycerin has a specific gravity of 1.249, how many mL of glycerin will be needed for this preparation?

10) You receive the following prescription for hand rolled lozenges:

Rx lidocaine HCl	100 mg
powdered sugar	10 g
acacia	700 mg
water	qs
food coloring and flavoring	gtt v aa

Disp: M et Ft 10 lozenges c 10 mg lidocaine each
Sig: dissolve in mouth prn mouth sores

The pharmacists suggests you do your calculations for 12 lozenges and when you're done you can simply discard the heaviest and the lightest lozenge.

a) How much of each ingredient should you measure?

b) If your stock lidocaine HCl solution has a concentration of 40 mg/mL how many milliliters will you need?

11) A prescription is written for: diphenhydramine 250 mg, Silica gel micronized 0.12 gm, Polyethylene glycol 4500 MW 6.5 g, Polyethylene glycol 300 MW 15 mL. How many 50 mg tablets of diphenhydramine are needed to prepare this compound?

12) The pharmacy needs to make a 150 mL of a 0.05 mg/mL alprazolam suspension. If each 2 mg alprazolam tablet weighs 80 mg, how many 2 mg alprazolam tablets will you need to pull out of stock, and after they are triturated how many grams of crushed alprazolam tablets will you weigh out?

13) Calculate the quantities required to make six cocoa butter based suppositories (each mold can hold 2.17 g of cocoa butter), each containing 100 mg aminophylline (aminophylline has a density factor of 1.1 when compared to 1 g of cocoa butter).

14) A physician has a patient with legitimate pain issues but also has a history of prescription drug abuse. Because of other medications the patient may be taking that contain acetaminophen, the physician does not want any additional acetaminophen in this product. After a brief discussion between the pharmacist and the patient's physician, they decide on dispensing 30 capsules with 5 mg of hydrocodone bitartrate, 325 mg of aspirin, and 5 mg of capsaicin each. What size capsule shells will you need to compound this prescription?

15) A pharmacist asks you to prepare a bottle of “Mile's Solution”. You find the following recipe in the book:

MILE'S SOLUTION

Prednisone Elixir 5 mg/5 mL	120 mL
Tetracycline 500 mg caps	3 caps
Cherry Syrup	60 mL
Nystatin Suspension 100,000 u/mL	30 mL
Diphenhydramine Elixir 12.5 mg/5 mL	75 mL
Sterile Water for Irrigation	240 mL
TOTAL VOLUME	525 mL

Auxiliary Labeling: Shake Well, Oral Use Only, Protect From Light

Expiration: Two Weeks

You gather all the supplies, and find that the only prednisone elixir you have in stock has a concentration of 5 mg/mL. You bring this to the pharmacist's attention and she tells you to adjust how much prednisone you add so that you end up with the correct concentration and change the amount of cherry syrup added to make up the difference in the total volume.

a) How much prednisone elixir 5 mg/mL are you going to add?

b) How much cherry syrup are you going to add?

Chapter 13

Calculations for Billing Compounds

You may have been thrilled
to help the folks who were ill,
but now you must bill
for the prescriptions you filled.
--Sean Parsons

Billing for compounded preparations can vary greatly from pharmacy to pharmacy, but the following formula is a common method for calculating the cost:

	Cost of ingredients
+	Dispensing fee
+	Cost of time
	Final Cost

Cost of ingredients – Just like the phrase implies, it is the cost of all the ingredients used to prepare a compound.

Dispensing fee – This represents the charge for the professional services provided by the pharmacy when dispensing a prescription and includes a distribution of the costs involved in running the pharmacy such as salaries, rent, utilities, costs associated with maintaining the computer system, etc. This is sometimes also referred to as a professional fee.

Cost of time – As compounding a prescription is often more time consuming then just filling a traditional prescription, the pharmacy will need to charge for the time required from the staff to make these preparations.

Let's look at a couple of example problems, one where we've already been given the cost of ingredients and another where we need to calculate the cost of ingredients.

Examples

1) Calculate the cost of 50 g of a 20% salicylic acid ointment that took 15 minutes to prepare if the cost of ingredients were only $1.03, your pharmacy has a standard dispensing fee of $5.00, and your cost of time is based on $35.00 per hour.

   Since we already know our cost of ingredients and our dispensing fee, the only thing we need to figure out before we add everything up is our cost of time.

$$\frac{15\,minutes}{1} \times \frac{\$35.00}{60\,minutes} = \$8.75$$

now we can simply add all of our costs together.

	Cost of ingredients	\$ 1.03
+	Dispensing fee	\$ 5.00
+	Cost of time	\$ 8.75
	Final Cost	**\$14.78**

2) You make the following compound in approximately 10 minutes:

Rx clindamycin phosphate 1200 mg in Cetaphil Lotion
Disp: 120 mL
Sig: aa hs ud

Use the following information to determine the final cost.

clindamycin phosphate 150 mg/mL, 4 mL vials, \$4.12/vial
Cetaphil Lotion 480 mL bottle, \$9.44/bottle
dispensing fee is \$5.00
cost of time is \$35.00/hour

We will need to determine the cost of each ingredient.

clindamycin phosphate

$$\frac{1200\,mg}{1} \times \frac{mL}{150\,mg} = 8\,mL$$

$$\frac{8\,mL}{1} \times \frac{vial}{4\,mL} \times \frac{\$4.12}{vial} = \$8.24$$

Cetaphil Lotion

$$120\,mL - 8\,mL = 112\,mL$$

$$\frac{112\,mL}{1} \times \frac{bottle}{480\,mL} \times \frac{\$9.44}{bottle} = \$2.20$$

You've been given the dispensing fee, but you'll need to determine the cost of time.

$$\frac{10\,minutes}{1} \times \frac{\$35.00}{60\,minutes} = \$5.83$$

Now you can add all your costs up.

	Cost of ingredients	$ 8.24
+		$ 2.20
+	Dispensing fee	$ 5.00
+	Cost of time	$ 5.83
	Final Cost	**$21.27**

Now we should look at some practice problems.

Practice Problems

1) Calculate the cost of 60 g of a 7.5% ibuprofen cream that took 20 minutes to prepare if the cost of ingredients were only $2.48, your pharmacy has a standard dispensing fee of $5.00, and your cost of time is based on $35.00 per hour.

2) You compound the following prescription in approximately 15 minutes:

    Rx metoprolol tartrate 6.25 mg/tsp in a 50:50 mixture of Ora-Plus and Ora-Sweet
    Disp: 300 mL
    Sig: i tsp po bid

    Use the following information to determine the final cost. (*Hint: expect the powder volume from the crushed up metoprolol tartrate tablets to be negligible*)

    metoprolol tartrate, 25 mg/tablet, $0.08/tablet
    Ora-Plus, 473 mL bottle, $33.02/bottle
    Ora-Sweet, 473 mL bottle, $33.02/bottle
    dispensing fee is $5.00
    cost of time is $35.00/hour

1) $19.15 2) $35.89

Worksheet 13-1

Name:

Date:

Solve the following problems using the formula:

	Cost of ingredients
+	Dispensing fee
+	Cost of time
	Final Cost

Use $5.00 for the dispensing fee and a rate of $35.00 per hour for the cost of time calculations.

1) Calculate the cost for 60 g of a 10% phenytoin in zinc oxide ointment that takes 20 minutes to prepare if the cost of ingredients is $3.17.

2) Calculate the cost for 120 g of a 2% testosterone and 4.33% menthol in hydrophilic petrolatum that takes takes 30 minutes to prepare if the cost of ingredients is $11.09.

3) Calculate the cost for 60 mL of celecoxib 100 mg/5 mL suspension if it took 10 minutes to prepare and the ingredients to make it cost $22.97.

4) Calculate the cost for 50 g of erythromycin ophthalmic ointment that requires 35 minutes to prepare if the cost of ingredients comes to $53.50.

5) Calculate the cost for 11 g of a 2% naproxen gel if it takes 20 minutes to prepare and the ingredients to compound it cost $21.86.

6) How much should you charge to dispense 300 mL of a mouthwash that has a recipe of 170 mL diphenhydramine elixir, 50 mL lidocaine viscous, 200 mL nystatin suspension, 52 mL erythromycin ethyl succinate suspension, and 28 mL cherry syrup to make 500 mL of mouthwash if it took you 20 minutes to prepare. (*Note: you will need to determine how much of each ingredient you actually needed to prepare this suspension in order to calculate your costs.*)

diphenhydramine elixir, 12.5 mg/5 mL, 120 mL/bottle, $7.36/bottle
lidocaine viscous, 2%, 100 mL/bottle, $3.42/bottle
nystatin suspension, 100,000 units/mL, 473 mL/bottle, $68.29/bottle
erythromycin ethyl succinate suspension, 200 mg/5 mL, 480 mL/bottle, $23.50/bottle
cherry syrup, 480 mL/bottle, $6.24/bottle

7) You compound the following prescription in approximately 15 minutes:

Rx allopurinol liquid 20 mg/mL in a 1:1 mixture of Ora-Plus and Ora-Sweet
Disp: 200 mL
Sig: 2/3 tsp po bid p meals

Use the following information to determine the final cost. (*Hint: expect the powder volume from the crushed up allopurinol tablets to be negligible*)

allopurinol, 100 mg/tablet, $0.08/tablet
Ora-Plus, 473 mL bottle, $33.02/bottle
Ora-Sweet, 473 mL bottle, $33.02/bottle

8) You compound a prescription for 30 g of equal parts triamcinolone 0.1% cream and Lamisil cream in approximately 15 minutes. How much should you charge for this compound?

triamcinolone, 0.1% cream, $3.75/80 g
Lamisil cream, $32.00/15 g

9) You compound a prescription for 60 g of 2.5% hydrocortisone in Eucerin cream in approximately 20 minutes. How much should you charge for this compound?

hydrocortisone powder, $175.00/100 g
Eucerin cream, $9.45/454 g

10) How much should the following compound cost if it took 10 minutes to prepare?

Rx Rifampin 600 mg/60 mL in Simple Syrup
Disp: 240 mL
Sig: 600 mg qd x 4 days

Rifampin, 300 mg/capsule, $1.89/capsule
Simple Syrup, 16 fl. oz./bottle, $16.45/bottle

Worksheet 13-2

Name:

Date:

Solve the following problems using the formula:

	Cost of ingredients
+	Dispensing fee
+	Cost of time
	Final Cost

Use $5.00 for the dispensing fee and a rate of $35.00 per hour for the cost of time calculations.

1) You compound a prescription for 60 g of equal parts hydrocortisone 2.5% cream and Lamisil cream in approximately 15 minutes. How much should you charge for this compound?

   hydrocortisone, 2.5% cream, $5.46/30 g
   Lamisil cream, $32.00/15 g

2) The pharmacy prepared 300 mL of diltiazem suspension 12 mg/mL in approximately 15 minutes using 90 mg diltiazem tablets and a 50:50 mixture of Ora-Plus and Ora-Sweet. How much should you charge based on the following information?

   diltiazem, 90 mg/tablet, $0.10/tablet
   Ora-Plus, 473 mL bottle, $33.02/bottle
   Ora-Sweet, 473 mL bottle, $33.02/bottle

3) You make the following compound in approximately 10 minutes:

    Rx tobramycin 800 mg in Cetaphil Lotion
    Disp: 60 mL
    Sig: aa hs ud

   Use the following information to determine the final cost.

   tobramycin, 40 mg/mL, 30 mL MDV, $27.50/vial
   Cetaphil Lotion, 480 mL bottle, $9.44/bottle

4) How much should you charge to dispense 120 mL of G.I. Cocktail that has a recipe of 120 mL Donnatal elixir, 120 mL lidocaine viscous, and 480 mL of Mylanta to make 720 mL of G.I. Cocktail if it took you 20 minutes to prepare. (*Note: you will need to determine how much of each ingredient you actually needed to prepare this suspension in order to calculate your costs.*)

   Donnatal elixir, 473 mL/bottle, $67.45/bottle
   lidocaine viscous, 2%, 100 mL/bottle, $3.42/bottle
   Mylanta, 720 mL/bottle, $8.89/bottle

5) How much should you charge to make the following medication stick if it took 30 minutes to prepare:

    Rx Acyclovir 1200 mg
    silica gel micronized 0.12 g
    PEG 4500 MW 6.5 g
    PEG 300 MW 15 mL
    Disp: tube i
    Sig: Apply to lips tid prn cold sores

   acyclovir, 200 mg/capsule, $0.15/capsule
   silica gel micronized, $30.00/100 g
   polyethylene glycol (PEG) 4500 MW, $37.10/500 g
   polyethylene glycol (PEG) 300 MW, $37.10/500 mL

6) The pharmacy prepared 200 mL of metformin 100 mg/mL suspension in approximately 15 minutes using 1000 mg metformin tablets and a 50:50 mixture of Ora-Plus and Ora-Sweet. How much should you charge based on the following information?

   metformin, 1000 mg/tablet, $1.44/tablet
   Ora-Plus, 473 mL bottle, $33.02/bottle
   Ora-Sweet, 473 mL bottle, $33.02/bottle

7) The pharmacy prepared 60 mL of a 5 mg/mL baclofen suspension in 20 minutes with 20 mg baclofen tablets, 5 mL of glycerin, and a sufficient quantity of simple syrup. Assuming that the powder volume from crushing the baclofen tablets was negligible, how much should you charge for this compounded prescription?

   baclofen, 20 mg/tablet, $0.09/tablet
   glycerin solution, $2.97/480 mL
   Simple Syrup, 16 fl. oz./bottle, $16.45/bottle

8) The pharmacy prepared 160 mL of amiodarone 5 mg/mL suspension in approximately 15 minutes using 200 mg amiodarone tablets and a 50:50 mixture of Ora-Plus and Ora-Sweet. How much should you charge based on the following information?

   amiodarone, 200 mg/tablet, $3.30/tablet
   Ora-Plus, 473 mL bottle, $33.02/bottle
   Ora-Sweet, 473 mL bottle, $33.02/bottle

9) The pharmacy prepared 120 mL of acetazolamide suspension 25 mg/mL in approximately 15 minutes using 250 mg acetazolamide tablets and a 50:50 mixture of Ora-Plus and Ora-Sweet. How much should you charge based on the following information?

acetazolamide, 250 mg/tablet, $0.44/tablet
Ora-Plus, 473 mL bottle, $33.02/bottle
Ora-Sweet, 473 mL bottle, $33.02/bottle

10) You make the following compound in approximately 10 minutes:

Rx atenolol suspension 2 mg/mL
in Cherry Syrup
Disp: 150 mL
Sig: i tsp po qd

Use the following information to determine the final cost.

atenolol, 25 mg/tablet, $0.10/tablet
cherry syrup, 480 mL/bottle, $6.24/bottle

Worksheet 13-3

Name:

Date:

Solve the following problems using the formula:

	Cost of ingredients
+	Dispensing fee
+	Cost of time
	Final Cost

Some pharmacies may use a sliding scale for their dispensing fees based on the cost of the ingredients used. Use the following chart to determine the dispensing fee based on the cost of the ingredients used:

Cost of Ingredients	*Dispensing Fee*
Less than $20.00	$3.50
$20.00 - $50.00	$5.00
Greater than $50.00	$7.50

Continue using a rate of $35.00 per hour for the cost of time calculations.

1) How much should you charge for the following medication stick if it took 30 minutes to prepare:

   Rx valacyclovir 1000 mg
   silica gel micronized 0.12 g
   PEG 4500 MW 6.5 g
   PEG 300 MW 15 mL
   Disp: tube i
   Sig: Apply to lips tid prn cold sores

   Valtrex (valacyclovir), 500 mg/tablet, $5.93/tablet
   silica gel micronized, $30.00/100 g
   polyethylene glycol (PEG) 4500 MW, $37.10/500 g
   polyethylene glycol (PEG) 300 MW, $37.10/500 mL

2) How much should be charged for a 1 liter SMOG enema if it takes 10 minutes to prepare? A SMOG enema is equal parts sorbitol solution, magnesium hydroxide suspension, mineral oil, and glycerin.

   70% sorbitol solution, $5.40/480 mL
   magnesium hydroxide suspension, 400 mg/5 mL, $2.06/480 mL
   mineral oil, $65.98/1000 mL
   glycerin solution, $2.97/480 mL

3) You make the following compound in approximately 10 minutes:

   Rx clindamycin phosphate 600 mg in Cetaphil Lotion
   Disp: 60 mL
   Sig: aa hs ud

   Use the following information to determine the final cost.

   clindamycin phosphate, 150 mg/mL, 4 mL vials, $4.12/vial
   Cetaphil Lotion, 480 mL bottle, $9.44/bottle

4) The pharmacy prepared 160 mL of a 1% hydrocortisone in Lubriderm Lotion in approximately 15 minutes. When calculating the cost for this prescription assume the powder volume from the hydrocortisone powder to be negligible.

   hydrocortisone powder, $175.00/100 g
   Lubriderm Lotion, $7.36/480 mL

5) You make the following compound in approximately 25 minutes:

> Rx promethazine 30 mg and codeine 18.75 mg per Tbs
> q.s. with cherry syrup
> Disp: 20 fl oz
> Sig: Tbs i po q4° prn

Use the following information to determine the final cost.

promethazine, 50 mg/mL, 1 mL/vial, $0.50/vial
codeine phosphate, 15 mg/mL, 2 mL/vial, $2.24/vial
cherry syrup, 480 mL/bottle, $6.24/bottle

6) The pharmacy prepared 4 fluid ounces of a 1 mg/mL amlodipine suspension in approximately 20 minutes using 5 mg amlodipine tablets and a 50:50 mixture of Ora-Plus and Ora-Sweet. How much should you charge based on the following information?

amlodipine, 5 mg/tablet, $1.85/tablet
Ora-Plus, 473 mL bottle, $33.02/bottle
Ora-Sweet, 473 mL bottle, $33.02/bottle

7) You make the following compound in approximately 20 minutes:

> Rx tetracycline HCl susp. 125 mg/tsp in
> 50:50 mixture of Ora-Plus and Ora-Sweet
> Disp: 300 mL
> Sig: tsp i po bid

Use the following information to determine the final cost.

tetracycline, 250 mg/tablet, $0.07/tablet
Ora-Plus, 473 mL bottle, $33.02/bottle
Ora-Sweet, 473 mL bottle, $33.02/bottle

8) The pharmacy prepared 120 mL of a 0.1 mg/mL clonazepam suspension in approximately 20 minutes using 1 mg clonazepam tablets and a 50:50 mixture of Ora-Plus and Ora-Sweet. How much should you charge based on the following information?

clonazepam, 1 mg/tablet, $0.90/tablet
Ora-Plus, 473 mL bottle, $33.02/bottle
Ora-Sweet, 473 mL bottle, $33.02/bottle

9) The pharmacy prepared 2 fluid ounces of a 5 mg/mL bethanechol suspension in approximately 20 minutes using 10 mg bethanechol tablets and a 50:50 mixture of Ora-Plus and Ora-Sweet. How much should you charge based on the following information?

bethanechol, 10 mg/tablet, $0.08/tablet
Ora-Plus, 473 mL bottle, $33.02/bottle
Ora-Sweet, 473 mL bottle, $33.02/bottle

10) The pharmacy compounded 8 fluid ounces of 1% glycopyrrolate topical solution with an appropriate quantity of glycopyrrolate powder, 1.7 mL of benzyl alcohol, and q.s. of purified water. How much should you charge if this medication took 10 minutes to prepare?

glycopyrrolate powder, $75.46/1 g
benzyl alcohol, $19.60/500 mL
Pharmacy maintains own water purifier and does not charge for it.

Worksheet 13-4

Name:

Date:

Solve the following problems using the formula:

	Cost of ingredients
+	Dispensing fee
+	Cost of time
	Final Cost

Some pharmacies may use a sliding scale for their dispensing fees based on the cost of the ingredients used. Use the following chart to determine the dispensing fee based on the cost of the ingredients used:

Cost of Ingredients	*Dispensing Fee*
Less than $20.00	$3.50
$20.00 - $50.00	$5.00
Greater than $50.00	$7.50

Continue using a rate of $35.00 per hour for the cost of time calculations.

1) How much should you charge to dispense a 4% diclofenac gel compounded with diclofenac powder, 4.8 mL of 200 proof ethanol, 28.8 mL of lipoil, qs ad 120 g with 20% Polox gel (this will require approximately 70 mL of 20% Polox gel). This preparation took 35 minutes to prepare and below are the average wholesale prices for each ingredient.

   diclofenac sodium, USP, $156.00/100 g
   denatured ethyl alcohol 200 proof, $13.65/118.25 mL
   lipoil, $18.20/473 mL
   Polox 20% gel, $21.35/473 mL

2) You need to charge for 2 ounces of ichthammol ointment that took 20 minutes to prepare. The recipe for 1 kilogram of ichthammol ointment is as follows: 100 g of ichthamol, 100 g of lanolin and 800 g of white petrolatum. Below are the AWPs for the ingredients.

   ichthammol powder, USP, $52.01/454 g
   lanolin powder, USP, $27.03/454 g
   white petrolatum, USP, $10.22/454 g

3) You make the following compound in 25 minutes:

   Rx Mudd Mixture
   Disp: 184 mL
   Sig: swish and swallow 23 mL q6h x 2 days

   For every 23 mL of Mudd mixture, you use 20 mL of nystatin (100,000 units/mL), 2 mL of gentamicin (40 mg/mL), and 1 mL of colistimethate (20 mg/mL). Calculate the charge for this extemporaneous compound.

   nystatin, 100,000 units/mL, 473 mL/bottle, $68.29/bottle
   gentamicin, 40 mg/mL, 20 mL/vial, $8.25/vial
   colistimethate, 150 mg/vial, $54.72/vial

4) You need to charge for the following compound that took 15 minutes to prepare:

   Rx diclofenac sodium 8% in Pentravan cream
   Disp: 60 grams
   Sig: aa bid ut dict

   Use the prices below to determine how much to charge.

   diclofenac sodium, USP, $156.00/100 g
   Pentravan cream, $33.29/454 g

5) The pharmacy prepared 30 mL of a 0.1 mg/mL clonidine suspension in approximately 15 minutes using 0.2 mg clonidine tablets and Simple Syrup. How much should you charge based on the following information?

   clonidine, 0.2 mg/tablet, $0.05/tablet
   Simple Syrup, 16 fl. oz./bottle, $16.45/bottle

6) How much should you charge for the following medication stick if it took 30 minutes to prepare:

   Rx vitamin E 1000 IU
   zinc oxide 100 mg
   silica gel micronized 0.12 g
   PEG 4500 MW 6.5 g
   PEG 300 MW 15 mL
   Disp: tube i
   Sig: Apply to lips tid prn cold sores

   vitamin E, 100 g/100 mL (1 mg = 1.1 IU), $70.70/100 mL
   zinc oxide powder, USP, $10.85/454 g
   silica gel micronized, $30.00/100 g
   polyethylene glycol (PEG) 4500 MW, $37.10/500 g
   polyethylene glycol (PEG) 300 MW, $37.10/500 mL

7) The pharmacy prepared 100 mL of a 2 mg/mL dapsone suspension in approximately 15 minutes using 25 mg dapsone tablets and a 50:50 mixture of Ora-Plus and Ora-Sweet. How much should you charge based on the following information?

   dapsone, 25 mg/tablet, $0.20/tablet
   Ora-Plus, 473 mL bottle, $33.02/bottle
   Ora-Sweet, 473 mL bottle, $33.02/bottle

8) The pharmacy compounded 4 fluid ounces of a 50 mg/tsp dipyridamole suspension in approximately 15 minutes using 50 mg dipyridamole tablets and a 50:50 mixture of Ora-Plus and Ora-Sweet. How much should you charge based on the following information?

   dipyridamole, 50 mg/tablet, $0.05/tablet
   Ora-Plus, 473 mL bottle, $33.02/bottle
   Ora-Sweet, 473 mL bottle, $33.02/bottle

9) The pharmacy prepared 100 mL of a 10 mg/mL disopyridamole suspension in approximately 10 minutes using 100 mg disopyridamole capsules and cherry syrup. How much should you charge based on the following information?

   disopyridamole, 100 mg/capsule, $0.06/capsule
   cherry syrup, 480 mL/bottle, $6.24/bottle

10) The pharmacy compounded 120 mL of a levodopa/carbidopa suspension with a concentration of 25 mg of levodopa and 6.25 mg of carbidopa per teaspoonful in approximately 15 minutes using levodopa 100 mg/carbidopa 25 mg tablets and a 50:50 mixture of Ora-Plus and Ora-Sweet. How much should you charge based on the following information?

    levodopa 100 mg/carbidopa 25 mg tablets, $0.05/tablet
    Ora-Plus, 473 mL bottle, $33.02/bottle
    Ora-Sweet, 473 mL bottle, $33.02/bottle

Chapter 14

Pharmacy Business Math

Money is always there, but the pockets change.
--unknown author

Being skilled in the art and practice of pharmacy is both exciting and important, but for the long term success of any pharmacy (even a non-profit pharmacy) it must achieve good cash flow and be able to meet its financial goals. This is an area where pharmacy technicians are rapidly expanding their roles in both community and institutional settings.

Today, technicians commonly handle inventory purchasing and receiving while assuring proper turnover and any necessary data maintenance involved with such responsibilities. They work with insurance companies on behalf of their pharmacy (sometimes technicians also work directly for insurance companies). Also, qualified technicians may have an opportunity to move into management positions where they need to give consideration to capital expenditures and justify professional fees. A pharmacy technician with solid business skills can achieve many things for the various practice settings they may work in.

In this chapter we will look at the following concepts:

- Terminology,
- Inventory management processes,
- Storage requirements,
- Markups/discounts,
- Gross profits and net profits,
- Third party reimbursement,
- Daily cash reports,
- Calculating dispensing fees, and
- Depreciation.

Terminology

To understand the concepts of this chapter you need to familiarize yourself with some basic terminology as it relates to pharmacy business math.

inventory - This is simply the entire stock on hand for sale at a given time.

inventory value - The total value of the drugs and merchandise in stock on a given day.

perpetual inventory - A system that maintains a continuous count of every item in inventory so that it always shows the stock on hand. Some pharmacies maintain perpetual inventories on all products while others only do this with their schedule II medications.

reorder point - Minimum and maximum stock levels which determine when a reorder is placed and for how much.

formulary - A list of medications available for use within a health care system. There are two major types of formularies, open formularies and closed formularies.

open formulary - The pharmacy must stock, or have ready access to, all drugs that may be written by the physicians in their practice area.

closed formulary - The drug inventory is limited to a list of approved medications.

purchasing - Purchasing is the ordering of products for use or sale by the pharmacy, and is usually carried out by either an independent or group process.

direct purchasing - This entails ordering medications directly from the original drug manufacturer. This typically requires completion of a purchase order, generally a preprinted form with a unique number, on which the product name(s), amount(s), and price(s) are entered.

wholesaler purchasing - Wholesaler purchasing enables the pharmacy to use a single source to purchase numerous products from numerous manufacturers. Most drug ordering of this fashion is done on-line.

prime vendor purchasing - This involves an agreement made by a pharmacy for a specified percentage or dollar volume of purchases in exchange for being given lower acquisition costs.

schedule II medications - Schedule II medications must be stocked separately in a secure place or distributed throughout your inventory and require a DEA 222 form for reordering. Their stock must be continually monitored and documented.

Occupational Safety and Health Administration (OSHA) - OSHA is a government agency within the United States Department of Labor responsible for maintaining safe and healthy work environments.

Material Safety Data Sheet (MSDS) - OSHA-required notices on hazardous substances which provide hazard, handling, clean-up, and first aid information.

gross profit - Gross profit is calculated as sales minus all costs directly related to those sales, or in simpler terms we can think of it as the difference between the selling price and the acquisition price.

net profit - Net profit is the difference between the gross profit and the sum of all the costs associated with filling the prescription. The costs associated with filling the prescription are accounted for with a dispensing fee or a professional fee. With that in mind, you can determine the net profit by subtracting a professional fee from the gross profit.

dispensing fee – A dispensing fee (also referred to as a professional fee) represents the charge for the professional services provided by the pharmacy when dispensing a prescription and includes a distribution of the costs involved in running the pharmacy such as salaries, rent, utilities, costs associated with maintaining the computer system, etc.

average wholesale price - This is the average price at which wholesalers typically sell medications to pharmacies.

usual and customary price – Usual and customary price is commonly known as the retail price and is the price paid for a prescription by a patient without insurance.

third party reimbursement - This is reimbursement for services rendered to a person in which an entity other than the receiver of the service is responsible for the payment. Third-party reimbursement for the cost of a subscriber's prescriptions is commonly paid in part by a health insurance plan or other prescription benefit manager, such as Highmark, Medco, Medicare, or Medicaid.

capitation fee - A method of payment for health services in which an individual or institutional provider is paid a fixed amount without regard to the actual number or nature of services provided to each patient. A common example would be a pharmacy in a long term care home receiving a fixed amount of money per month regardless of whether the patient required no pharmaceuticals or if their prescriptions exceeded the money allotted by a capitation fee.

capital expenditures - Money spent to acquire or upgrade physical assets such as property, fixtures, or machinery (i.e., a building, shelving, and computers). Capital expenditures are not for day-to-day operations such as payroll, inventory, maintenance, and advertising. This is also called capital spending or capital expense.

depreciation - This is the decline in value of assets. As an example, a pharmacy delivery car declines in value as it becomes older and obtains more mileage.

Worksheet 14-1

Name:

Date:

Fill in the following crossword puzzle using the terminology presented in this chapter. The clues are on the back of this worksheet.

1 2 3 4 5 6 7 8 9 10 11 12 13 14 15 16 17 18 19 20 21 22

Across

4) This is simply the entire stock on hand for sale at a given time.
6) These must be stocked separately in a secure place or distributed throughout your inventory and require a DEA 222 form for reordering. Their stock must be continually monitored and documented.
7) The total value of the drugs and merchandise in stock on a given day.
10) This involves an agreement made by a pharmacy for a specified percentage or dollar volume of purchases in exchange for being given lower acquisition costs.
14) OSHA-required notices on hazardous substances which provide hazard, handling, clean-up, and first aid information.
16) A system that maintains a continuous count of every item in inventory so that it always shows the stock on hand.
17) The drug inventory is limited to a list of approved medications.
18) A method of payment for health services in which an individual or institutional provider is paid a fixed amount without regard to the actual number or nature of services provided to each patient.
20) This is reimbursement for services rendered to a person in which an entity other than the receiver of the service is responsible for the payment.
21) Minimum and maximum stock levels which determine when a reorder is placed and for how much.
22) This is the decline in value of assets.

Down

1) A government agency within the United States Department of Labor responsible for maintaining safe and healthy work environments.
2) This is the difference between the gross profit and the sum of all the costs associated with filling the prescription.
3) This entails ordering medications directly from the original drug manufacturer.
5) This is commonly known as the retail price and is the price paid for a prescription by a patient without insurance.
8) The pharmacy must stock, or have ready access to, all drugs that may be written by the physicians in their practice area.
9) This represents the charge for the professional services provided by the pharmacy when dispensing a prescription, and includes a distribution of the costs involved in running the pharmacy such as salaries, rent, utilities, costs associated with maintaining the computer system, etc.
11) This enables the pharmacy to use a single source to purchase numerous products from numerous manufacturers. Most drug ordering of this fashion is done on-line.
12) This is the average price at which wholesalers typically sell medications to pharmacies.
13) Money spent to acquire or upgrade physical assets such as property, fixtures, or machinery.
15) This is calculated as sales minus all costs directly related to those sales, or in simpler terms, we can think of it as the difference between the selling price and the acquisition price.
16) This is the ordering of products for use or sale by the pharmacy and is usually carried out by either an independent or group process.
19) A list of medications available for use within a health care system.

Inventory management processes

The primary purpose of inventory management is the timely purchase and receipt of pharmaceuticals and to establish and maintain appropriate levels of materials in stock. Purchasing, receiving, and inventory should be as uncomplicated as possible so as not to disrupt or to interfere with the other activities of the pharmacy. Community pharmacies usually maintain an open formulary, which means they must stock, or have ready access to, all drugs that may be written by the physicians in their practice area. In the community pharmacy, most pharmaceutical products are purchased through a local wholesaler. Specialty pharmacies and hospital pharmacies use a closed formulary, which means the drug inventory is limited to a list of approved medications. To minimize the cost of doing business, inventory levels must be adequate, but not excessive, with a rapid turnover of drug stock on the shelf. The purchasing, receipt, and inventory of controlled-drug substances requires special procedures and record-keeping requirements.

Purchasing

Purchasing is the ordering of products for use or sale by the pharmacy and is usually carried out by either an independent or group process. In independent purchasing, the pharmacist or technician deals directly with a drug wholesaler (or rarely the pharmaceutical manufacturer) regarding matters such as price and terms. In group purchasing, a number of pharmacies work together to negotiate a discount for high-volume purchases and more favorable contractual terms. Several purchasing methods are used in pharmacy including Direct Purchasing, Wholesaler Purchasing, and Prime Vendor Purchasing.

Direct purchasing entails ordering medications directly from the original drug manufacturer. This typically requires completion of a purchase order, generally a preprinted form with a unique number on which the product name(s), amount(s), and price(s) are entered.

Advantages to direct purchasing

- lower cost
- lack of add-on fees

Disadvantages to direct purchasing

- a commitment of time as it will take longer than other methods to receive drugs
- a commitment of staff since there are multiple requisitions to be completed and mailed to multiple pharmaceutical companies

Wholesaler purchasing enables the pharmacy to use a single source to purchase numerous products from numerous manufacturers. Most drug ordering of this fashion is done on-line, although gathering information may be done in a number of different ways, such as writing items on a 'want book', walking the shelves and scanning items that need reordered into a portable bar code scanner (an example is pictured to the right), or many pharmacy management software programs will automatically populate a reorder list when the pharmacy stock reaches a predetermined reorder point. A system that maintains a continuous inventory record is known as perpetual inventory.

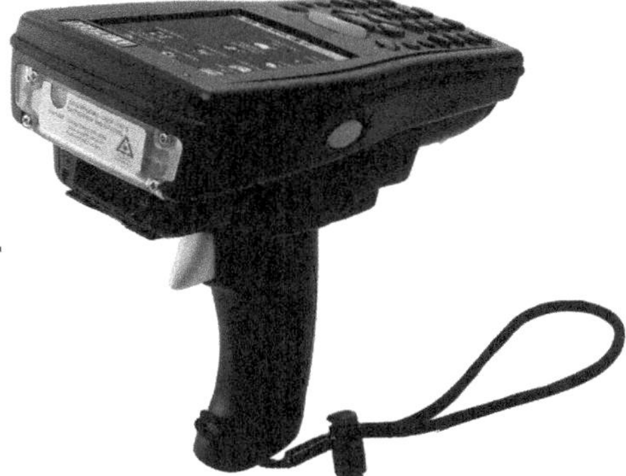

Advantages to wholesaler purchasing

- reduced turnaround time for orders
- lower inventory and associated costs
- reduced commitment of time and staff

Disadvantages to wholesaler purchasing

- a higher purchase cost
- supply difficulties
- loss of control provided by in-house purchase orders
- unavailability of some pharmaceuticals

Sometimes to offset some of the increased costs associated with using a wholesaler, as opposed to direct purchasing, pharmacies will establish a contract identifying a particular wholesaler as a prime vendor purchaser. Prime vendor purchasing involves an agreement made by a pharmacy for a specified percentage or dollar volume of purchases in exchange for being given lower acquisition costs.

Advantages to prime vendor purchasing

- lower acquisition costs
- competitive service fees
- electronic order entry
- often, emergency delivery services
- promotes just in time (JIT) purchasing

Disadvantages to prime vendor purchasing

- limits ability to use other wholesalers
- in term of JIT, it can only be used when supplies are readily available and needs can be accurately predicted

Controlled substances, as expected, require special consideration when it comes to purchasing. The Controlled Substance Act (CSA) defines procedures for purchasing and receiving and requirements for inventory and record keeping.

Schedule III – V drugs (see Chapter 8 for information on drug schedules) may be ordered by a pharmacy or other appropriate dispensary on a general order from a wholesaler, and you should check the delivery in against the original order.

Schedule II drugs have much more stringent requirements. A pharmacy must register with the Drug Enforcement Administration (DEA) to purchase Schedule II medications. The purchase of such controlled substances must be authorized by a pharmacist and executed on either a triplicate DEA 222 order form or an electronic 222 form through a controlled substances ordering system (CSOS)

The DEA form 222 (pictured to the right) is a triplicate form. The pharmacy retains the third sheet while sending the first and second pages to the wholesaler. The wholesaler is responsible for sending the second page to the DEA, while retaining the first page for its own records. When the Schedule II medications arrive in the pharmacy, they should be checked in against the DEA form.

On the next page is a chart explaining how an electronic 222 form works using CSOS.

Sample DEA Form 222

See Reverse of PURCHASER's Copy for Instructions | No order form may be issued for Schedule I and II substance unless completed application form has been received. (21 CFR 1305.04) | OMB APPROVAL NO. 1117-0010

TO: | STREET ADDRESS

CITY and STATE | DATE | TO BE FILLED IN BY SUPPLIER

SUPPLIER DEA REGISTRATION No.

	TO BE FILLED IN BY PURCHASER					
	No. of Packages	Size of Package	Name of Item	National Drug Code	Packaging Shipped	Date Shipped
1						
2						
3						
4						
5						
6						
7						
8						
9						
10						

◄ LAST LINE COMPLETED (MUST BE 10 OR LESS) | SIGNATURE OF PURCHASER OR ATTORNEY OR AGENT

Date Issued | DEA Registration No. | Name and Address of Registrant

Schedules

Registered as a | No. of this Order Form

DEA Form 222 (Oct. 1992)

US OFFICIAL ORDER FORMS - SCHEDULES I & II

DRUG ENFORCEMENT ADMINISTRATION

SUPPLIER'S Copy 1

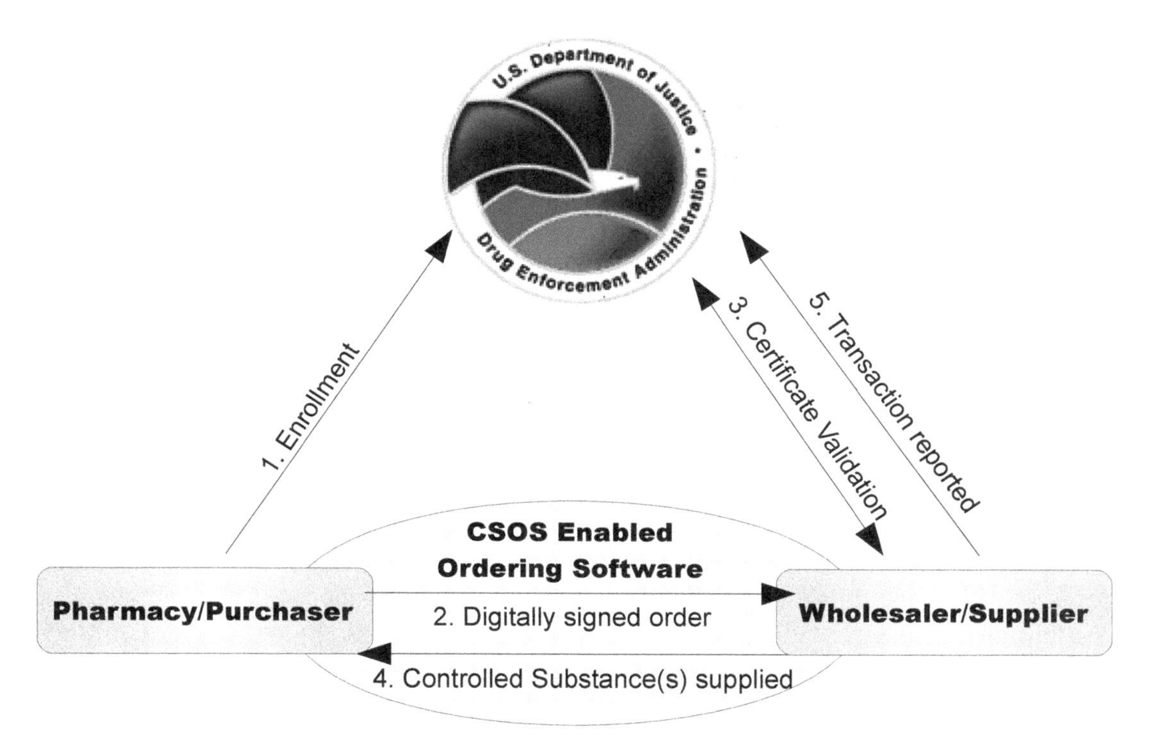

1) An individual enrolls with the DEA and, once approved, is issued a personal CSOS Certificate.
2) The purchaser creates an electronic 222 order using an approved ordering software. The order is digitally signed using the purchaser's personal CSOS Certificate and then transmitted to the suppliers. The paper 222 is not required for electronic ordering.
3) The supplier receives the purchase order and verifies that the purchaser's certificate is valid with the DEA. Additionally, the supplier validates the electronic order information just like it would a paper order.
4) The supplier completes the order and ships to the purchaser. Any communications regarding the order are sent electronically.
5) The order is reported by the supplier to the DEA within two business days.

Receiving

The pharmaceutical received should be carefully checked in against the purchase order including product name, quantity, strength, and package size. Controlled substances are shipped in separate containers and should be checked in by a pharmacist, although pharmacy technicians may assist with this process under the direct supervision of a pharmacist. Schedule II medications need to be checked in against your DEA 222 form (whether the paper triplicate form, or the electronic form on your CSOS enabled software).

If something is damaged in shipment or improperly shipped, it must be reported to the pharmacist and vendor immediately.

Inventory

Inventory, simply put, is the entire stock on hand for sale at a given time. Inventory typically includes prescription drugs, over the counter medications, dietary supplements, and front end merchandise.

Traditionally, inventory is the single largest expense in pharmacy, and so proper management of it is essential to the success of any pharmacy. As such, inventory value is the total value of the drugs and merchandise in stock on a given day.

Several things a technician should keep in mind while working with their inventory are:

- appropriate maximum and minimum levels on products,
- turnover rates on inventory, and
- days' supply of inventory.

Most pharmacies must borrow and pay interest in order to keep an adequate quantity of medications on the shelves in order to meet anticipated needs. Most pharmacies establish maximum and minimum inventory levels to try and assure an adequate, but not excessive, quantity of medicines are on the shelf so that they do not need to borrow too much money (or even if they aren't using loans, they don't want too much money tied up in non moving inventory). Often, max and min levels are monitored by computer systems, but pharmacy staff will need to initially establish those levels and may occasionally need to intervene and adjust those levels. Furthermore, some pharmacies still do not have automated inventory systems. When you establish a max level, that is the most you want to have on the shelves and you should not exceed that quantity. When you establish a minimum level (or par level), that is the point when you order more (commonly called your reorder point), but not till you reach that minimum quantity. Based on this information fill in the chart below with the appropriate number of vials you should purchase.

Practice Problems

	Medications	***Package Size***	***Minimum # Units***	***Maximum # Units***	***Current Inventory***	***Re-order***
1)	tetracycline 250 mg cap	500	120	700	80	
2)	metronidazole 500 mg tab	50	30	120	12	
3)	doxycycline 50 mg cap	50	30	150	42	

1) 1 bottle 2) 2 bottles 3) 0 bottles

Determining the inventory turnover rate is a good method of measuring the overall effectiveness of the purchasing and inventory control programs. The inventory turnover rate is calculated by dividing the total dollars spent to purchase drugs for one year by the actual value of the pharmacy inventory at any point in time. The number produced by this calculation offers an indication of how many times a year the inventory may have been used or replaced.

Example

A pharmacy does a quarterly inventory count and has an average inventory value of $100,000.00. Annual inventory purchases are $500,000.00. What is the turnover rate, and what does that number mean?

$$\frac{\$500,000.00}{\$100,000.00} = 5\ \boldsymbol{turnovers}$$

Looking at that number, we can say that this pharmacy turns over all its inventory approximately 5 times per year. Most community pharmacies usually have a goal of turning over their inventory 12 times a year and institutional pharmacies may push for even higher turnover rates. Solve the practice problem below.

Practice Problem

1) If Bidwell Pharmacy's average inventory for the last year was $132,936.00, and the annual cost total was $1,612,000.00, what was the turnover rate?

1) Bidwell Pharmacy has a turnover rate of 12.1

Often, pharmacies have $75,000.00 to $200,000.00 (or more) in inventory sitting on the shelves as drug products. Pharmacies often set goals for lowering the inventory to improve cash flow. A common method used is to set inventory goals called days' supply of inventory, which refers to making the value of the inventory approximately equal to the cost to the pharmacy of the products sold in a certain number of days. A common number for pharmacies to shoot for is in the 25-35 day range. Let's look at an example problem.

Example

Bidwell Pharmacy has a total inventory value of $132,936.00. Last week it had sales of $45,813.00, and the cost to the pharmacy of the products sold came to $36,592.00. Bidwell's current goal is 25 days' supply. How close is Bidwell to its goal, and how much inventory value does this difference represent? (*Hint: Inventory value is based on how much you paid for it, not how much you can sell it for.*)

$$\frac{\$132,936.00}{1} \times \frac{7\ days}{\$36,592.00} = \mathbf{25.43}\ \boldsymbol{days}$$

The pharmacy is very close to achieving its goal, but it is a little over where it wants to be. Let's calculate how much more money is invested in inventory than it wants.

$$\$\,132,936.00 - \left(\frac{25\ days}{1} \times \frac{\$\,36,592.00}{7\ days}\right) = \mathbf{\$\,2,250.29}$$

Therefore the pharmacy has $2,250.29 more dollars tied up in inventory than it wants.

Let's go to the next page where you can attempt a practice problem with days' supply of inventory.

Practice Problem

1) Adam's Pharmacy has a total inventory value of $183,445.00. Last week the pharmacy had sales of $47,293.00, and the cost to the pharmacy for the products sold came to $38,207.00. Adam's goal is to have a 28 days' supply of inventory. How many days' supply does Adam have? How much is he over or under his goal in dollars?

1) Adam's Pharmacy has 33.6 days' supply of inventory which is $30,617.00 over Adam's goal.

Inventory requirements

While individual pharmacies may have various frequencies in which they verify their inventory levels, it is worth mentioning the requirements that the DEA sets for controlled substances. A pharmacy is required by the DEA to take an inventory of controlled substances every 2 years (biennially). This inventory must be done on any date that is within 2 years of the previous inventory date. The inventory record must be maintained at the registered location in a readily retrievable manner for at least 2 years for copying and inspection by the Drug Enforcement Administration. An inventory record of all Schedule II controlled substances must be kept separate from those of other controlled substances. Submission of a copy of any inventory record to the DEA is not required unless requested.

When taking the inventory of Schedule II controlled substances, an actual physical count must be made. For the inventory of Schedule III, IV, and V controlled substances, an estimate count may be made. If the commercial container holds more than 1000 dosage units and has been opened, however, an actual physical count must be made.

State law may strengthen this requirement with annual actual physical counts of all controlled substances. It also may require such an inventory be submitted before reregistration by the board of pharmacy.

Worksheet 14-2

Name:

Date:

Determine how much of each medication to reorder based on the package size and the minimum and maximum quantities the pharmacy wants to stock. Reorder medications when they reach the minimum.

	Medications	*Package Size*	*Minimum # Units*	*Maximum # Units*	*Current Inventory*	*Re-order*
1)	albuterol inh sol vial-neb	25	50	150	25	
2)	carisoprodol 350 mg tab	20	24	100	48	
3)	E.E.S. 400 mg tab	100	60	240	57	
4)	fluoxetine 20 mg cap	100	2000	6000	1870	
5)	glyburide 5 mg tab	100	200	800	400	
6)	levothyroxine 100 mcg tab	100	180	720	140	
7)	metformin 500 mg tab	500	200	800	250	
8)	metoprolol tar. 50 mg tab	100	120	480	90	
9)	ondansetron 4 mg/5 mL	50	100	400	100	
10)	quetiapine 200 mg tab	100	200	800	100	
11)	risperidone 2 mg tab	100	120	480	60	
12)	sertraline 50 mg tab	100	200	800	120	
13)	sildenafil cit. 100 mg tab	30	60	240	69	
14)	tramadol 50 mg tab	100	200	800	150	
15)	zidovudine 300 mg tab	60	20	80	24	

Calculate how many times each of the following pharmacies turnover their inventory on an annual basis.

16) If Epocrates's Apothecary has an average inventory for the last year was $112,936.87, and the annual cost total was $1,298,774.13, what was the turnover rate?

17) If Frank's Pharmacy has an average inventory for the last year was $187,639.11, and the annual cost total was $2,251,669.30, what was the turnover rate?

18) If Lou's Legend Drug has an average inventory for the last year was $97,812.50, and the annual cost total was $1,222,656.33, what was the turnover rate?

Perform the following days' supply of inventory calculations.

19) Epocrates's Apothecary has a total inventory value of $112,937.00. Last week the pharmacy had sales of $29,971.71, and the cost to the pharmacy for the products sold came to $24,976.43. Epocrates's goal is to have a 35 days' supply of inventory. How many days' supply does Epocrates have? How much is he over or under his goal in dollars?

20) Frank's Pharmacy has a total inventory value of $187,638.87. Last week the pharmacy had sales of $51,061.60, and the cost to the pharmacy for the products sold came to $43,301.33. Frank's goal is to have a 30 days' supply of inventory. How many days' supply does Frank have? How much is he over or under his goal in dollars?

21) Lou's Legend Drug has a total inventory value of $97,812.50. Last week the pharmacy had sales of $28,215.14, and the cost to the pharmacy for the products sold came to $23,512.62. Lou's goal is to have a 25 days' supply of inventory. How many days' supply does Lou have? How much is he over or under his goal in dollars?

Answer the following questions.

22) Do you think most pharmacies use wholesaler purchasing or direct purchasing and why?

23) How are Schedule III – V medications ordered?

24) How are Schedule II medications ordered?

25) What should be done if something is damaged in shipment or improperly shipped?

26) Whom is allowed to check in controlled substances from the medication delivery?

Storage requirements

Now that we've discussed how to purchase and receive various pharmaceuticals, we need to start looking at how to properly store our inventory, as this is another area in which pharmacy technicians have great responsibility. There are three main concepts to discuss in this section:

- environmental considerations,
- security issues, and
- safety requirements.

Environmental considerations

Environmental considerations include proper temperature, ventilation, humidity, light and sanitation. Specific storage conditions are required to be printed in product literature, on drug packaging, and drug labels to ensure proper storage and product integrity. The conditions are defined by the following terms[1]:

Cold: any temperature not exceeding 8° C (45° F)
Freezer: -25° to -10° C (-13° to 14° F)
Refrigerator: 2° to 8° C (36° to 46° F)
Cool: 8° to 15° C (46° to 59° F)
Room temperature: the temperature prevailing in a working area
Controlled room temperature: 15° to 30° C (59° to 86° F)
Warm: 30° to 40° C (86° to 104° F)
Excessive heat: any temperature above 40° C (104° F)

The temperatures that you will need to be most concerned with are freezer, refrigerator, and controlled room temperature. Pharmacies should maintain some sort of daily log for the refrigerators and freezers that medications are stored within.

Volatile or flammable substances such as the alcohols that a pharmacy may use for compounding or other purposes must be stored in an area with proper ventilation to prevent build up of fumes in case of accidental spill or damaged storage container.

Humidity can cause a tablet to become moist and powdery. While all medications should not be exposed to excessive levels of humidity, some medications, such as acyclovir, mycophenolate, and zidovudine, seem to be more sensitive to degradation from humidity.

There are more than 200 different medications which are light sensitive. The chemical composition of these medications can be altered by exposure to direct light. As an example, when nitroprusside is exposed to direct sun light it will breakdown into cyanide. Some common light sensitive medications include acetazolamide, doxycycline, linezolid, and zolmatriptan. While many drugs need to be protected from light while in storage, their original package from the manufacturer should suffice. If you need to repackage any medications, always be sure to consult the manufacturers recommendations to determine if you need to place the medication in light resistant packaging or not.

---

1 These important standards are contained in a combined publication that is recognized as the official compendium, the *United States Pharmacopeia* (USP) and the *National Formulary* (NF).

Sanitation standards are usually set by the state that your pharmacy practices in. Below is an example of what various states request as a sanitation standard. The standard quoted comes from the Commonwealth of Pennsylvania[2].

> § 27.15. Sanitary standards.
>
> (a) The pharmacy and equipment shall be maintained in a clean and orderly condition and in good repair.
>
> (b) The pharmacy shall comply with the health and sanitation statutes of the Commonwealth and of the municipality and county in which the pharmacy is located.
>
> (c) Waste material may not be permitted to collect upon the floor, counter or other area of the pharmacy. The pharmacy shall have a waste removal system adequate to maintain clean and sanitary conditions.
>
> (d) The prescription area shall be dry and well ventilated, free from rodents, insects, dirt and foreign material, and well lighted.
>
> (e) Plumbing shall be in good repair and working order.
>
> (f) The prescription area shall contain only appliances, instruments, equipment, materials, drugs, medicines, chemicals and supplies necessary for the practice of pharmacy, as set forth in section 2(11) of the act (63 P. S. § 390-2(11)), and other equipment and supplies deemed reasonable for the operation and management of a pharmacy as established by the Board.
>
> (g) Persons working in the prescription area shall be required to keep themselves and their apparel in a clean, sanitary and professional manner.

Security issues

Security requirements that restrict access to medications to "authorized personnel only" are often the result of legal requirements, institutional policy, and established standards of practice. All drugs in an institutional setting must be maintained in restricted locations so that they are only accessible to professional staff who are authorized to receive, store, prepare, dispense, distribute, or administer such products. Whereas, in a community pharmacy, the public has ready access to various over the counter medications.

Prescription (legend) drugs require a prescription and are otherwise restricted to "authorized personnel only" such as pharmacists and pharmacy technicians in all pharmacy settings.

As should be expected, there are additional security measures with respect to controlled substances. Schedule III – V medications must either be stored in a secured vault or be distributed throughout the pharmacy stock. By dispersing your controlled substances throughout your inventory, you effectively prevent someone from being able to steal all your scheduled medications. Schedule II medications must also either be stored in a secured vault or be distributed throughout the pharmacy stock; although,

2 The Pennsylvania Code, Chapter 27. State Board of Pharmacy, The provisions of this § 27.15 amended September 4, 1998, effective September 5, 1998, http://www.pacode.com/secure/data/049/chapter27/s27.15.html

some states specifically require Schedule II medications to be stored in a secured vault.

Safety requirements

The safety requirements include everything from the proper inventory rotation to avoid dispensing expired products, to material safety data sheets to provide the necessary information for safe clean up after accidental spills, to appropriate handling of oncology materials, and proper storage of chemicals and flammable items.

Proper rotation of inventory and periodic checking of expirations help to reduce the potential for dispensing expired medications. It also maximizes the utilization of inventory before medications become outdated. When looking at expirations on medication vials it is important to note that if a medication only mentions the month and year but not the day, then you are to treat it as expiring at the end of the month. As an example, if a medication is marked as expiring on 02/2020, then you would treat it as expiring on February 29, 2020.

The Occupational Safety and Health Administration (OSHA) requires all work places, including pharmacies, to carry material safety data sheets (MSDS) for all hazardous substances that are stored on the premises. This includes oncology drugs and volatile chemicals along with other hazardous chemicals. The MSDS provide handling, clean-up, and first-aid information.

Segregating inventory by drug categories helps to prevent potentially harmful errors. The Joint Commission (TJC), formerly known as the Joint Commission on Accreditation of Healthcare Organizations (JCAHO), requires that internal and external medications must be stored separately. This reduces the potential that someone will dispense or administer an external product for internal use. The Joint Commission also has requirements for separate storage of oncology drugs and volatile or flammable substances.

Hazardous drugs (i.e., oncology drugs) should have a separate space on the shelves and be labeled in such a way that it will alert staff of the hazardous potential of these medications. Oncology drugs are often cytotoxic themselves and must be handled with extreme care. They should be received in a sealed protective outer bag that restricts dissemination of the drug if the container leaks or is broken. When potential exists for exposure to hazardous drugs, all personnel involved must wear appropriate personal protective equipment while following a hazardous materials cleanup procedure. All exposed materials must be properly disposed of in hazardous waste containers.

Volatile or flammable substances (including tax free alcohol) require careful storage. They must have a cool location that is properly ventilated. Their storage area must be designed to reduce fire and explosion potential.

Worksheet 14-3

Name:

Date:

Fully answer the following questions.

1) What are the major environmental considerations for medications?

2) List the storage ranges in both Celsius and Fahrenheit for freezer, refrigerator, and controlled room temperature.

3) Approximately how many medications are considered light sensitive?

4) What happens to nitroprusside when it is exposed to direct sun light?

5) Whom usually sets the sanitation standards for your pharmacy?

6) What is a legend drug and who should have access to it?

7) How should Schedule III – V medications be stored in the pharmacy?

8) How should Schedule II medications be stored in the pharmacy?

9) Create a short list of safety requirements as pertaining to medication storage.

10) If a medication had an expiration of 10/2020, what day does it expire?

Markups and discounts

Most pharmacies need to provide a markup on their products in order to be profitable. Usually a product will be given a percent markup although an item may be given a flat rate markup instead. Even when the markup is given as a flat rate, it is a common practice to want to determine what kind of percent markup it equals. Let's look at an example of each.

Examples

1) A tube of ointment that cost the pharmacy $10 is given a flat rate markup of $2.25. What is the selling price?

$$purchase\ price + flat\ rate\ markup = selling\ price$$

$$\$10.00 + \$2.25 = \mathbf{\$12.25}$$

2) What is the percent markup in the previous problem?

$$\frac{(selling\ price - purchase\ price)}{purchase\ price} = \frac{\%\,markup}{100}$$

$$\frac{(\$12.25 - \$10.00)}{\$10.00} = \frac{N}{100}$$

$$N = \boldsymbol{a\,markup\,of}\ \mathbf{22.5\,\%}$$

3) An unmedicated lip balm cost the pharmacy $0.83, and it wants to sell it for a 20% markup. What is the unmedicated lip balm's selling price?

$$purchase\ price + (purchase\ price \times percent\ markup) = selling\ price$$

$$\$0.83 + \left(\$0.83 \times \frac{20}{100}\right) = \mathbf{\$1.00}$$

As percent markups are more common, we will focus on those. Please attempt the following practice problems working with percent markups.

Practice Problems

Several items are listed below with the price that Bidwell Pharmacy paid for each item. Perform the appropriate markup calculations below.

1) Bidwell Pharmacy purchased a 120 mL bottle of acetaminophen elixir 160 mg/5 mL $3.15. What would its selling price be if it were marked up 200%?

2) Bidwell Pharmacy purchased 2 packets of Goody's extra strength headache powder for $0.39. What would its selling price be if it were marked up 200%?

3) Bidwell Pharmacy purchased a 120 mL bottle of guaifenesin syrup 100 mg/5 mL for $2.50 and sold it for $7.50. What is the percent markup on this product?

1) $9.45 2) $1.17 3) 200% markup

Another common concept in a pharmacy is to add a percent markup and a professional fee for prescription medications. The professional fee (or dispensing fee) is intended to displace the pharmacy's costs and then the markup would provide the pharmacy with its profit. Typically if a professional fee is included, then the percent markup tends to be lower. Let's look at an example problem.

Example

1) If a pharmacy purchases a product for $10, adds a 50% markup, and applies a $3.50 professional fee, how much will the product be sold for?

$$\text{purchase price} + (\text{purchase price} \times \text{percent markup}) + \text{professional fee} = \text{selling price}$$

$$\$10.00 + \left(\$10.00 \times \frac{50}{100}\right) + \$3.50 = \mathbf{\$18.50}$$

Let's look at a couple of practice problems.

Practice Problems

Calculate the selling price if the pharmacy is charging a 50% markup and a $5 professional fee.

1) The pharmacy purchases 500 tablets of 25 mg metoprolol tartrate for $40 and is dispensing 60 tablets to a patient. How much should the pharmacy charge?

2) The pharmacy purchases 30 capsules of 300 mg rifampin capsules for $56.70 and is dispensing 8 capsules to a patient. How much should the pharmacy charge?

1) $12.20 2) $27.68

Additionally, pharmacies may receive and/or offer various special discounts. There may be a reduced price offered by a wholesaler if a purchase exceeds a particular amount, or some pharmacies may have a sale or reduced prices for various groups such as a discount for senior citizens. Maybe there is a discount for patients that transfer their prescriptions into a particular pharmacy. Let's look at several practice discount problems.

Examples

1) The pharmacy has switched wholesalers for its over the counter medications and it wants to get rid of its old inventory. Normally the pharmacy sells its generic ibuprofen for $11.99 but is offering a flat rate discount of $3.00. What is the selling price?

$$\textit{retail price} - \textit{flat rate discount} = \textit{discount price}$$

$$\$11.99 - \$3.00 = \mathbf{\$8.99}$$

2) For marketing purposes, the pharmacy wants to state what the percent discount is on the ibuprofen in the previous problem.

$$\frac{(\textit{retail price} - \textit{discount price})}{\textit{retail price}} = \frac{\%\,\textit{discount}}{100}$$

$$\frac{(\$11.99 - \$8.99)}{\$11.99} = \frac{N}{100}$$

$$N = \mathbf{25\%\ \textit{discount}}$$

3) The pharmacy offers a 10% discount to senior citizens on Wednesdays. On Wednesday a senior citizen purchases $32.87 worth of products. How much will they pay after their 10% discount?

$$\textit{retail price} - (\textit{retail price} \times \textit{percent discount}) = \textit{discount price}$$

$$\$32.87 - \left(\$32.87 \times \frac{10}{100}\right) = \mathbf{\$29.58}$$

Percent discounts tend to be very common and are therefore worth focusing on. Perform the necessary discount calculations below as we look at several practice problems.

Practice Problems

Several items are listed below with the price that Bidwell Pharmacy normally charges for each item. Perform the appropriate discount calculations below.

1) At Bidwell Pharmacy, a 120 mL bottle of acetaminophen elixir 160 mg/5 mL retails for $9.45. What would the sale price be if it were given a 5% discount?

2) At Bidwell Pharmacy, 2 packets of Goody's extra strength headache powder retail for $1.17. What would the sale price be if it were given a 5% discount?

3) Ordinarily Bidwell Pharmacy charges $7.50 for a 120 mL bottle of guaifenesin syrup 100 mg/5 mL, but it is on sale for $7.13. What is the percent discount?

1) $8.98 2) $1.11 3) 5% discount

Worksheet 14-4

Name:

Date:

Perform the following markups for problems 1-10 with a 200% markup rate.

1) The purchase price for a medication is $4.99. After a 200% markup the selling price is _________

2) The purchase price for a medication is $12.47. After a 200% markup the selling price is _________

3) The purchase price for a medication is $35.20. After a 200% markup the selling price is _________

4) The purchase price for a medication is $89.90. After a 200% markup the selling price is _________

5) The purchase price for a medication is $120.47. After a 200% markup the selling price is _________

6) The purchase price for a medication is $27.50. After a 200% markup the selling price is _________

7) The purchase price for a medication is $68.50. After a 200% markup the selling price is _________

8) The purchase price for a medication is $153.50. After a 200% markup the selling price is _________

9) The purchase price for a medication is $67.12. After a 200% markup the selling price is _________

10) The purchase price for a medication is $127.30. After a 200% markup the selling price is _________

Perform the following markups for problems 11-20 with a 75% markup rate.

11) The purchase price for a medication is $25.50. After a 75% markup the selling price is _________

12) The purchase price for a medication is $118.70. After a 75% markup the selling price is _________

13) The purchase price for a medication is $195.20. After a 75% markup the selling price is _________

14) The purchase price for a medication is $162.50. After a 75% markup the selling price is _________

15) The purchase price for a medication is $188.95. After a 75% markup the selling price is _________

16) The purchase price for a medication is $135.20. After a 75% markup the selling price is _________

17) The purchase price for a medication is $199.41. After a 75% markup the selling price is _________

18) The purchase price for a medication is $207.33. After a 75% markup the selling price is _________

19) The purchase price for a medication is $211.65. After a 75% markup the selling price is _________

20) The purchase price for a medication is $472.50. After a 75% markup the selling price is _________

Determine the selling price of the prescriptions in problems 21-25 if the pharmacy is charging a 50% markup and a $5 professional fee.

21) The pharmacy purchases 100 tablets of diltiazem 90 mg for $10 and is dispensing 90 tablets to a patient. How much should the pharmacy charge?

22) The pharmacy purchases 90 capsules of acyclovir 200 mg for $13.50 and is dispensing 90 capsules to a patient. How much should the pharmacy charge?

23) The pharmacy purchases 1000 tablets of metformin 1000 mg for $50 and is dispensing 180 tablets to a patient. How much should the pharmacy charge?

24) The pharmacy purchases 30 tablets of amiodarone 200 mg for $99 and is dispensing 30 tablets to a patient. How much should the pharmacy charge?

25) The pharmacy purchases 100 tablets of atenolol 25 mg for $10 and is dispensing 30 tablets to a patient. How much should the pharmacy charge?

Perform the following discounts for problems 26-35 with a 5% discount rate.

26) The retail price for a medication is $14.97. After a 5% discount the discount price is

____

27) The retail price for a medication is $37.41. After a 5% discount the discount price is

____

28) The retail price for a medication is $105.60. After a 5% discount the discount price is

____

29) The retail price for a medication is $269.70. After a 5% discount the discount price is

____

30) The retail price for a medication is $361.41. After a 5% discount the discount price is

____

31) The retail price for a medication is $82.50. After a 5% discount the discount price is

____

32) The retail price for a medication is $205.50. After a 5% discount the discount price is

____

33) The retail price for a medication is $460.50. After a 5% discount the discount price is

____

34) The retail price for a medication is $201.36. After a 5% discount the discount price is

____

35) The retail price for a medication is $381.90. After a 5% discount the discount price is

____

Perform the following discounts for problems 36-45 with a 10% discount rate.

36) The retail price for a medication is $44.63. After a 10% discount the discount price is

____

37) The retail price for a medication is $207.73. After a 10% discount the discount price is

____

38) The retail price for a medication is $341.60. After a 10% discount the discount price is

____

39) The retail price for a medication is $284.38. After a 10% discount the discount price is

____

40) The retail price for a medication is $330.66. After a 10% discount the discount price is

____

41) The retail price for a medication is $236.60. After a 10% discount the discount price is

____

42) The retail price for a medication is $348.97. After a 10% discount the discount price is

____

43) The retail price for a medication is $362.83. After a 10% discount the discount price is

____

44) The retail price for a medication is $370.39. After a 10% discount the discount price is

____

45) The retail price for a medication is $826.88. After a 10% discount the discount price is

____

Solve the following problems.

46) If a medication is purchased for $25 and sold for $30 what was the percent markup?

47) If the purchase price for 100 tablets of a medication was $49.25, and the percent markup is 15%, then what is the selling price for 30 tablets?

48) If the discount on a wholesaler invoice for $9,300 was $310, what was the percent discount?

49) If your wholesaler offers you a 5% discount on orders over $5,000, how much would you expect your invoice to be for if you purchased $5,500 worth of product?

50) A senior citizen is paying for a prescription for penicillin VK 250 mg #30. The usual and customary price is $8.49. However this patient qualifies for a 10% discount. How much will the patient pay?

51) A wholesaler offers the pharmacy a 10% discount on orders of $14,400 or more, or a 15% discount on orders of $28,800. If the pharmacy makes an $18,000 purchase, how much should its invoice be for?

Gross profits and net profits

In this section we will look at gross profits and net profits. Gross profit is calculated as sales minus all costs directly related to those sales, or in simpler terms we can think of it as the difference between the selling price and the acquisition price.

$$selling\ price - acquisition\ price = \mathbf{gross\ profit}$$

Net profit is the difference between the gross profit and the sum of all the costs associated with filling the prescription. The costs associated with filling the prescription are accounted for with a dispensing fee or a professional fee. With that in mind, you can determine the net profit by subtracting a professional fee from the gross profit.

$$gross\ profit - dispensing\ fee = \mathbf{net\ profit}$$

While many pharmacies will have a standard dispensing fee on all prescriptions, others may have a tiered dispensing fee placing a lower fee on less expensive medications and a higher fee on more expensive medications. Let's look at a couple of examples of each examples of each.

Examples

1) The pharmacy purchases 30 tablets of amiodarone 200 mg for $99 and is dispensing 30 tablets to a patient for $153.50. What is the gross profit? If the pharmacy charged a $5.00 dispensing fee what is the net profit?

$$selling\ price - acquisition\ price = \mathbf{gross\ profit}$$

$$\$153.50 - \$99 = \mathbf{\$54.50\ gross\ profit}$$

$$gross\ profit - dispensing\ fee = \mathbf{net\ profit}$$

$$\$54.50 - \$5.00 = \mathbf{\$49.50\ net\ profit}$$

2) The pharmacy purchases 100 tablets of diltiazem 90 mg for $10 and is dispensing 90 tablets to a patient for 18.50. What is the gross profit? If the pharmacy charged a $5.00 dispensing fee what is the net profit?

$$selling\ price - acquisition\ price = \mathbf{gross\ profit}$$

$$\$18.50 - \left(\frac{90\ tablets}{1} \times \frac{\$10}{100\ tablets}\right) = \mathbf{\$9.50\ gross\ profit}$$

$$gross\ profit - dispensing\ fee = \mathbf{net\ profit}$$

$$\$9.50 - \$5.00 = \mathbf{\$4.50\ net\ profit}$$

Some pharmacies may use a sliding scale for their dispensing fees based on the average wholesale price of a medication. Let's do a couple of example problems using this chart:

AWP	*Dispensing Fee*
Less than $20.00	$3.50
$20.00 - $50.00	$5.00
Greater than $50.00	$7.50

3) The pharmacy purchases 30 tablets of amiodarone 200 mg for $99 (its AWP is $112.00) and is dispensing 30 tablets to a patient for $153.50. What is the gross profit? If the pharmacy charges a tiered dispensing fee based on the average wholesale price, what is the net profit?

$$\text{selling price} - \text{acquisition price} = \textbf{gross profit}$$

$$\$153.50 - \$99 = \textbf{\$54.50 gross profit}$$

$$\text{gross profit} - \text{dispensing fee} = \textbf{net profit}$$

The AWP for 30 tablets is $112.00 so the dispensing fee is $7.50.

$$\$54.50 - \$7.50 = \textbf{\$47.00 net profit}$$

4) The pharmacy purchases 100 tablets of diltiazem 90 mg for $10 (its AWP is $11 for 100 tablets) and is dispensing 90 tablets to a patient for 18.50. What is the gross profit? If the pharmacy charges a tiered dispensing fee based on the average wholesale price, what is the net profit?

$$\text{selling price} - \text{acquisition price} = \textbf{gross profit}$$

$$\$18.50 - \left(\frac{90\ \text{tablets}}{1} \times \frac{\$10}{100\ \text{tablets}}\right) = \textbf{\$9.50 gross profit}$$

$$\text{gross profit} - \text{dispensing fee} = \textbf{net profit}$$

The AWP for is $\frac{90\ \text{tablets}}{1} \times \frac{\$11}{100\ \text{tablets}} = \$9.90$ so the dispensing fee is $3.50.

$$\$9.50 - \$3.50 = \textbf{\$6.00 net profit}$$

Now that we've looked at some examples, you should try several practice problems.

Practice Problems

1) The pharmacy purchases 90 capsules of acyclovir 200 mg for $13.50 and is dispensing 90 capsules to a patient for $25.25. What is the gross profit? If the pharmacy charged a $5.00 dispensing fee, what is the net profit?

2) The pharmacy purchases 1000 tablets of a medication for $50 and is dispensing 180 tablets to a patient for $18.50. What is the gross profit? If the pharmacy charged a $5.00 dispensing fee, what is the net profit?

Some pharmacies may use a sliding scale for their dispensing fees based on the average wholesale price of a medication. Let's do a couple of practice problems using this chart:

AWP	*Dispensing Fee*
Less than $20.00	$3.50
$20.00 - $50.00	$5.00
Greater than $50.00	$7.50

3) The pharmacy purchases 90 capsules of acyclovir 200 mg for $13.50 (its AWP is $15.00) and is dispensing 90 capsules to a patient for $25.25. What is the gross profit? If the pharmacy charges a tiered dispensing fee based on the average wholesale price, what is the net profit?

4) The pharmacy purchases 1000 tablets of a medication for $50 (its AWP is $55.00 for 1000 tablets) and is dispensing 180 tablets to a patient for $18.50. What is the gross profit? If the pharmacy charges a tiered dispensing fee based on the average wholesale price, what is the net profit?

1) gross profit = $11.75; net profit = $6.75 2) gross profit = $9.50; net profit = $4.50
3) gross profit = $11.75; net profit = $8.25 4) gross profit = $9.50; net profit = $6.00

Worksheet 14-5

Name:

Date:

Solve for the gross profit and the net profit for problems 1-20 using a set dispensing fee of $5.

1) The pharmacy purchased 25 nebulizer vials of albuterol inhalation solution for $27.90 and sold 25 vials for $46.85. What is the gross profit? What is the net profit?

2) The pharmacy purchased 20 tablets of carisoprodol 350 mg for $37.26 and sold 20 tablets for $60.89. What is the gross profit? What is the net profit?

3) The pharmacy purchased 100 tablets of E.E.S. 400 mg for $21.06 and sold 28 tablets for $12.35. What is the gross profit? What is the net profit?

4) The pharmacy purchased 100 capsules of fluoxetine 20 mg for $223.56 and sold 30 capsules for $108.10. What is the gross profit? What is the net profit?

5) The pharmacy purchased 100 tablets of glyburide 5 mg for $25.48 and sold 30 tablets for $16.47. What is the gross profit? What is the net profit?

6) The pharmacy purchased 100 tablets of levothyroxine 100 mcg for $26.87 and sold 30 tablets for $15.59. What is the gross profit? What is the net profit?

7) The pharmacy purchased 1000 tablets of metformin 500 mg for $633.40 and sold 60 tablets for $62.01. What is the gross profit? What is the net profit?

8) The pharmacy purchased 100 tablets of metoprolol tartrate 25 mg for $6.48 and sold 60 tablets for $9.33. What is the gross profit? What is the net profit?

9) The pharmacy purchased a 50 mL bottle of ondansetron oral solution 4 mg/5 mL for $242.77 and sold 1 fl. oz. for $225.99. What is the gross profit? What is the net profit?

10) The pharmacy purchased 100 tablets of quetiapine 200 mg for $631.49 and sold 30 tablets for $291.67. What is the gross profit? What is the net profit?

11) The pharmacy purchased 100 tablets of risperidone 2 mg for $638.96 and sold 30 tablets for $295.03. What is the gross profit? What is the net profit?

12) The pharmacy purchased 100 tablets of sertraline 50 mg for $244.11 and sold 30 tablets for $117.35. What is the gross profit? What is the net profit?

13) The pharmacy purchased 30 tablets of sildenafil citrate 100 mg for $324.81 and sold 4 tablets for $72.46. What is the gross profit? What is the net profit?

14) The pharmacy purchased 100 tablets of tramadol 50 mg for $27.61 and sold 60 tablets for $28.34. What is the gross profit? What is the net profit?

15) The pharmacy purchased 60 tablets of zidovudine 300 mg for $197.12 and sold 60 tablets for $303.18. What is the gross profit? What is the net profit?

16) The pharmacy purchased 100 tablets of atenolol 50 mg for $9.52 and sold 30 tablets for $7.78. What is the gross profit? What is the net profit?

17) The pharmacy purchased 100 capsules of benzonatate 100 mg for $39.48 and sold 30 capsules for $21.27. What is the gross profit? What is the net profit?

18) The pharmacy purchased 100 tablets of cyclobenzaprine 5 mg for $147.81 and sold 30 tablets for $71.51. What is the gross profit? What is the net profit?

19) The pharmacy purchased 100 tablets of furosemide 20 mg for $5.07 and sold 30 tablets for $5.78. What is the gross profit? What is the net profit?

20) The pharmacy purchased 100 tablets of hydrochlorothiazide 25 mg for $1.58 and sold 60 tablets for $4.92. What is the gross profit? What is the net profit?

Solve for the gross profit and the net profit for problems 21-40 using the tiered dispensing fees on the chart below based on the average wholesale price.

AWP	*Dispensing Fee*
Less than $20.00	$3.50
$20.00 - $50.00	$5.00
Greater than $50.00	$7.50

21) The pharmacy purchased 25 nebulizer vials of albuterol inhalation solution for $27.90 (the AWP is $31.00 for 25 vials) and sold 25 vials for $46.85. What is the gross profit? What is the net profit?

22) The pharmacy purchased 20 tablets of carisoprodol 350 mg for $37.26 (the AWP for 20 tablets is $41.40) and sold 20 tablets for $60.89. What is the gross profit? What is the net profit?

23) The pharmacy purchased 100 tablets of E.E.S. 400 mg for $21.06 (the AWP for 100 tablets is $23.46) and sold 28 tablets for $12.35. What is the gross profit? What is the net profit?

24) The pharmacy purchased 100 capsules of fluoxetine 20 mg for $223.56 (the AWP for 100 capsules is $248.40) and sold 30 capsules for $108.10. What is the gross profit? What is the net profit?

25) The pharmacy purchased 100 tablets of glyburide 5 mg for $25.48 (the AWP for 100 tablets is $28.31) and sold 30 tablets for $16.47. What is the gross profit? What is the net profit?

26) The pharmacy purchased 100 tablets of levothyroxine 100 mcg for $26.87 (the AWP for 100 tablets is $29.85) and sold 30 tablets for $15.59. What is the gross profit? What is the net profit?

27) The pharmacy purchased 1000 tablets of metformin 500 mg for $633.40 (the AWP for 1000 tablets is $703.78) and sold 60 tablets for $62.01. What is the gross profit? What is the net profit?

28) The pharmacy purchased 100 tablets of metoprolol tartrate 25 mg for $6.48 (the AWP for 100 tablets is $7.20) and sold 60 tablets for $9.33. What is the gross profit? What is the net profit?

29) The pharmacy purchased a 50 mL bottle of ondansetron oral solution 4 mg/5 mL for $242.77 (the AWP for 50 milliliters is $269.74) and sold 1 fl. oz. for $225.99. What is the gross profit? What is the net profit?

30) The pharmacy purchased 100 tablets of quetiapine 200 mg for $631.49 (the AWP for 100 tablets is $701.65) and sold 30 tablets for $291.67. What is the gross profit? What is the net profit?

31) The pharmacy purchased 100 tablets of risperidone 2 mg for $638.96 (the AWP for 100 tablets is $709.96) and sold 30 tablets for $295.03. What is the gross profit? What is the net profit?

32) The pharmacy purchased 100 tablets of sertraline 50 mg for $244.11 (the AWP for 100 tablets is $271.23) and sold 30 tablets for $117.35. What is the gross profit? What is the net profit?

33) The pharmacy purchased 30 tablets of sildenafil citrate 100 mg for $324.81 (the AWP for 30 tablets is $360.90) and sold 4 tablets for $72.46. What is the gross profit? What is the net profit?

34) The pharmacy purchased 100 tablets of tramadol 50 mg for $27.61 (the AWP for 100 tablets is $30.68) and sold 60 tablets for $28.34. What is the gross profit? What is the net profit?

35) The pharmacy purchased 60 tablets of zidovudine 300 mg for $197.12 (the AWP for 60 tablets is $219.02) and sold 60 tablets for $303.18. What is the gross profit? What is the net profit?

36) The pharmacy purchased 100 tablets of atenolol 50 mg for $9.52 (the AWP for 100 tablets is $10.58) and sold 30 tablets for $7.78. What is the gross profit? What is the net profit?

37) The pharmacy purchased 100 capsules of benzonatate 100 mg for $39.48 (the AWP for 100 capsules is $43.87) and sold 30 capsules for $21.27. What is the gross profit? What is the net profit?

38) The pharmacy purchased 100 tablets of cyclobenzaprine 5 mg for $147.81 (the AWP for 100 tablets is $164.23) and sold 30 tablets for $71.51. What is the gross profit? What is the net profit?

39) The pharmacy purchased 100 tablets of furosemide 20 mg for $5.07 (the AWP for 100 tablets is $5.63) and sold 30 tablets for $5.78. What is the gross profit? What is the net profit?

40) The pharmacy purchased 100 tablets of hydrochlorothiazide 25 mg for $1.58 (the AWP for 100 tablets is $1.75) and sold 60 tablets for $4.92. What is the gross profit? What is the net profit?

Third party reimbursement

Currently, approximately 80% of Americans have health care insurance including some kind of prescription benefits, and this number is expected to increase in coming years due to the *Affordable Health Care for America Act*. With this in mind, we need to look at how we should be billing prescriptions for patients with insurance. Insurance companies know that for most medications the average wholesale price (AWP) is less than the typical usual and customary (U&C) price, but they are also aware that some pharmacies offer certain prescriptions below their AWP (such as $4.00 generics offered by some of the chains). Most insurances are going to reimburse the pharmacy based on whichever is less.

Every pharmacy and insurance company may negotiate their own reimbursement rates, so there is no singular set formula for all insurances. A typical formula may be something along the lines of:

$$87\%\, AWP\, or\, 100\%\, U\&C\,(whichever\, is\, less) + a\,\$\,3.50\, dispensing\, fee = \mathbf{reimbursement\, rate}$$

Let's use the above formula for an example.

Example

1) A patient receives a prescription for thirty hydrochlorothiazide 25 mg. The retail price for 30 tablets is $0.72, whereas the AWP for 100 tablets is $1.75. Determine the charge for this medication based on the aforementioned insurance calculation.

   87% of AWP

   $$\frac{30\,tablets}{1} \times \frac{\$\,1.75}{100\,tablets} \times \frac{87}{100} = \$\,0.46$$

   100% of U&C

   $$\frac{30\,tablets}{1} \times \frac{\$\,0.72}{30\,tablets} = \$\,0.72$$

   87% of the AWP is less so then we can add our $3.50 dispensing fee.
   $\$\,0.46 + \$\,3.50 = \mathbf{\$\,3.96}$

Now you should attempt a couple of practice problems.

Practice Problems

Continue using the following sample equation for the practice problems:

$$87\%\, AWP\, or\, 100\%\, U\&C\,(whichever\, is\, less) + a\,\$\,3.50\, dispensing\, fee = \mathbf{reimbursement\, rate}$$

1) A patient receives a prescription for sixty 500 mg metformin tablets. The pharmacy offers certain medications on their $4.00 prescription plan including these 60 tablets. The AWP for this medication is $703.78 for 1000 tablets. How much will the patient's insurance reimburse for this medication?

2) A patient receives a prescription for thirty furosemide 20 mg tablets. The U&C on this product before a dispensing fee is added is $2.28. The AWP for 100 tablets is $5.07. How much should the patient's insurance reimburse for this prescription?

1) $7.50 2) $4.82

Capitation fee

A capitation fee is another concept to give consideration to for third-party reimbursement. A capitation fee is a method of payment for health services in which an individual or institutional provider is paid a fixed amount without regard to the actual number or nature of services provided to each patient. A common example would be a pharmacy in a long term care home receiving a fixed amount of money per month regardless of whether the patient required no pharmaceuticals or if their prescriptions exceeded the money allotted by a capitation fee.

To demonstrate this concept, attempt the practice problems below.

Practice Problems

1) Bidwell Senior Care Pharmacy receives a monthly capitation fee of $250.00/month for Theophyllus Monk. Last month Mr. Monk's prescriptions totaled $198.75. How much did the pharmacy make off of his capitation fee?

2) Bidwell Senior Care Pharmacy receives a monthly capitation fee of $250.00/month for Alopecia Allen. Last month Mr. Allen's prescriptions totaled $301.25. How much did the pharmacy make off of his capitation fee?

1) The pharmacy had a profit margin of $51.25 on Theophyllus Monk's prescriptions.
2) The pharmacy had a loss of $51.25 on Alopecia Allen's prescriptions.

Worksheet 14-6

Name:

Date:

Solve problems 1-20 using the sample formula below for third-party reimbursement:

$87\%\, AWP\, or\, 100\%\, U\&C\,(whichever\, is\, less) + a\, \$\, 3.50\, dispensing\, fee = \textbf{reimbursement rate}$

1) A pharmacy receives a prescription for 25 nebulizer vials of albuterol inhalation solution. The U&C on this product before a dispensing fee is added is $41.85. The AWP for 25 vials is $31.00. How much should the patient's insurance reimburse for this prescription?

2) A pharmacy receives a prescription for 20 tablets of carisoprodol 350 mg. The U&C on this product before a dispensing fee is added is $55.89. The AWP for 20 tablets is $41.40. How much should the patient's insurance reimburse for this prescription?

3) A pharmacy receives a prescription for 40 erythromycin 400 mg tablets. This product is part of this pharmacy's $4.00 plan. The AWP for 100 tablets is $23.46. How much should the patient's insurance reimburse for this prescription?

4) A pharmacy receives a prescription for 30 capsules of fluoxetine 20 mg. The U&C on this product before a dispensing fee is added is $100.60. The AWP for 100 capsules is $248.40. How much should the patient's insurance reimburse for this prescription?

5) A pharmacy receives a prescription for 30 tablets of 5 mg glyburide. This product is part of this pharmacy's $4.00 plan. The AWP for 100 tablets is $28.31. How much should the patient's insurance reimburse for this prescription?

6) A pharmacy receives a prescription for 30 tablets of levothyroxine 100 mcg tablets. The U&C on this product before a dispensing fee is added is $12.09. The AWP for 100 tablets is $29.85.

How much should the patient's insurance reimburse for this prescription?

7) A pharmacy receives a prescription for 30 tablets of metoprolol succinate 25 mg tablets. The U&C on this product before a dispensing fee is added is $35.29. The AWP for 1000 tablets is $871.40. How much should the patient's insurance reimburse for this prescription?

8) A pharmacy receives a prescription for 60 tablets of metoprolol tartrate 25 mg tablets. The U&C on this product before a dispensing fee is added is $5.83. The AWP for 100 tablets is $7.20. How much should the patient's insurance reimburse for this prescription?

9) A pharmacy receives a prescription for 1 fl. oz. of ondansetron oral solution 4 mg/5 mL. The U&C on this product before a dispensing fee is added is $218.49. The AWP for 50 mL is $269.74. How much should the patient's insurance reimburse for this prescription?

10) A pharmacy receives a prescription for thirty quetiapine 200 mg tablets. The U&C on this product before a dispensing fee is added is $284.17. The AWP for 100 tablets is $701.65. How much should the patient's insurance reimburse for this prescription?

11) A pharmacy receives a prescription for 30 tablets of risperidone 2 mg. The U&C on this product before a dispensing fee is added is $295.03. The AWP for 100 tablets is $709.96. How much should the patient's insurance reimburse for this prescription?

12) A pharmacy receives a prescription for 30 tablets of sertraline 50 mg. The U&C on this product before a dispensing fee is added is $109.85. The AWP for 100 tablets is $271.23. How much should the patient's insurance reimburse for this prescription?

13) A pharmacy receives a prescription for 4 tablets of sildenafil citrate 100 mg. The U&C on this product before a dispensing fee is added is $64.96. The AWP for 30 tablets is $72.46. How much should the patient's insurance reimburse for this prescription?

14) A pharmacy receives a prescription for 60 tablets of tramadol 50 mg. The U&C on this product before a dispensing fee is added is $23.34. The AWP for 100 tablets is $30.68. How much should the patient's insurance reimburse for this prescription?

15) A pharmacy receives a prescription for 60 tablets of zidovudine 300 mg. The U&C on this product before a dispensing fee is added is $295.68. The AWP for 60 tablets is $219.02. How much should the patient's insurance reimburse for this prescription?

16) A pharmacy receives a prescription for 30 tablets of atenolol 50 mg. This product is part of this pharmacy's $4.00 plan. The AWP for 100 tablets is $10.58. How much should the patient's insurance reimburse for this prescription?

17) A pharmacy receives a prescription for 30 capsules of benzonatate 100 mg. The U&C on this product before a dispensing fee is added is $17.77. The AWP for 100 capsules is $43.87. How much should the patient's insurance reimburse for this prescription?

18) A pharmacy receives a prescription for 30 tablets of cyclobenzaprine 5 mg. The U&C on this product before a dispensing fee is added is $64.01. The AWP for 100 tablets is $164.23. How much should the patient's insurance reimburse for this prescription?

19) A pharmacy receives a prescription for 30 tablets of furosemide 40 mg. This product is part of this pharmacy's $4.00 plan. The AWP for 100 tablets is $5.99. How much should the patient's insurance reimburse for this prescription?

20) A pharmacy receives a prescription for 60 tablets of hydrochlorothiazide 25 mg. The U&C on this product before a dispensing fee is added is $1.42. The AWP for 100 tablets is $1.75. How much should the patient's insurance reimburse for this prescription?

Determine the profit margin based on the capitation fees in problems 21 and 22.

21) Fred's Pharmacy receives a capitation fee of $500.00 for filling Senile Sally's prescriptions each month. Last month she required six medications with the following associated costs:

Aricept 10 mg = $200
Lipitor 40 mg = $125
Digoxin 250 mcg = $7
Furosemide 20 mg = $4
Celebrex 200 mg = $115
Nexium 40 mg = $170

What was the profit margin for Sally's prescriptions last month?

22) The Grainger House pays a capitation fee of $275.00 per month per patient to Epocrates's Apothecary. Epocrates's Apothecary services 10 patients for the Grainger House. The pharmacy filled prescriptions for five clients last month (the other five had no prescriptions). The costs for these patients' prescriptions are as follows:

Patient ABK: $89.63
Patient BTC: $126.54
Patient MBC: $420.45
Patient SEP: $170.85
Patient BRS: $39.90

a) What is the total capitation received if the pharmacy is paid for all 10 patients?

b) What is the cost to the Epocrates's Apothecary?

c) What was the loss or gain to Epocrates's Apothecary for servicing the Grainger House?

Daily cash reports

In many community pharmacy settings you will be expected to assist with the the daily cash reports for the store. The process of counting the money, reconciling the receipts and balancing the cash drawer creates an accountability of the day's transactions.

In order to do the cash report, you will need to collect the following data:

- opening and closing readings for each register
- cash and checks
- bank charges and credit cards
- other charges
- paid outs
- coupons
- discounts
- voids
- refunds
- and over-rings

Whenever you are balancing a cash report, you should always double check your vertical and horizontal totals. Below is an example cash report. Please finish filling it in and compare it to the cash report on the next page to see if you filled everything out correctly. This report include operators to help.

	Register 1	Register 2	Register 3	Total
+ Cash and Checks	1513.12	45.12	2002.02	
+ Bank Charges	120.00	---	350.44	
+ House Charges	---	---	---	
+ Paid Outs	---	---	---	
Total				
+ Closing Reading	1,760.02	95.12	2402.50	
- Opening Reading	50.00	50.00	50.00	
= Difference				
- Coupons	1.10	---	---	
- Discounts	8.90	---	---	
- Voids	10.00	---	---	
- Refunds	---	---	---	
- Over-Rings	56.65	---	---	
Total				
Over + or Short -				

The following table is the completed version of the cash report on the previous page.

	Register 1	**Register 2**	**Register 3**	**Total**
+ Cash and Checks	1,513.12	45.12	2002.02	*3,560.26*
+ Bank Charges	120.00	---	350.44	*470.44*
+ House Charges	---	---	---	*0.00*
+ Paid Outs	---	---	---	*0.00*
Total	*1,633.12*	*45.12*	*2,352.46*	*4,030.70*
+ Closing Reading	1,760.02	95.12	2,402.50	*4,257.64*
- Opening Reading	50.00	50.00	50.00	*150.00*
= Difference	*1,710.02*	*45.12*	*2,352.50*	*4,107.64*
- Coupons	1.10	---	---	*1.10*
- Discounts	8.90	---	---	*8.90*
- Voids	10.00	---	---	*10.00*
- Refunds	---	---	---	*0.00*
- Over-Rings	56.65	---	---	*56.65*
Total	*1,633.37*	*45.12*	*2,352.50*	*4,030.99*
Over + or Short -	*-0.25*	*0.00*	*-0.04*	*-0.29*

All the answers that you should have filled in on the previous page are italicized above. You can achieve all of your vertical numbers by following the operators listed in the first column and then you can determine if the register is over or short by subtracting the total on the bottom half from the total on the top half. Horizontally all your numbers are achieved by adding across. The highlighted points on the table are excellent places to check your numbers both vertically and horizontally to make sure you didn't make any simple mistakes.

Something worth noting is that two of the registers are off (register 1 is short $0.25 and register 3 is short $0.04). Most stores are forgiving up to some small amount (typically $2.00 per register). If the register is off by more than that, someone will have to figure out, "*Why?*", which could range from a missing credit card receipt, to human error or even employee theft!

Worksheet 14-7

Name:

Date:

Balance the following cash reports.

1)	Register 1	Register 2	Register 3	Total
+ Cash and Checks	513.12	---	300.44	
+ Bank Charges	120.00	90.00	---	
+ House Charges	---	---	52.02	
+ Paid Outs	---	5.12	---	
Total				
+ Closing Reading	760.02	145.12	402.50	
- Opening Reading	50.00	50.00	50.00	
= Difference				
- Coupons	---	---	---	
- Discounts	8.90	---	---	
- Voids	10.00	---	---	
- Refunds	---	---	---	
- Over-Rings	56.65	---	---	
Total				
Over + or Short -				

2)

	Register 1	Register 2	Register 3	Total
+ Cash and Checks	3,897.65	12.00	1,533.12	
+ Bank Charges	2,111.05	96.67	140.00	
+ House Charges	25.00	---	---	
+ Paid Outs	---	---	---	
Total				
+ Closing Reading	6,121.47	158.70	1,800.02	
- Opening Reading	50.00	50.00	50.00	
= Difference				
- Coupons	14.25	---	1.10	
- Discounts	23.47	---	8.90	
- Voids	---	---	10.00	
- Refunds	---	---	---	
- Over-Rings	---	---	56.65	
Total				
Over + or Short -				

3)	Register 1	Register 2	Register 3	Total
+ Cash and Checks	1,713.12	45.12	2,002.18	
+ Bank Charges	240.00	20.00	350.44	
+ House Charges	---	---	---	
+ Paid Outs	---	---	10.00	
Total				
+ Closing Reading	2,172.67	115.12	2,429.66	
- Opening Reading	50.00	50.00	50.00	
= Difference				
- Coupons	---	---	7.00	
- Discounts	130.00	---	---	
- Voids	27.50	---	---	
- Refunds	12.00	---	---	
- Over-Rings	---	---	10.00	
Total				
Over + or Short -				

4)	Register 1	Register 2	Register 3	Total
+ Cash and Checks	1234.56	789.01	234.56	
+ Bank Charges	789.01	234.56	78.90	
+ House Charges	---	---	25.00	
+ Paid Outs	20.00	10.00	---	
Total				
+ Closing Reading	2,116.45	1,091.03	390.43	
- Opening Reading	50.00	50.00	50.00	
= Difference				
- Coupons	2.50	1.75	---	
- Discounts	12.34	5.67	---	
- Voids	---	---	---	
- Refunds	8.00	---	---	
- Over-Rings	---	---	---	
Total				
Over + or Short -				

Calculating dispensing fees

Pharmacies need to offset the costs of doing business through dispensing fees. This represents the charge for the professional services provided by the pharmacy when dispensing a prescription, and includes a distribution of the costs involved in running the pharmacy such as salaries, rent, utilities, costs associated with maintaining the computer system, etc. This is sometimes also referred to as a professional fee. While some pharmacies will create tiered dispensing fees (like we did in some of our earlier worksheets), most pharmacies will use just a flat rate dispensing fee for all prescriptions as it requires the same amount of labor to fill a low cost medication as it does to fill a higher cost medication.

A common formula for calculating a pharmacy's dispensing fee is:

$$\frac{Labor\ Expenses + Direct\ Expenses + Indirect\ Expenses}{Quantity\ of\ Prescriptions\ Filled\ Annually} = dispensing\ fee$$

- *Labor Expenses* = These are the wages paid to all pharmacy staff.
- *Direct Expenses* = These include items like the cost of the vials and the costs involved in maintaining the pharmacy computer system.
- *Indirect Expenses* = These are fixed items such as rent and utilities.

Let's use this formula for an example problem.

Example

1) Calculate the dispensing fee for a pharmacy with the following annual expenses if they filled 54,000 prescriptions annually.

Labor Expenses:

Wages for 1.6 pharmacists = $129,600.00
Wages for 1.0 technician = $43,200.00
5% match into 401k = $8,640.00
Healthcare contribution = $29,664.00
Total = $129,600.00 + $43,200.00 + $8,640.00 + $29,664.00 = **$211,104.00**

Direct Expenses:

Vials and caps = $\frac{\$0.12}{prescription} \times \frac{54{,}000\ prescriptions}{1}$ = $6,480.00
Labels, patient information, bags = $5,614.39
Computer system = $2,000.00
Total = $6,480.00 + $5,614.39 + $2,000.00 = **$14,094.39**

Indirect Expenses:

Rent = $33,600.00
Utilities = $9,000.00
Property insurance = $1,800.00
Total = $33,600.00 + $9,000.00 + $1,800.00 = **$44,400.00**

$$\frac{Labor\ Expenses + Direct\ Expenses + Indirect\ Expenses}{Quantity\ of\ Prescriptions\ Filled\ Annually} = dispensing\ fee$$

$$\frac{\$211{,}104.00 + \$14{,}094.39 + \$44{,}400.00}{54{,}000} = \mathbf{\$5.00\ dispensing\ fee}$$

Now that the calculation for determining a dispensing fee has been demonstrated, use the same method to calculate the dispensing fee in a practice problem below.

Practice Problem

1) Calculate the dispensing fee for a pharmacy with the following annual expenses if they filled 79,800 prescriptions annually.

Labor Expenses:
- Wages for pharmacists = $132,192.00
- Wages for technicians = $58,752.00
- Cost of benefits = $50,088.00

Direct Expenses:
- Vials and caps = $\frac{\$0.12}{prescription} \times \frac{79{,}800\ prescriptions}{1}$ = $9,576.00
- Labels, patient information, bags = $8,296.83
- Computer system = $2,000.00

Indirect Expenses:
- Rent = $33,600.00
- Utilities = $9,000.00
- Property insurance = $1,800.00

1) $3.83 dispensing fee

Depreciation

Depreciation is the decline in value of assets over time. Some items in a pharmacy lose value and eventually need to be replaced due to use, obsolescence, and the passage of time. The straight-line method of calculating depreciation uses the total cost, the estimated life of the property (in years), and the disposal value. Below is the formula you will need for this method:

$$\frac{total\ cost - disposal\ value}{estimated\ life} = \mathbf{annual\ depreciation}$$

Let's look at an example using the aforementioned formula.

Example

1) A pharmacy purchases a 4 cylinder sedan for drug deliveries to customers. The cost of the car is $11,965.00. Its estimated useful life is five years, and the resale value after five years is expected to be $2,632.30. What is the annual depreciation?

$$\frac{\$11{,}965.00 - \$2{,}632.30}{5} = \mathbf{\$1{,}866.54} \text{ } \boldsymbol{is} \textbf{ the } \boldsymbol{annual\ depreciation}$$

Now you should attempt a practice problem below.

Practice Problem

1) A Pharmacy purchases 3 workstations, a server, a printer, and various scanners in order to process prescriptions, maintain medication inventories, and various other functions related to running the pharmacy. These items cost the pharmacy approximately $6,000.00. The pharmacy's expectation is that this equipment will last five years and have negligible value after those five years are up (use $0.00 for the disposal value). What is the annual depreciation?

1) $1,200.00 is the annual depreciation of this computer equipment

Worksheet 14-8

Name:

Date:

Solve the following problems.

1) Calculate the dispensing fee for a pharmacy with the following annual expenses if they filled 54,900 prescriptions annually.

Labor Expenses:
Wages for pharmacists = $108,000.00
Wages for technicians = $48,000.00
Cost of benefits = $35,772.00
Direct Expenses:
Vials and caps = $\frac{\$0.12}{prescription} \times \frac{54{,}900\ prescriptions}{1}$ = $6,588.00
Labels, patient information, bags = $5,707.97
Computer system = $2,000.00
Indirect Expenses:
Rent = $12,600.00
Utilities = $2,400.00
Insurances = $5,520.00

2) Calculate the dispensing fee for a pharmacy with the following annual expenses if they filled 79,800 prescriptions annually.

Labor Expenses:
Wages for pharmacists = $132,192.00
Wages for technicians = $58,752.00
Cost of benefits = $50,088.00
Direct Expenses:
Vials and caps = $\frac{\$0.12}{prescription} \times \frac{79{,}800\ prescriptions}{1}$ = $9,576.00
Labels, patient information, bags = $8,296.83
Computer system = $2,000.00
Indirect Expenses:
Rent = $12,600.00
Utilities = $2,400.00
Insurances = $5,520.00

3) The pharmacy has a new point of sale (POS) system. The system cost $8,294.00 and should last six years. Its disposal value is $2,138.00. What is the annual depreciation?

4) A pharmacy purchased a new barrier isolator for $9,175.00 and is expected to last 12 years if properly maintained. Its disposal value is $1,567.00. What is the annual depreciation?

Worksheet 14-9

Name:

Date:

Match the definitions with their proper terms.

_____ 1) This represents the charge for the professional services provided by the pharmacy when dispensing a prescription.
_____ 2) The pharmacy must stock, or have ready access to, all drugs that may be written by the physicians in their practice area.
_____ 3) This is reimbursement for services rendered to a person in which another entity is responsible for the payment.
_____ 4) OSHA-required notices on hazardous substances.
_____ 5) This is calculated as sales minus all costs directly related to those sales.
_____ 6) A system that maintains a continuous count of every item in inventory so that it always shows the stock on hand.
_____ 7) This involves an agreement for a specified percentage or dollar volume of purchases.
_____ 8) The ordering of products for use or sale by the pharmacy and is carried out by an independent or group process.
_____ 9) This enables the pharmacy to use a single source to purchase products from numerous manufacturers.
_____ 10) This is the difference between the gross profit and the sum of all the costs associated with filling the prescription.
_____ 11) A government agency within the U.S. Department of Labor responsible for maintaining safe & healthy work places.
_____ 12) The total value of the drugs and merchandise in stock on a given day.
_____ 13) This is commonly known as the retail price and is the price paid for a prescription by a patient without insurance.
_____ 14) This is the decline in value of assets.
_____ 15) These require a DEA 222 form for reordering. Their stock must be continually monitored and documented.
_____ 16) This is the average price at which wholesalers typically sell medications to pharmacies.
_____ 17) Money spent to acquire or upgrade physical assets such as property, fixtures, or machinery
_____ 18) The drug inventory is limited to a list of approved medications.
_____ 19) A list of medications available for use within a health care system.
_____ 20) This entails ordering medications directly from the original drug manufacturer.
_____ 21) This is simply the entire stock on hand for sale at a given time.
_____ 22) Minimum and maximum stock levels which determine when a reorder is placed and for how much.
_____ 23) A method of payment for health services in which an individual or institutional provider is paid a fixed amount without regard to the actual number or nature of services provided to each patient.

a) average wholesale price
b) capital expenditures
c) capitation fee
d) closed formulary
e) depreciation
f) direct purchasing
g) dispensing fee
h) formulary
i) gross profit
j) inventory
k) inventory value
l) Material Safety Data Sheet (MSDS)
m) net profit
n) Occupational Safety and Health Administration (OSHA)
o) open formulary
p) perpetual inventory
q) prime vendor purchasing
r) purchasing
s) reorder point
t) schedule II medications
u) third party reimbursement
v) usual and customary price
w) wholesaler purchasing

Determine how much of each medication to reorder based on the package size and the minimum and maximum quantities the pharmacy wants to stock. Reorder medications when they reach the minimum.

	Medications	*Package Size*	*Minimum # Units*	*Maximum # Units*	*Current Inventory*	*Re-order*
24)	ramipril 5 mg cap	100	120	240	64	
25)	captopril 50 mg tab	100	150	300	76	
26)	doxazosin 2 mg tab	100	80	240	83	
27)	simvastatin 20 mg tab	100	360	500	190	
28)	simvastatin 40 mg tab	100	180	440	304	

Calculate how many times each of the following pharmacies turnover their inventory on an annual basis.

29) If April's Apothecary has an average inventory for the last year was $101,643.18, and the annual cost total was $1,168,896.70, what was the turnover rate?

30) If Drummond's Drug Store has an average inventory for the last year was $168,875.20, and the annual cost total was $2,026,402.40, what was the turnover rate?

Perform the following days' supply of inventory calculations.

31) April's Apothecary has a total inventory value of $101,643.30. Last week the pharmacy had sales of $26,974.54, and the cost to the pharmacy for the products sold came to $22,478.79. April's goal is to have a 35 days' supply of inventory. How many days' supply does April have? How much is she over or under her goal in dollars?

32) Drummond's Drug Store has a total inventory value of $168,874.98. Last week the pharmacy had sales of $45,955.44, and the cost to the pharmacy for the products sold came to $38,971.20. Drummond's goal is to have a 30 days' supply of inventory. How many days' supply does Drummond have? How much is he over or under his goal in dollars?

Answer the following questions.

33) How are Schedule II medications ordered?

34) How are Schedule III-V medications ordered?

35) How are controlled substances checked into the pharmacy?

36) How are controlled substances to be stored?

37) What are the major environmental considerations with respect to medications?

38) Define freezer, refrigerated, and controlled room temperature.

39) If two medications had an expiration of 02/2020 and 02/28/2020, which medication would expire first and why?

Solve the following markup and discount questions.

40) If a medication is purchased for $25 and sold for $35 what was the percent markup?

41) If the purchase price for 100 tablets of a medication was $49.25 and the percent markup is 25%, then what is the selling price for 30 tablets?

42) If the discount on a wholesaler invoice for $12,300 was $369, what was the percent discount?

43) If your wholesaler offers you a 5% discount on orders over $5,000, how much would you

expect your invoice to be for if you purchased $6,500 worth of product?

44) A senior citizen is paying for a prescription for furosemide 30 mg #30. The usual and customary price is $5.00. However, this patient qualifies for a 10% discount. How much will the patient pay?

45) A wholesaler offers the pharmacy a 10% discount on orders of $14,400 or more, or a 15% discount on orders of $28,800. If the pharmacy makes a $19,800 purchase, how much should its invoice be for?

Solve for the gross profit and the net profit for problems 46-51 using a set dispensing fee of $5.

46) The pharmacy purchased 25 nebulizer vials of albuterol inhalation solution for $25.90 and sold 25 vials for $46.85. What is the gross profit? What is the net profit?

47) The pharmacy purchased 20 tablets of carisoprodol 350 mg for $35.26 and sold 20 tablets for $60.89. What is the gross profit? What is the net profit?

48) The pharmacy purchased 100 tablets of E.E.S. 400 mg for $19.06 and sold 28 tablets for $12.35. What is the gross profit? What is the net profit?

49) The pharmacy purchased 100 capsules of fluoxetine 20 mg for $213.56 and sold 30 capsules for $108.10. What is the gross profit? What is the net profit?

50) The pharmacy purchased 100 tablets of glyburide 5 mg for $22.48 and sold 30 tablets for $16.47. What is the gross profit? What is the net profit?

51) The pharmacy purchased 100 tablets of levothyroxine 100 mcg for $24.87 and sold 30 tablets for $15.59. What is the gross profit? What is the net profit?

Solve the following problems using the sample formula below for third-party reimbursement:

$$87\%\, AWP\ or\ 100\%\, U\&C\,(whichever\ is\ less)+a\,\$\,3.50\, dispensing\ fee=\mathbf{reimbursement\ rate}$$

52) A pharmacy receives a prescription for 125 nebulizer vials of albuterol inhalation solution. The U&C on this product is $129.00. The AWP for 25 vials is $31.00. How much should the patient's insurance reimburse for this prescription?

53) A pharmacy receives a prescription for 20 tablets of carisoprodol 350 mg. The U&C on this product before a dispensing fee is added is $38.12. The AWP for 20 tablets is $41.40. How much should the patient's insurance reimburse for this prescription?

54) A pharmacy receives a prescription for 40 erythromycin 400 mg tablets. This product's usual and customary price is 12.51. The AWP for 100 tablets is $23.46. How much should the patient's insurance reimburse for this prescription?

55) A pharmacy receives a prescription for 30 capsules of fluoxetine 20 mg. The U&C on this product before a dispensing fee is added is $64.62. The AWP for 100 capsules is $248.40. How much should the patient's insurance reimburse for this prescription?

56) A pharmacy receives a prescription for 30 tablets of 5 mg glyburide. This product's usual and customary price is $11.79. The AWP for 100 tablets is $28.31. How much should the patient's insurance reimburse for this prescription?

Determine the profit margin for the following capitation fee problem.

57) April's Apothecary receives a monthly capitation fee of $210.00/month for Nauseous Nancy. Last month, Nancy's prescriptions totaled $188.75. How much did the pharmacy make off of her capitation fee?

Solve the following depreciation problem using the straight-line method.

58) The pharmacy uses a 1 year old compact car for drug deliveries. The car cost $9,420.00 and is expected to last four years from date of purchase. Its disposal value is $4,867.00. What is the annual depreciation?

Balance the following cash report.

59)	Register 1	Register 2	Register 3	Total
+ Cash and Checks	1264.56	790.01	233.56	
+ Bank Charges	789.03	235.56	78.90	
+ House Charges	---	---	25.00	
+ Paid Outs	20.00	10.00	---	
Total				
+ Closing Reading	2,151.47	1,095.03	389.47	
- Opening Reading	50.00	50.00	50.00	
= Difference				
- Coupons	7.50	1.75	---	
- Discounts	12.34	7.67	---	
- Voids	---	---	---	
- Refunds	8.00	---	---	
- Over-Rings	---	---	---	
Total				
Over + or Short -				

Solve the following dispensing fee problem.

60) Calculate the dispensing fee for a pharmacy with the following annual expenses if they filled 80,000 prescriptions annually.

Labor Expenses:

Wages for pharmacists = $138,801.60

Wages for technicians = $61,689.60

Cost of benefits = $51,340.20

Direct Expenses:

Vials and caps = $\frac{\$0.12}{perscription} \times \frac{80,000\ prescriptions}{1}$ = $9,600.00

Labels, patient information, bags = $8,317.63

Computer system = $2,000.00

Indirect Expenses:

Rent = $13,860.00

Utilities = $2,400.00

Insurances = $5,520.00

UNIT 4

INSTITUTIONAL PHARMACY MATH

What does institutional pharmacy math consist of?

In order to explain what institutional pharmacy math is, we should first define what an institutional pharmacy is. An institutional pharmacy is a pharmacy practice that provides drugs, devices, and other materials used in the diagnosis and treatment of patients in any of the following settings: hospitals, long term care facilities, convalescent homes, nursing homes, extended care facilities, mental health facilities, rehabilitation centers, psychiatric centers, developmental disability centers, drug abuse treatment centers, family planning clinics, penal institutions, hospice, public health facilities, and athletic facilities. The technician in this practice setting may need to use math for anything from compounding sterile products and chemotherapy to calculating radioactive decay of an isotope for a stress test.

What are the specific learning objectives in this unit?

- Parenteral routes of administration,
- Parenteral dosage calculations,
- Working with insulins,
- MilliMoles, milliEquivalents, millicuries, and international units,
- Reconstituting lyophilized powders,
- Percentage strength,
- Ratio strength,
- Reducing and enlarging formulas,
- Dosage calculations based on body weight,
- Dosage calculations based on body surface area,
- Infusion rates,
- Dilutions and alligations,
- Parenteral nutrition,
- Aliquots, and
- Drug desensitization therapy.

Chapter 15

Parenteral Dosage Calculations

"What is it he keeps saying about their parents?"
"He's not talking about parents, he's saying 'parenteral'."
--A discussion between one of my students and her spouse.

To get started, we should define the term parenteral. In order to define this, we should look at its root words:

para/o = despite, other than, or beside
enteron = meaning the alimentary canal, more commonly referred to as the GI tract
-al = a suffix meaning pertaining to

So, based on that we can define parenteral as a route of administration other than the GI tract. Technically this includes everything from topical medications and inhalation therapies to ear drops and injections, but today the term parenteral is intended to mean various kinds of injections and infusions and generally excludes all other routes of administration.

In this chapter our goals are:

- to learn about parenteral routes of administration,
- perform basic dosage calculations using dimensional analysis and/or ratio proportions, and
- use medication labels to perform necessary calculations.

Parenteral routes of administration

The following is a short list of parenteral routes of administration and is by no means considered comprehensive, but is instead intended to make you start thinking about these various routes[1].

- Intravenous - (IV) into a vein
- Intramuscular - (IM) into a muscle
- Subcutaneous - (SC, SQ) under the skin
- Intraarterial - (IA) into an artery
- Intracardiac - (IC) into the heart
- Intrathecal - (IT) into the spinal canal
- Intradermal - (ID) into the skin itself
- Intraperitoneal - infusion or injection into the peritoneum
- Epidural - infusion or injection into the epidural space (the outermost part of the spinal canal)
- Intravitreal - through the eye
- Intraosseus infusion - through the bone marrow
- Intrahepatic - into the liver
- Intracerebral - into the cerebrum
- Intracerebroventricular - into the cerebral ventricles
- Intravesical infusion - into the urinary bladder
- Intracavernosal injection - into the base of the penis

1 Some references will also include inhalation and ophthalmic as they also need to be sterile.

Additional precautions need to be kept in mind when preparing parenterals because they are able to avoid many of a patient's barriers to absorption due to how they are administered. These special considerations are that:

- solutions for injection must be sterile – i.e., free from bacteria and other microorganisms,
- solutions must be free of all visible particulate material,
- all parenteral solutions must be pyrogen-free,
- the solution must be stable for its intended use,
- the pH of an intravenous solution should not vary significantly from physiological pH (approximately 7.4), and
- intravenous solutions should be formulated to have an osmotic pressure similar to that of blood (isotonic).

Basic dosage calculations

Often, as a pharmacy technician, you will receive a label for a medication you will need to make in the IV room. The medication will request a specific patient dose in milligrams, grams, units, milliEquivalents, etc. You will need to use information on the vial or in the literature to determine how many milliliters you will need to draw up in order to fulfill the requested dose. Conveniently, you already know the problem solving methods you will need to employ to solve these kinds of problems. The challenge is filtering through all the information on the label to decide what you need to use and what you don't need. Let's look at an example.

Example

Magnesium sulfate 2 g in 100 mL of 5% dextrose in water (D5W) is ordered by the physician. How many milliliters magnesium sulfate will be added to the bag that patient receives if the magnesium sulfate vial provides the following information: 50% magnesium sulfate (500 mg/mL), 4.06 mEq/mL, 10 mL single dose vial?

QUESTION

How many milliliters will be added to the bag?

DATA

2 g of magnesium sulfate wanted

medication is being added to a 100 mL bag of D5W

50% magnesium sulfate = $\frac{50\,g\,magnesium\,sulfate}{100\,mL}$

500 mg magnesium sulfate/mL

4.06 mEq magnesium sulfate/mL

10 mL single dose vial

MATHEMATICAL METHOD/FORMULA

dimensional analysis or ratio-proportion

DO THE MATH

dimensional analysis

$$\frac{2\,g}{1} \times \frac{1000\,mg}{1\,g} \times \frac{mL}{500\,mg} = \mathbf{4\,mL}$$

ratio-proportion

$$\frac{2\,g}{N} = \frac{50\,g}{100\,mL}$$

$N = \mathbf{4\,mL}$

DOES THE ANSWER MAKE SENSE?

Yes

When you look at the example problem, you can identify multiple ways to solve it (even more than the two ways demonstrated). It is also important to note that there were many aspects of the example problem that could be completely ignored, such as the diluent (100 mL of D5W), the size of the vial being used (10 mL), and even some of the information on its concentration (4.06 mEq/mL).

Try and solve the practice problem below using the images presented with the problem. Notice that there is much more information presented then what you need for the practice problem below.

Practice Problem

A 250 mL bag of Sterile Water For Injection (SWFI) needs the addition of 19.4 mEq of sodium chloride. How many milliliters should be added to the SWFI bag? Draw a line on the syringe pictured below to demonstrate how much of the stock sodium chloride you will need.

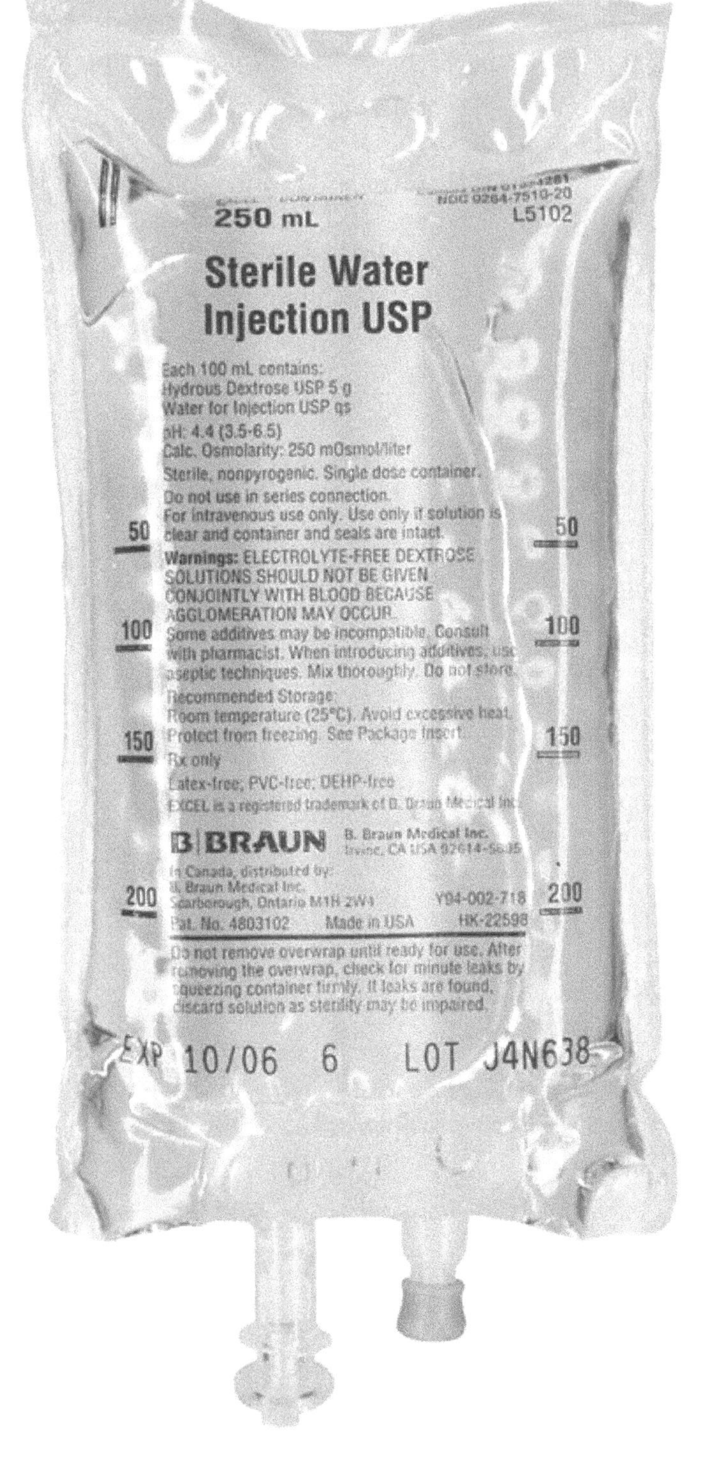

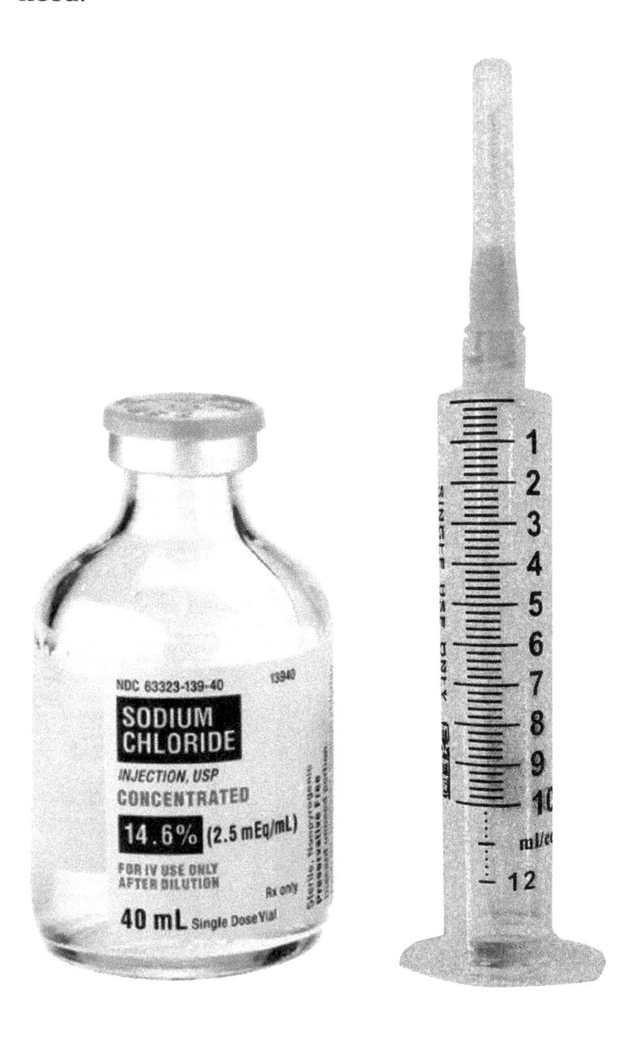

The syringe should be marked at approximately 7.8 mL.

Worksheet 15-1

Name:

Date:

Solve the following problems.

1) A 1,500 mL TPN being infused intravenously through a central line needs the addition of 99 mEq of sodium chloride. The label on the vial of concentrated sodium chloride injection has the following information: 30 mL single dose vial, 234 mg/mL, 4 mEq/mL, and 23.4%. How many milliliters should be added to the TPN bag?

2) Digoxin injection is available in a concentration of 0.5 mg/2 mL. The physician orders a 250 mcg dose in 250 mL of D5W. How many milliliters will the patient need?

3) Tobramycin injection is available in a concentration of 80 mg/2 mL. The patient received 1.25 mL in 100 mL of 0.9% sodium chloride. What was the dose in mg that the patient received?

4) Twelve units of Humulin R are to be added to a 2 liter TPN. The 10 mL vial of Humulin R has a U-100 concentration (100 units/mL). How many milliliters of Humulin R are required?

5) Morphine sulfate 12 mg is ordered by the physician. The label on the morphine sulfate vial reads 15 mg/mL. How many milliliters will the patient receive?

6) Atropine sulfate injection 0.4 mg per mL is available in the pharmacy. The doctor orders 1 mg. How many milliliters will complete this order?

7) A patient requires potassium chloride 25 mEq in a 1000 mL bag of lactated ringers solution.

The pharmacy has on hand potassium chloride for injection 40 mEq in 20 mL vials. How many milliliters will be needed in the IV bag?

8) Aminophylline injection is available in a 20 mL vial containing 500 mg (25 mg/mL). The physician orders a dose of 350 mg. How many milliliters will be needed to fill this order?

9) A 500 mL bag of D5W with 16,000 units of heparin is ordered for a patient. A 5 mL vial of heparin contains 10,000 units per mL. How many milliliters of heparin are needed for this patient?

10) A physician orders 25 mg of theophylline to be given orally to a pediatric patient. If the elixir of theophylline contains 80 mg per tablespoonful, how many milliliters of the elixir should be administered?

11) A 1,000 ml bag of Sterile Water For Injection needs the addition of 77.5 mEq of sodium

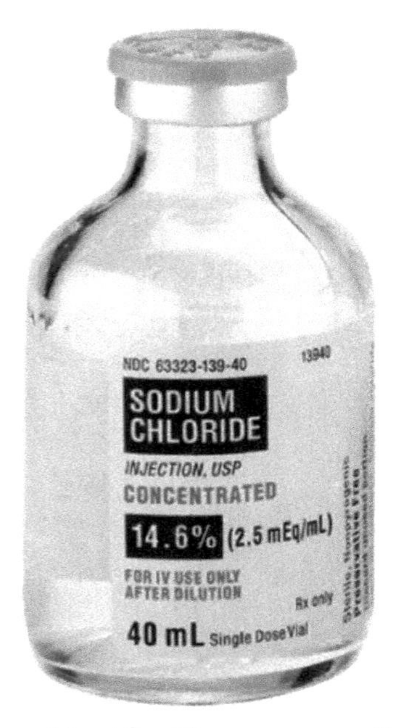

chloride. How many milliliters should be added to the SWFI bag?

12) A patient is to receive gentamicin in 50 ml of 0.9% sodium chloride. Look at the syringe and check what the dose is in mg that the patient is to receive.

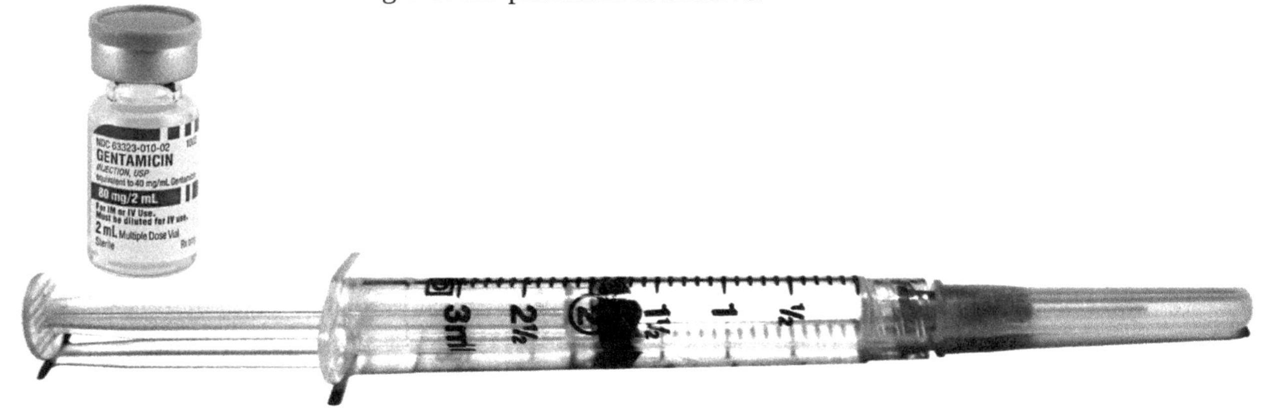

13) Humulin R 50 units is to be added to a 50 mL bag of NS. How many milliliters of Humulin R are required to make this infusion?

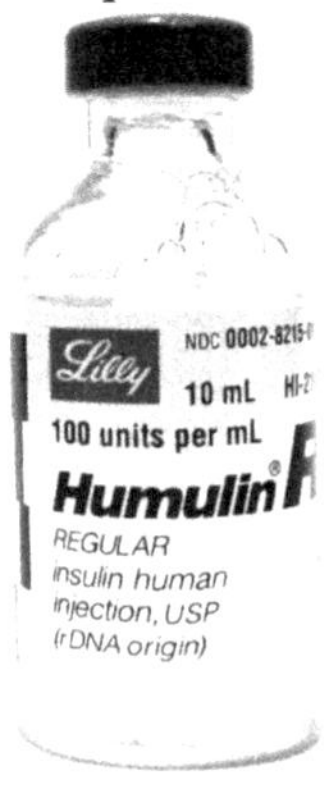

14) Magnesium sulfate 2 g in 100 mL of D5W is ordered by the physician. How many milliliters magnesium sulfate will be added to the bag that patient receives?

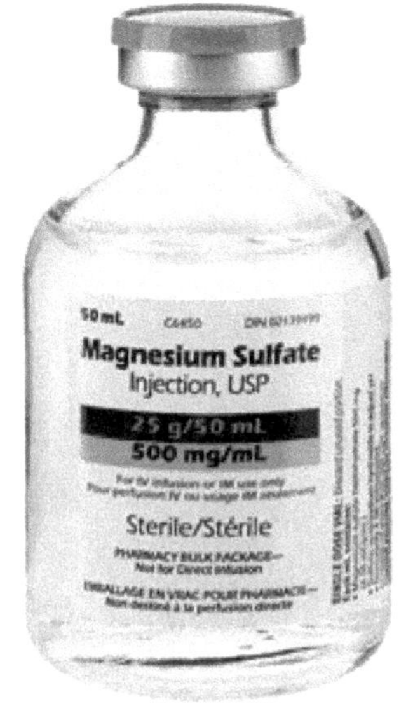

15) A patient requires potassium chloride 15 mEq in a 5,000 mL bag of prismasate solution. How many milliliters will need to be withdrawn from the 10 mL vial below?

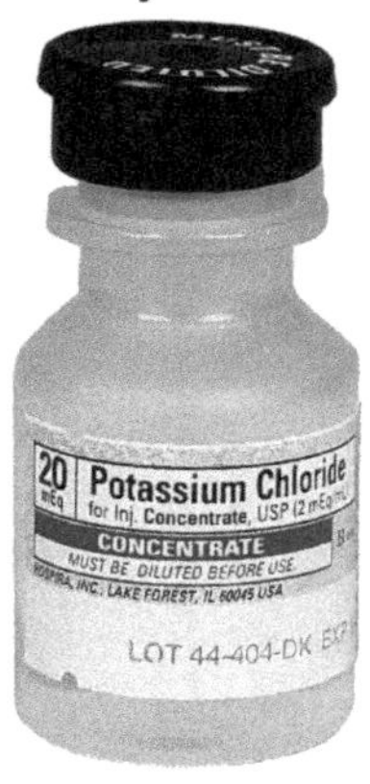

16) The physician orders 300 mg of aminophylline. How many milliliters will be needed to fill this order?

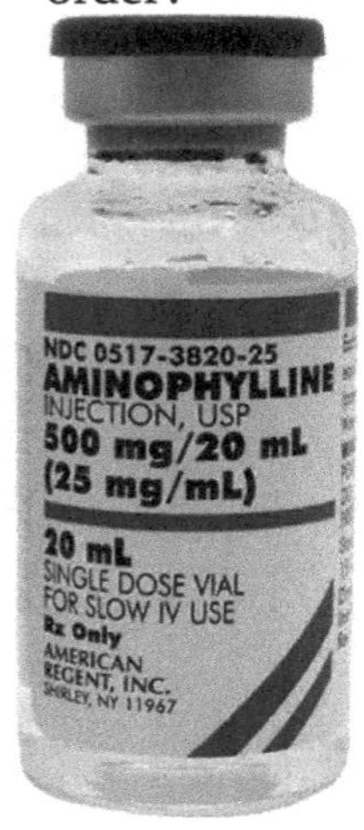

17) 600 units of Heparin is ordered for a patient. How many milliliters of heparin are needed for this patient?

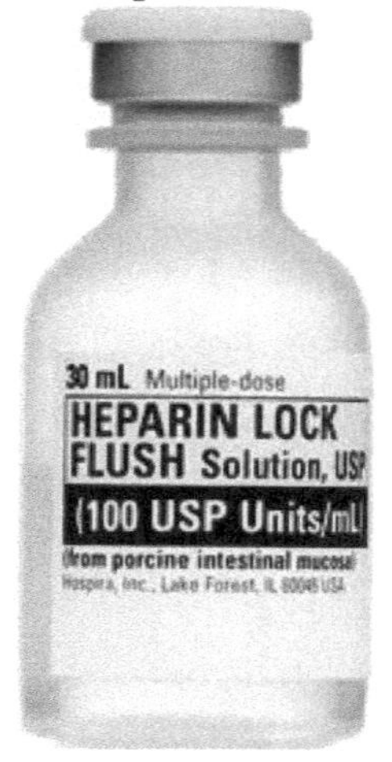

Worksheet 15-2

Name:

Date:

Answer the following questions.

1) What does the word '*parenteral*' mean?

2) Correctly match the following terms to their meaning.

_____ Intravenous (IV)

_____ Intramuscular (IM)

_____ Subcutaneous (SC, SQ)

_____ Intraarterial (IA)

_____ Intracardiac (IC)

_____ Intrathecal (IT)

_____ Intradermal (ID)

_____ Intraperitoneal

_____ Epidural

_____ Intravitreal

_____ Intraosseus infusion

_____ Intrahepatic

_____ Intracerebral

_____ Intracerebroventricular

_____ Intravesical infusion

_____ Intracavernosal injection

a) infusion or injection into the epidural space (the outermost part of the spinal canal)
b) infusion or injection into the peritoneum
c) into a muscle
d) into a vein
e) into an artery
f) into the base of the penis
g) into the cerebral ventricles
h) into the cerebrum
i) into the heart
j) into the liver
k) into the skin itself
l) into the spinal canal
m) into the urinary bladder
n) through the bone marrow
o) through the eye
p) under the skin

3) Name two other routes of administration that require sterile products.

4) Make a short list of precautions/considerations when dealing with compounded sterile preparations (CSPs).

5) You receive an order for heparin 12,500 units in 250 mL of D5W. If the strength of the heparin available is 5,000 units/mL, how many mL of heparin will you need?

6) A TPN requires the addition of 15 units of regular insulin. If you are using the insulin vial pictured below, how many mL of insulin will you need to add to the TPN?

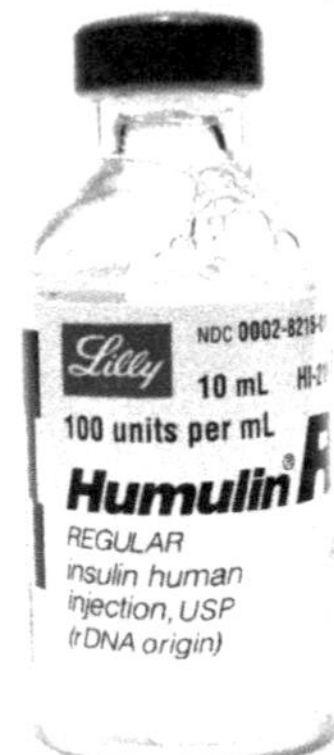

7) Calculate the number of milliliters required to prepare the following concentrations:
   a) 25 mEq potassium chloride

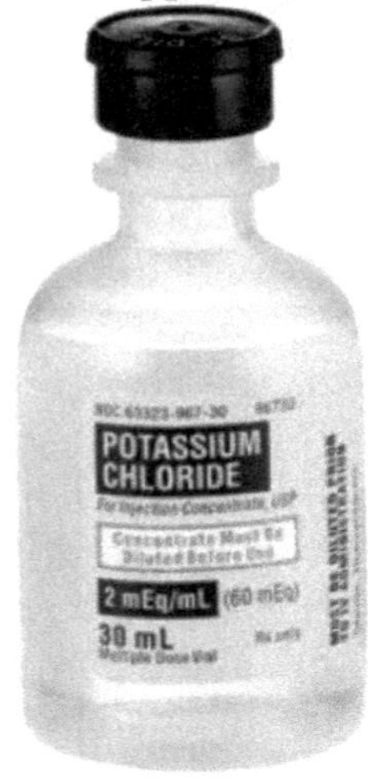

   b) 37.5 mg methotrexate

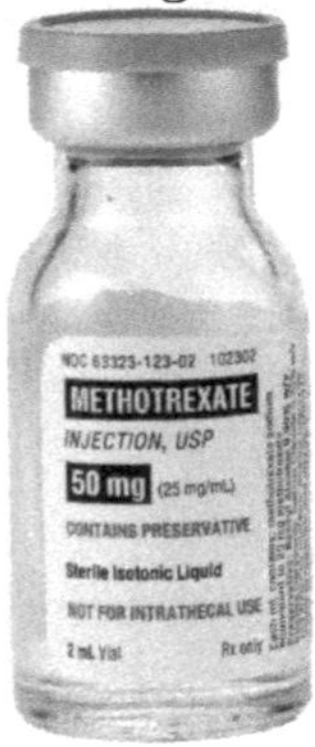

c) 1050 mg fluorouracil

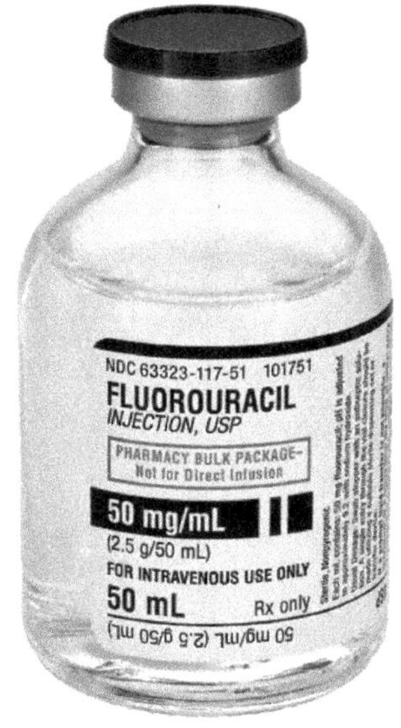

d) 62.5 mg doxorubicin

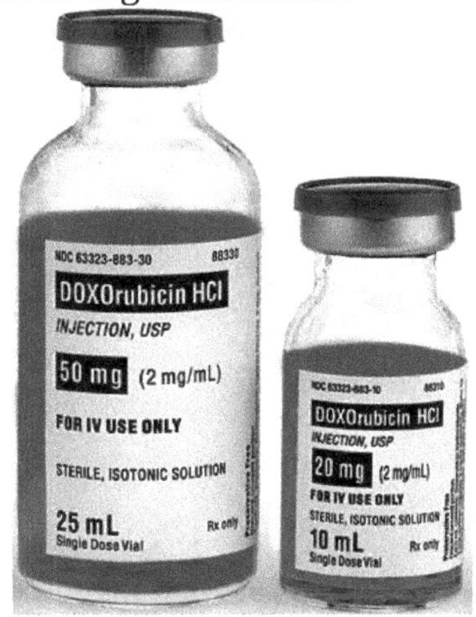

e) 30 units Novolin N

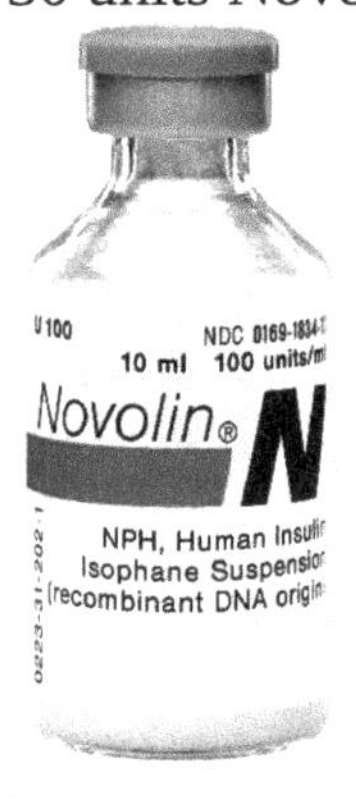

f) 200 mcg scopolamine

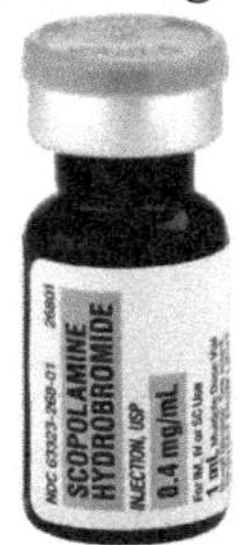

g) 17.6 mEq potassium phosphate

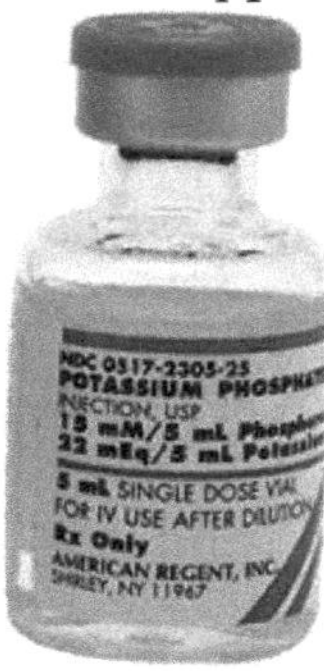

8) Levothyroxine comes in 500 mcg vials. If the powder is diluted with 10 mL of sterile water (the medication has negligible powder volume), how many mL are required to provide 0.1 mg? Draw a line on the syringe below showing what volume you would draw up in the syringe.

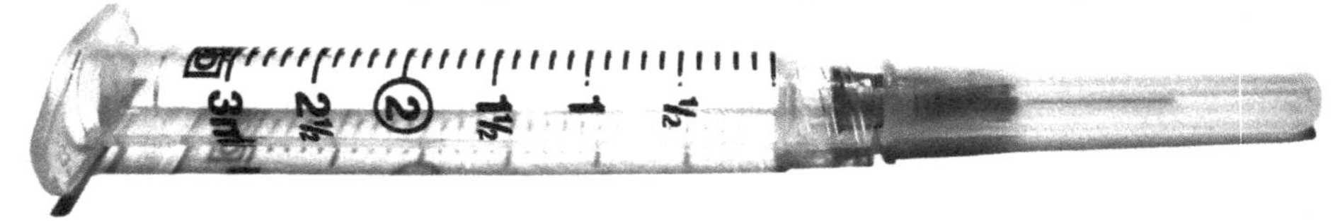

9) Clindamycin phosphate comes as 600 mg/4 mL. How many mL are needed to make an IVPB of 750 mg in 100 mL of NS?

10) A patient requires 30 units of oxytocin by IV infusion in a liter of D5W. Oxytocin is available as 10 units per mL. How many mL should be added to the IV bag?

11) Tobramycin is available in a concentration of 80 mg/2 mL. The patient received 2.5 mL in 100 mL of NS infused over 1 hour. How many mg did the patient receive?

12) A physician orders 6.5 MU of penicillin G potassium. The stock vial has three different possible concentrations depending as to how much volume it is reconstituted with. How many mL would you need to draw up for each possible concentration?

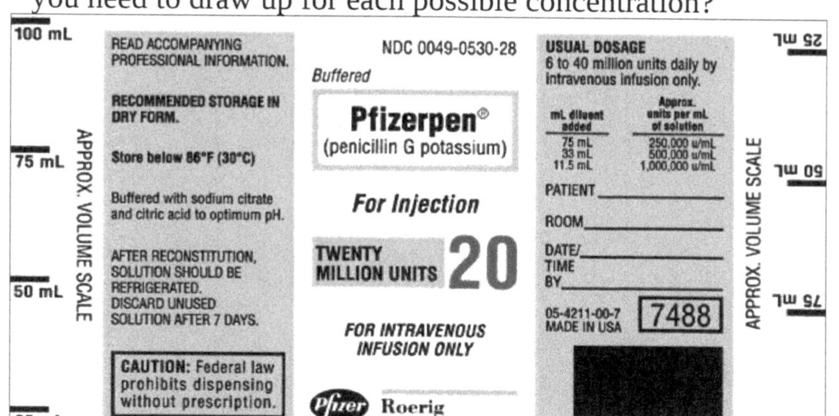

13) A physician orders 16 mg of norepinephrine in 234 mL of D5W. On the syringes below mark how many mL will need to be removed from the D5W bag and how many mL of norepinephrine will need to be added to it.

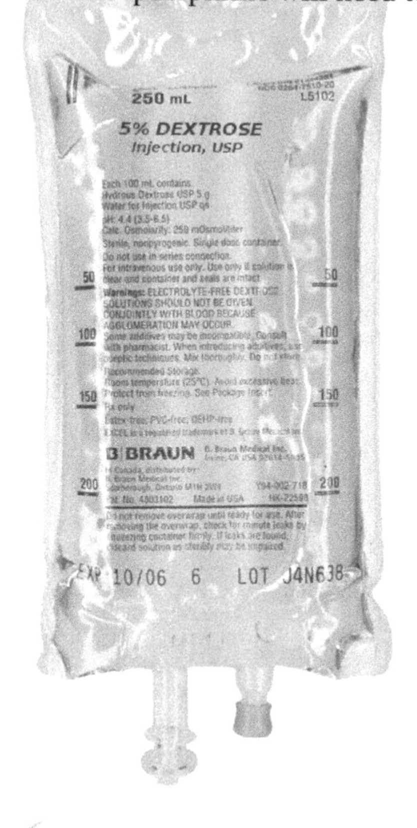

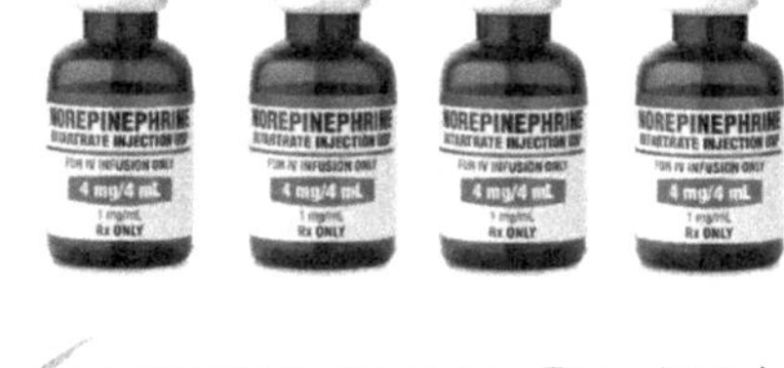

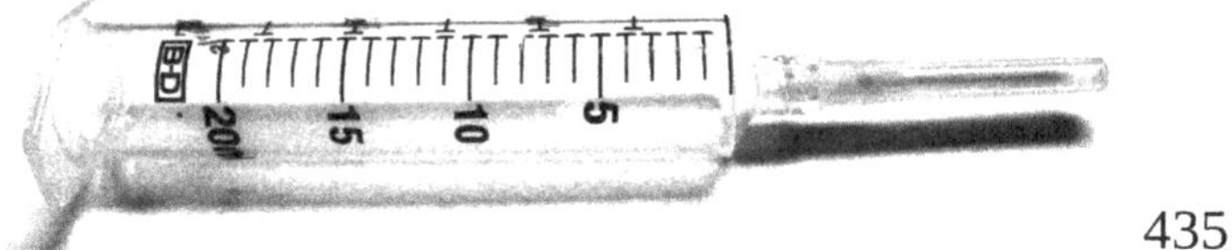

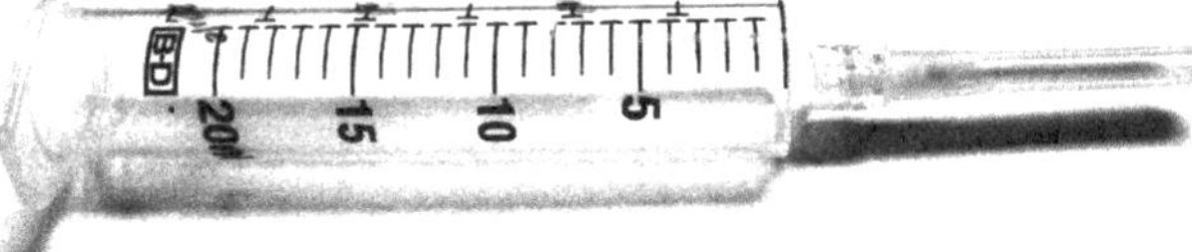

CHAPTER 16

INSULIN

"How did she get the nickname of 'Sweet Pea'?"
"When her classmates discovered the Greek and Latin roots for diabetes mellitus, they thought it would be a good pet name for her as a diabetic."
--A discussion between two instructors.

Ideally, the human body will effectively create its own insulin and make proper utilization of it in order to regulate its blood glucose level. When the human body fails to regulate it effectively, a patient may end up receiving some form of insulin therapy from a pharmacy.

In this chapter we will look at the following concepts with respect to insulin therapy:

- Categories of patients receiving insulin therapy,
- Definitions,
- Dosing and types of insulins,
- Insulin syringes, and
- Mixing insulins.

Categories of patients receiving insulin therapy

Insulin therapy is required in type 1 diabetes (insulin dependent diabetes mellitus, IDDM), and may be necessary in some individuals with type 2 diabetes (non-insulin dependent diabetes mellitus). The general objective of insulin replacement therapy is to approximate the physiological pattern of insulin secretion. This requires a basal insulin throughout the day, supplemented by prandial insulin at mealtime. Insulin injections are intended to mimic the natural process shown in the image above:

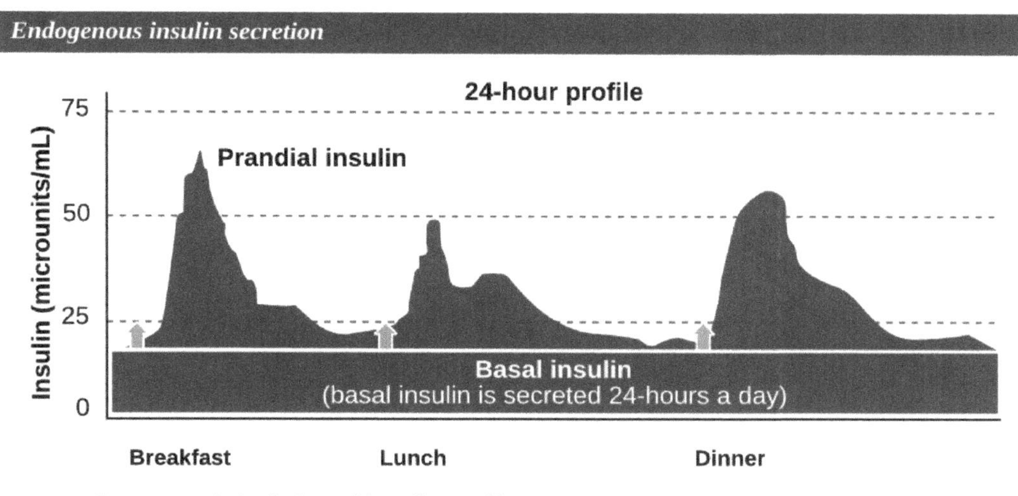

Conceptual depiction of insulin profiles

Definitions

Insulin is a hormone central to regulating carbohydrate and fat metabolism in the body. Insulin causes cells in the liver, muscle, and fat tissue to take up glucose from the blood, storing it as glycogen in the liver and muscle. Ideally it will be created endogenously in the pancreas and effectively used by the body's cells, but when either one or both of those situations are not occurring, an exogenous source of insulin may be required.

Basal insulin may also sometimes be called "background" insulin, that is, the insulin working behind the scenes. Basal insulin may be covered with a long-acting insulin like glargine insulin, or an intermediate-acting insulin like isophane (NPH) insulin.

Prandial insulin, also known as nutritional insulin, is the insulin used to cover the spike in blood sugar from consuming food. Prandial insulin may be covered with regular insulin (a short-acting insulin), or a rapid-acting insulin like lispro insulin.

Dosing and types of insulin

Although some patients may receive a continuous infusion of insulin in an institutional setting, multiple daily doses guided by blood glucose monitoring are the standard of diabetes care. Combinations of insulins are commonly used. The number and size of daily doses, time of administration, and diet and exercise require continuous medical supervision. In addition, specific formulations may require distinct administration procedures/timing.

There is solid scientific documentation of the benefit of tight glucose control, either by insulin pump or multiple daily injections (4-6 times daily). However, the benefits must be balanced against the risk of hypoglycemia, the patient's ability to adhere to the regimen, and other issues regarding the complexity of management. Diabetic education and nutritional counseling are essential to maximize the effectiveness of therapy. Patients should also be instructed in administration techniques, timing of administration, and sick day management.

Type 1 diabetics (IDDM) will traditionally use either a rapid or short-acting insulin prior to each meal simulating prandial insulin and they will often take an intermediate or long-acting insulin either once or twice a day simulating basal insulin. The image to the right demonstrates this concept.

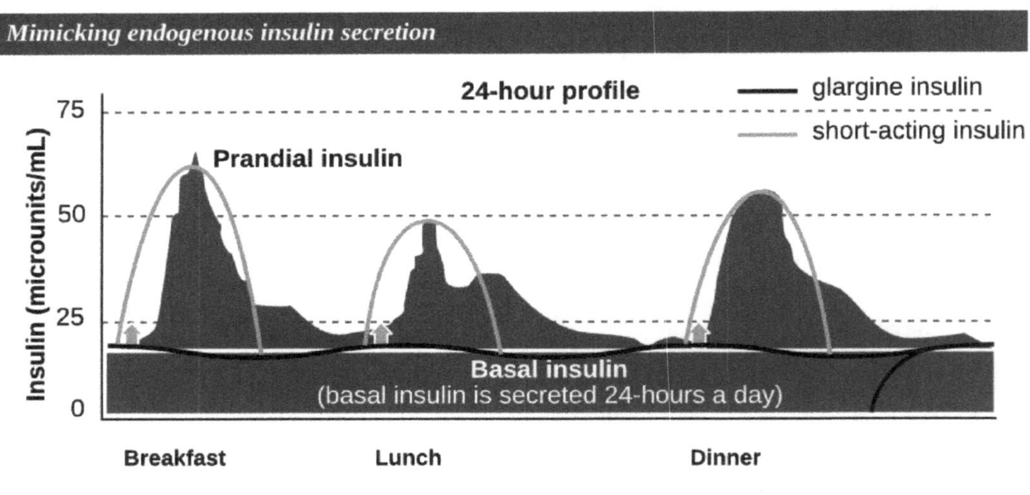

Conceptual depiction of insulin profiles

The chart on the following page lists the onset, peak, and duration of various insulin products. Combination products include either rapid or short-acting insulin in combination with an intermediate-acting insulin.

Types of Insulin	*Onset (h)*	*Peak (h)*	*Duration (h)*
Rapid-acting			
lispro insulin - Humalog	0.2-0.5	0.5-1.5	3-4
aspart insulin - NovoLog	0.2-0.5	1-3	3-5
glulisine insulin - Apidra	0.2-0.5	0.5-1.5	3-4
Short-acting			
regular insulin (clear) - Humulin R, NovolinR	0.5-1	2-4	6-8
Intermediate-acting			
isophane (NPH) insulin (cloudy) - Humulin N, Novolin N	1-2	6-12	18-24
Intermediate to long-acting			
detemir insulin - Levemir	3-4	6-8	6-23
Long-acting			
glargine insulin - Lantus	3-4	*	24
Combinations			
70% isophane (NPH) insulin & 30% regular insulin - Humulin 70/30, Novolin 70/30	0.5	2-12	18-24
aspart protamine insuline & aspart insulin - Humalog Mix 50/50, Humalog Mix 75/25	0.2-0.5	1-4	18-24
lispro protamine insulin & lispro insulin - NovoLog Mix 70/30	0.2-0.5	2-12	18-24

* glargine insulin has no pronounced peak

Insulin syringes

Insulin dosing is very customized for various patients, and therefore, flexible delivery systems allowing for a wide range of required doses make it common place for the use of insulin syringes and vials. Insulin vials are marked with their concentration in two ways: 100 units per mL and U-100. Most insulins have this same concentration, but occasionally, pharmacies need to carry a concentration of 500 units per mL / U-500 regular insulin for patients that need very high doses of insulin. Pictured below are both concentrations.

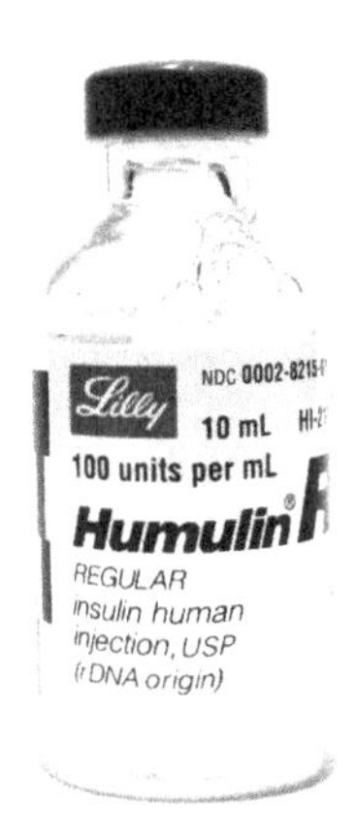

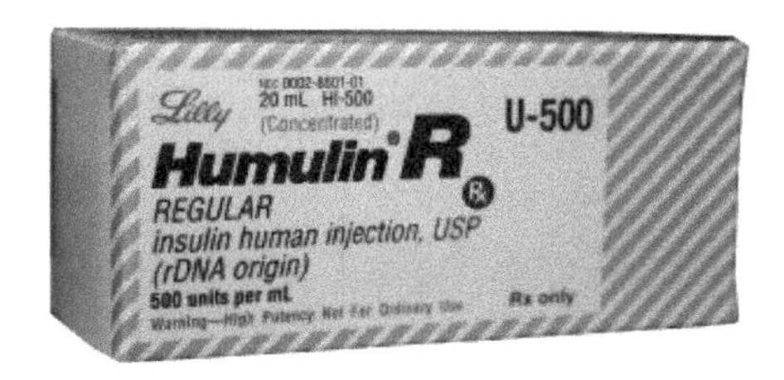

Insulin syringes (easily identifiable by their orange caps) are intended for the much more common dosing of U-100 insulins. Note, insulin syringes should only be used for insulins with a concentration of U-100 as the higher concentration of U-500 insulin will lead to a medication error. The following are images providing examples of various insulin syringes and a 1 cc syringe for dosing a U-500 insulin.

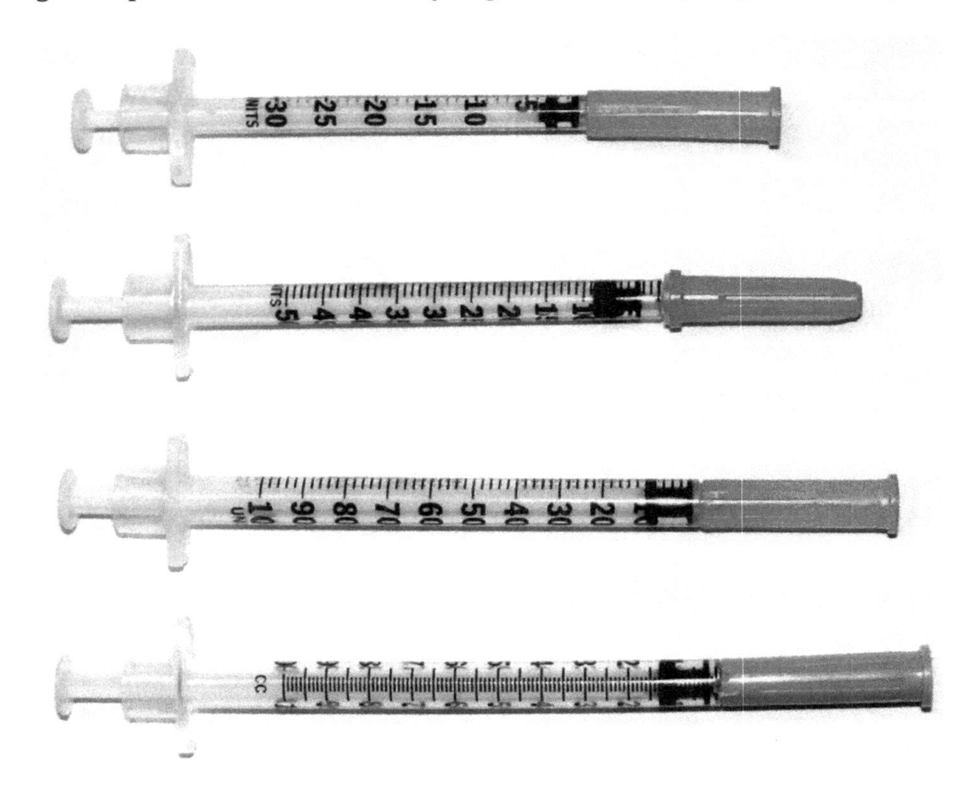

When selecting a syringe, choose the one that will most closely compliment the dose being given to the patient.

Practice Problems

Mark on the syringes near each problem how many units or mL you should draw up for each scenario.

1) 66 units of Novolin N in is ordered for a patient.

2) 150 units of U-500 concentrated Humulin R is ordered for a patient.

3) 8 units of Humalog is ordered for a patient.

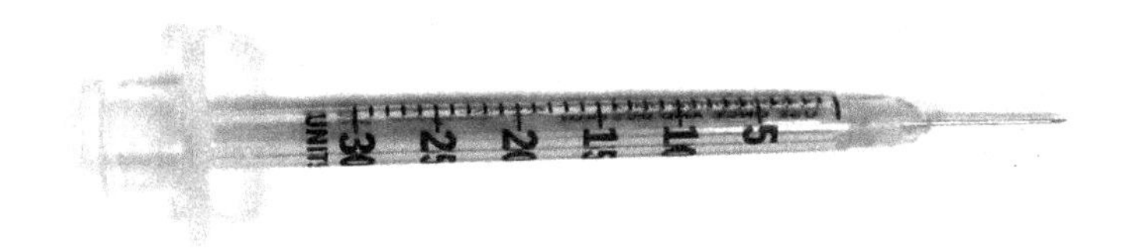

1) The syringe should be marked at 66 units
2) The syringe should be marked at 0.3 cc
3) The syringe should be marked at 8 units

Mixing insulins

As previously mentioned, a physician will often order two kinds of insulin. In order to decrease the number of injections a patient might need to take in a day, a physician may order a combination vial such as Novolin 70/30 or Humalog Mix 75/25 in order to allow a patient to receive both kinds at the same time. If a patient's needs can not be met that way with premixed combination vials, some solutions can be mixed together by the patient immediately before administration and injected. Isophane (NPH) insulin may be mixed with the following insulins: regular insulin, lispro insulin, aspart insulin, or glulisine insulin. Below is a list explaining the procedure for drawing up two different insulins into the same syringe.

1) Calculate the total dose of both insulins combined.
2) Draw up a volume of air equivalent to the volume of isophane (NPH) insulin desired and inject the air into the isophane (NPH) insulin vial, but do not draw up the dose. Withdraw the needle from the vial.
3) Draw up a volume of air equivalent to the volume of the rapid or short-acting insulin and inject it into the rapid or short-acting insulin vial. Draw up the appropriate quantity of this insulin.
4) Carefully insert the needle through the stopper of the isophane (NPH) insulin vial. Invert the vial without injecting any of the rapid or short-acting insulin into the vial.
5) Slowly draw up the isophane (NPH) insulin until the syringe reaches the appropriate dose for both insulins combined.

A good memory trick to help with memorization of this pattern is *'clear before cloudy'* as the clear rapid or short-acting insulin is actually drawn into the syringe prior to adding the cloudy isophane (NPH) insulin.

Example

A patient requires 42 units of Humulin N and 10 units of Humulin R to be given at the same time. To minimize the number of needle sticks the patient needs to endure, they should be drawn up at the same time.

First, calculate the total dose.

$$42\,U + 10\,U = 52\,\text{units}$$

Then draw up 42 units of air and inject it into the Humulin N vial, but do not draw up any solution yet. Withdraw the needle from the vial.

Next, draw up 10 units of air, inject it into the Humulin R, and draw 10 units of regular insulin.

Next, insert the needle into the Humulin N vial and carefully invert the vial without injecting any solution into the Humulin N.

Lastly, slowly withdraw insulin from the Humulin N vial until the vial contains a total of 52 units of insulin.

Practice Problem

Using the above example as a guide, explain how to prepare an insulin syringe with 43 units of Novolin N and 22 units of NovoLog.

First, calculate the total dose. (43 units + 22 units = 65 units) Then draw up 43 units of air and inject it into the Novolin N vial, but do not draw up any solution yet. Withdraw the needle from the vial. Next, draw up 22 units of air, inject it into the NovoLog, and draw 22 units of aspart insulin. Next, insert the needle into the Novolin N vial and carefully invert the vial without injecting any solution into the isophane (NPH) insulin. Lastly, slowly withdraw insulin from the Novolin N vial until the vial contains a total of 65 units of insulin.

Worksheet 16-1

Name:

Date:

Select the vial or box that corresponds with each order in problems 1-14 and 16-17. Also mark the syringes pictured in problems 1-14 with the correct volume for the ordered dose.

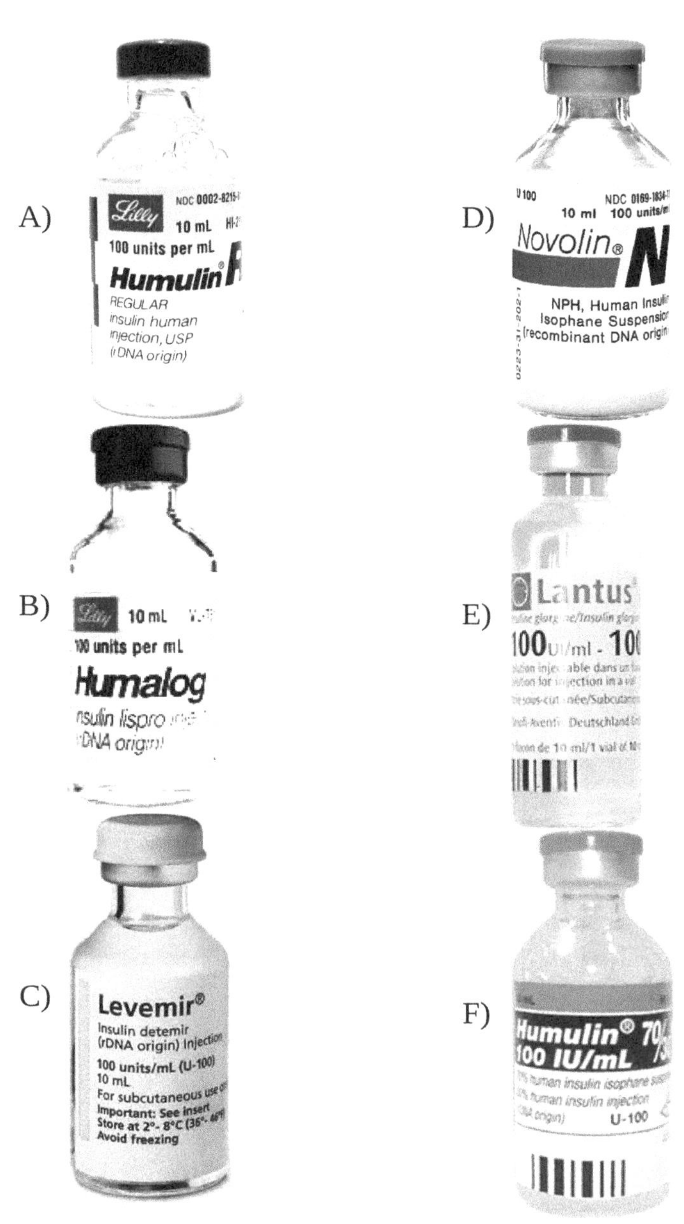

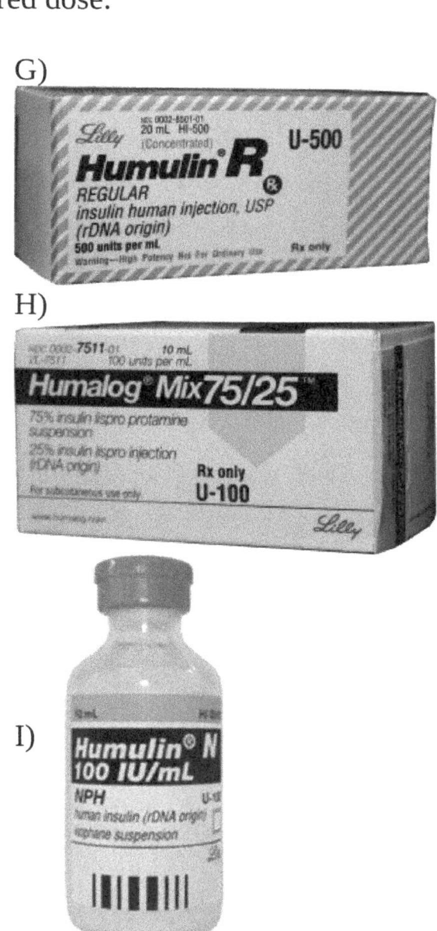

1) A patient is ordered 12 units of Humulin R SQ before breakfast.
Selected vial: _____

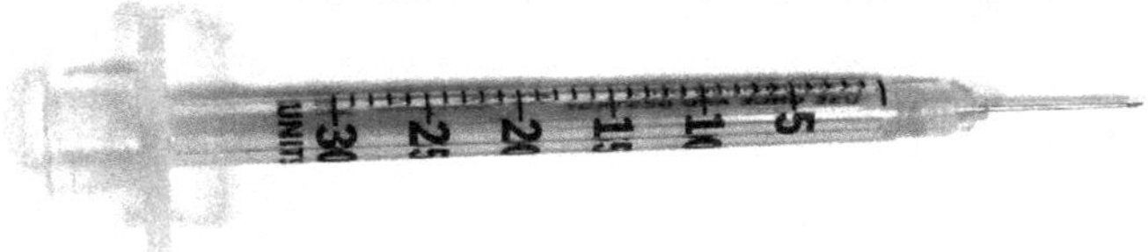

2) A patient is ordered 5 units of Humalog SQ 15 minutes before lunch.
Selected vial: _____

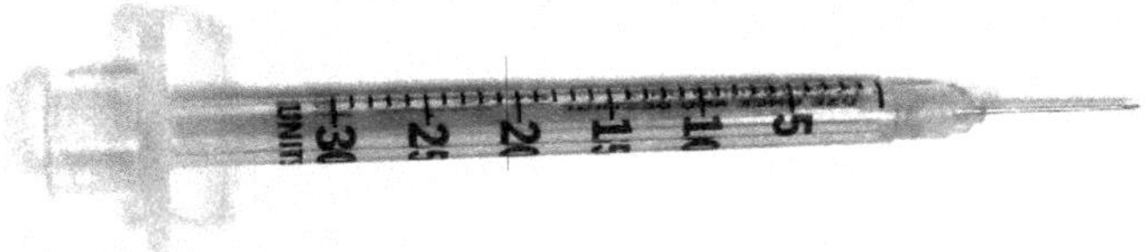

3) A patient is ordered 35 units of Novolin N SQ every morning.
Selected vial: _____

4) A patient is ordered 72 units of Lantus SQ every evening.
Selected vial: _____

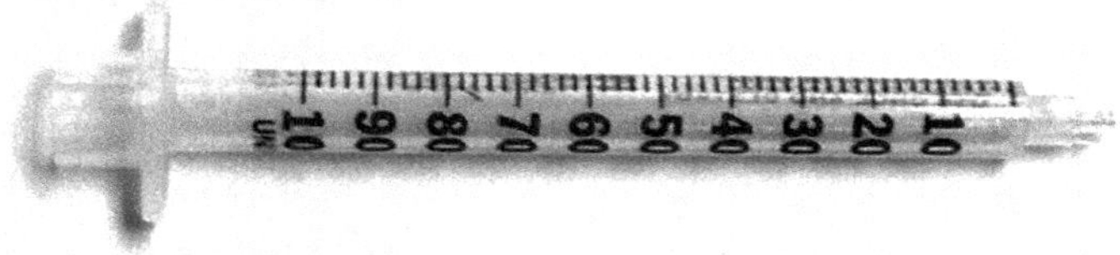

5) A patient is ordered 55 units of Humulin N SQ every evening.
Selected vial: _____

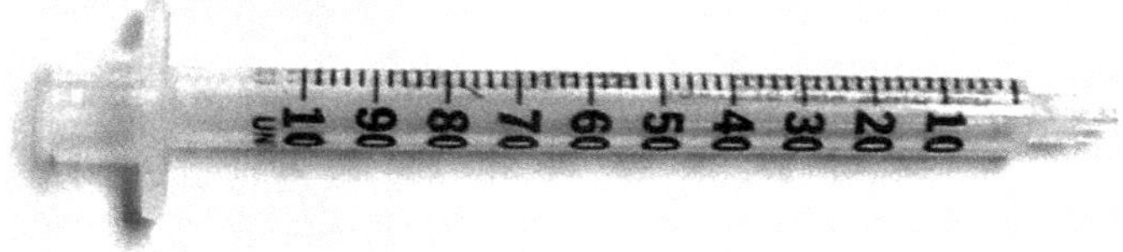

6) A patient is ordered 22 units of Levemir SQ every morning.
Selected vial: _____

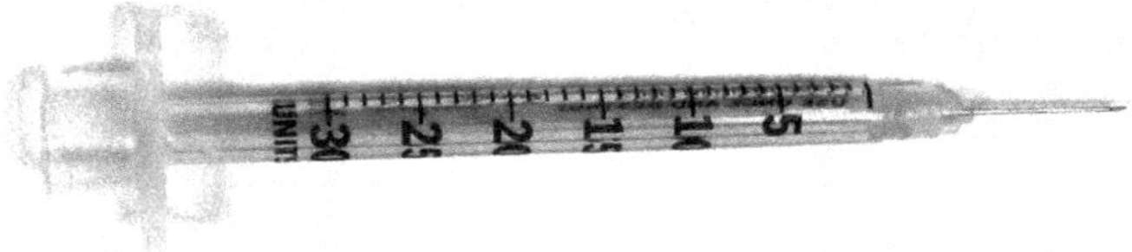

7) A patient is ordered 80 units of Humulin 70/30 SQ every morning.
Selected vial: _____

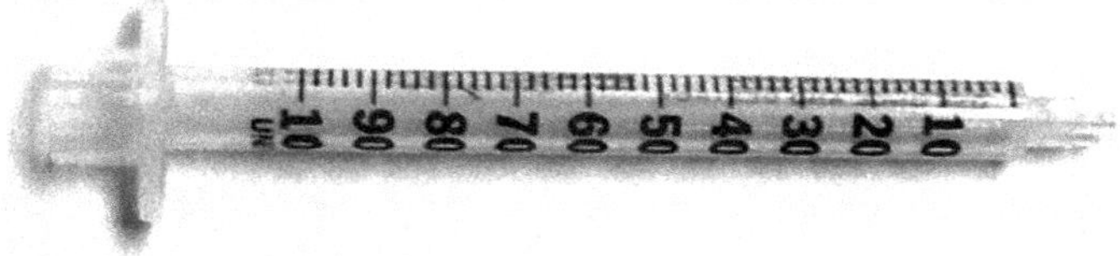

8) A patient is ordered 45 units of Humalog Mix 75/25 SQ every evening.
Selected vial: _____

9) A patient is ordered 25 units of Humalog SQ daily before lunch.
Selected vial: _____

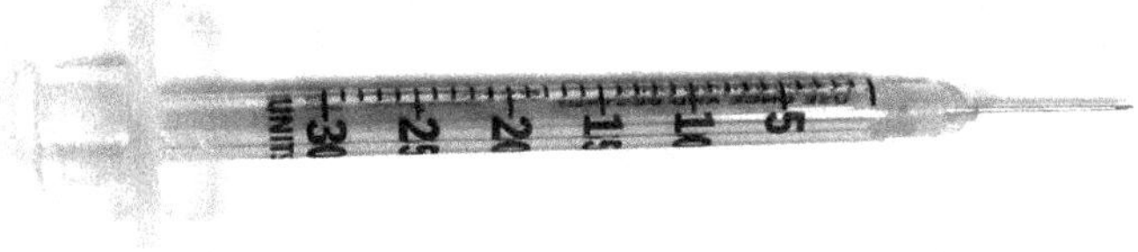

10) A patient is ordered 170 units of U-500 Humulin R every morning.
Selected vial: _____

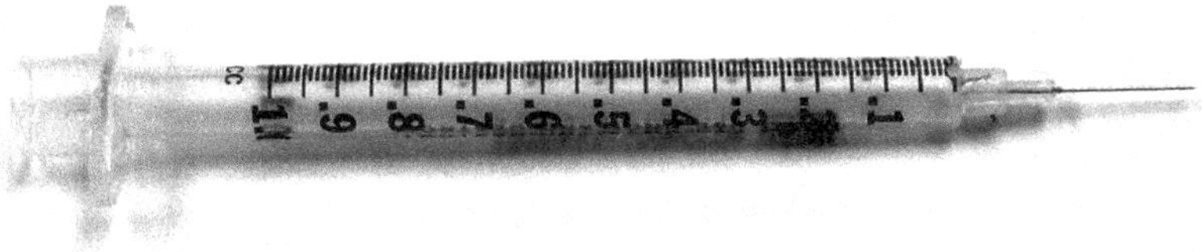

11) A patient is ordered 40 units of Novolin N SQ every evening.
Selected vial: _____

12) A patient is ordered 77 units of Humulin 70/30 SQ every morning 30 minutes before breakfast.
Selected vial: _____

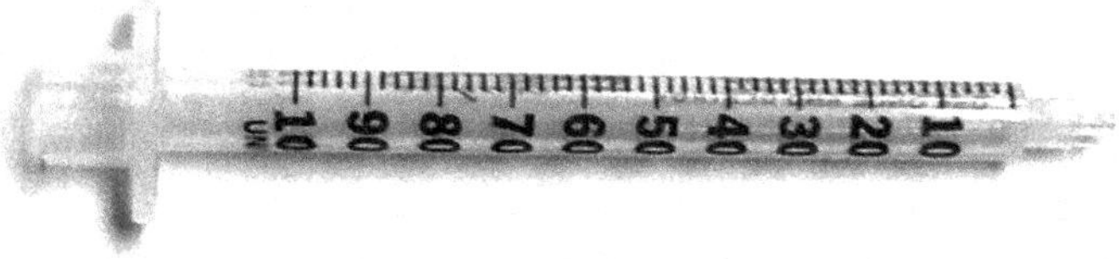

13) A patient is ordered 21 units of Lantus SQ every evening.
Selected vial: _____

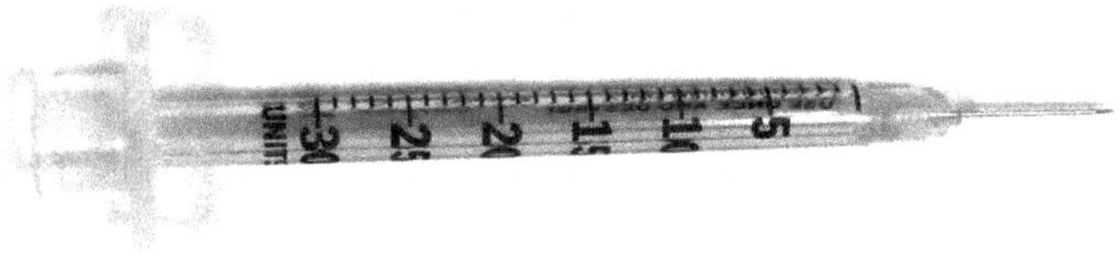

14) A patient is ordered 50 units of Humulin R in a 50 mL bag of normal saline. Use sliding scale to maintain appropriate blood glucose levels. (*Note: It's easier to use a 1 cc syringe rather than an insulin syringe for this, as you need a needle long enough to pierce the injection port on an IV bag.*)
Selected vial: _____

15) When mixing insulins, explain what the phrase 'clear before cloudy' means.

16) A patient is to receive 15 units of Humulin R and 30 units of Humulin N. Explain how the two insulins can be drawn up at the same time.
Selected vials: _____ & _____

17) A patient is to receive 20 units of Humalog and 45 units of Humulin N. Explain how the two insulins can be drawn up and injected at the same time.
Selected vials: _____ & _____

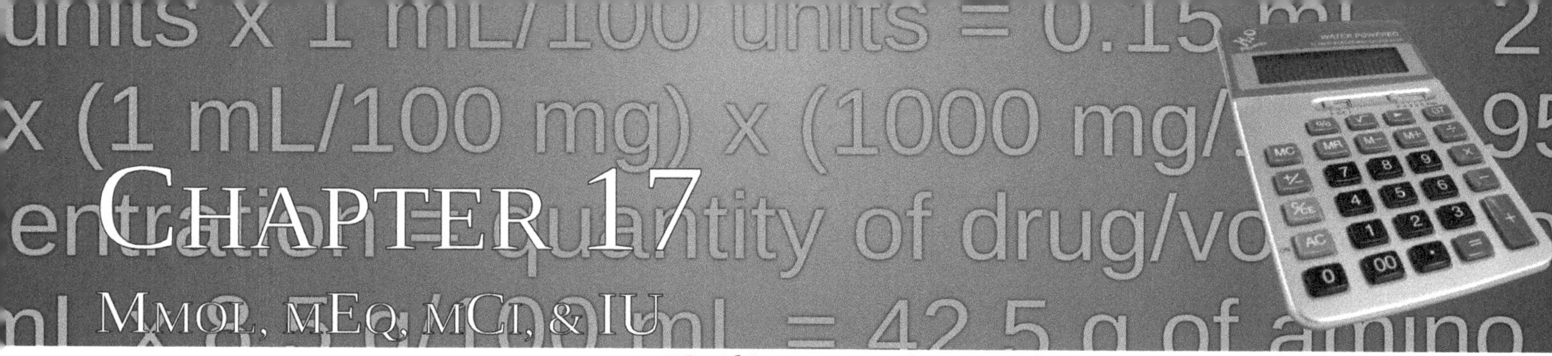

Chapter 17

mmol, mEq, mCi, & IU

"So if I get exposed to enough radiation at our field trip to a nuclear pharmacy, could I turn into the Hulk?"
"Probably not, but if you did, it would be fairly short-lived since they primarily work with technetium-99 and that only has a 6 hour half-life."
--A discussion between a student and his instructor.

Many systems of measures for drugs are easily understandable, such as mass of drug (30 grams of ointment), volume of a solution (250 mL of NS), concentration as mass of drug per quantity of volume (400 mg E.E.S./tsp), percentage strength (50% Magnesium Sulfate = 50 g/100 mL), etc. Other forms require more manipulation to arrive at how they are calculated such as

- millimoles,
- milliEquivalents,
- millicuries, and
- international units.

Due to the concepts in this chapter, we need to take a brief moment and explain how to read a chemical formula. A chemical formula (or molecular formula) is a way of expressing information about the atoms that constitute a particular chemical compound. The chemical formula identifies each constituent element by its chemical symbol and indicates the number of atoms of each element found in each discrete molecule of that compound. If a molecule contains more than one atom of a particular element, this quantity is indicated using a subscript after the chemical symbol. Therefore, if you were to look at the chemical formula for water, H_2O, you would know that there are 2 hydrogen atoms and one oxygen atom in a single water molecule. (If you are not sure of the proper abbreviations for various elements, you may reference the periodic table that appears on the next page.)

Practice Problems

To ensure that you are comfortable with reading chemical formulas, please attempt the following practice problems. Determine how many of each type of atom are present in the following molecules.

1) Salt - NaCl

2) Dextrose - $C_6H_{12}O_6$

3) Dibasic sodium phosphate - Na_2HPO_4

4) Calcium hydroxide - $Ca(OH)_2$

1) 1 sodium, 1 chloride 2) 6 carbon, 12 hydrogen, 6 oxygen
3) 2 sodium, 1 hydrogen, 1 phosphorous, 4 oxygen 4) 1 calcium, 2 oxygen, 2 hydrogen

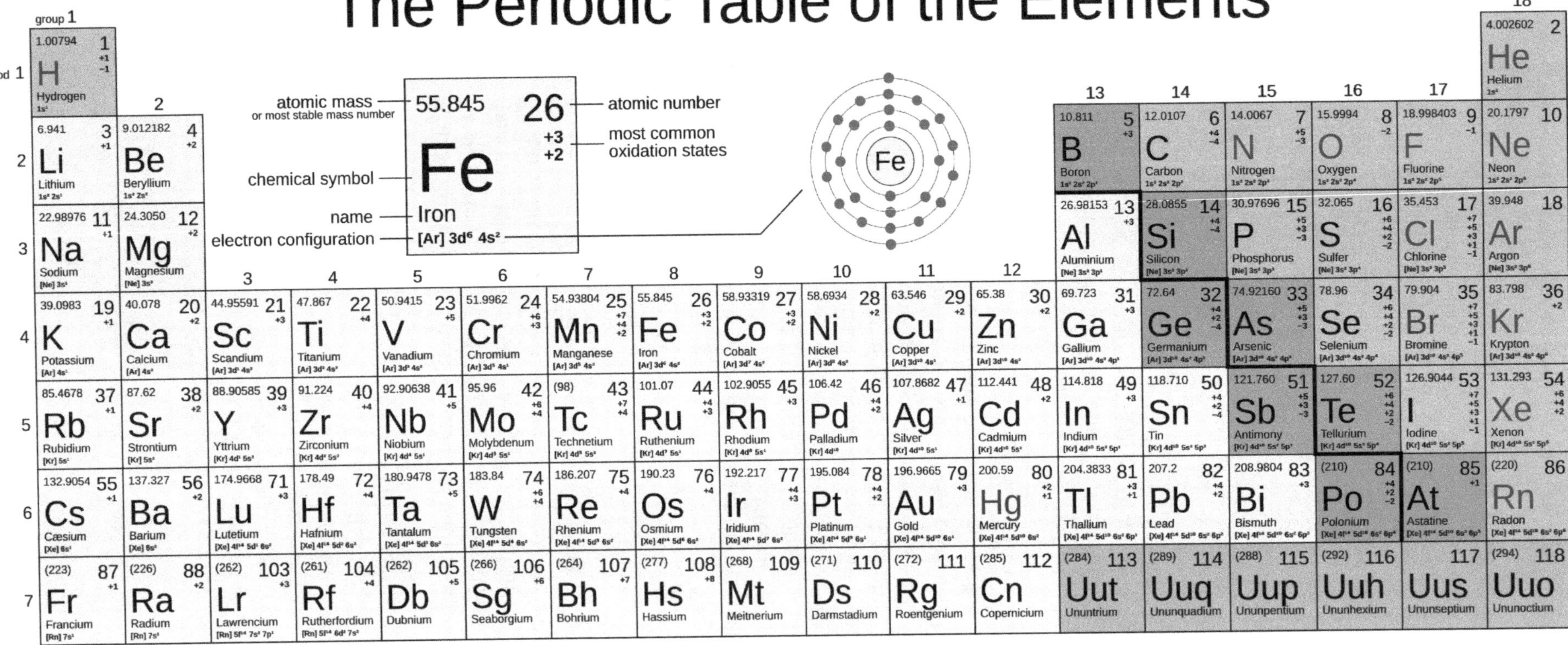

Series														
Lanthanide Series	138.9054 57 La Lanthanum $[Xe]\,5d^1 6s^2$ +3	140.116 58 Ce Cerium $[Xe]\,4f^1 5d^1 6s^2$ +4 +3	140.9076 59 Pr Praseodymium $[Xe]\,4f^3 6s^2$ +3	144.242 60 Nd Neodymium $[Xe]\,4f^4 6s^2$ +3	(145) 61 Pm Promethium $[Xe]\,4f^5 6s^2$ +3	150.36 62 Sm Samarium $[Xe]\,4f^6 6s^2$ +3	151.964 63 Eu Europium $[Xe]\,4f^7 6s^2$ +3 +2	157.25 64 Gd Gadolinium $[Xe]\,4f^7 5d^1 6s^2$ +3	158.9253 65 Tb Terbium $[Xe]\,4f^9 6s^2$ +3	162.500 66 Dy Dysprosium $[Xe]\,4f^{10} 6s^2$ +3	164.9303 67 Ho Holmium $[Xe]\,4f^{11} 6s^2$ +3	167.259 68 Er Erbium $[Xe]\,4f^{12} 6s^2$ +3	168.9342 69 Tm Thulium $[Xe]\,4f^{13} 6s^2$ +3	173.054 70 Yb Ytterbium $[Xe]\,4f^{14} 6s^2$ +3
Actanide Series	(227) 89 Ac Actinium $[Rn]\,6d^1 7s^2$ +3	232.0380 90 Th Thorium $[Rn]\,6d^2 7s^2$ +4	231.0358 91 Pa Protactinium $[Rn]\,5f^2 6d^1 7s^2$ +5	238.0289 92 U Uranium $[Rn]\,5f^3 6d^1 7s^2$ +6	(237) 93 Np Neptunium $[Rn]\,5f^4 6d^1 7s^2$ +5	(244) 94 Pu Plutonium $[Rn]\,5f^6 7s^2$ +4	(243) 95 Am Americium $[Rn]\,5f^7 7s^2$ +3	(247) 96 Cm Curium $[Rn]\,5f^7 6d^1 7s^2$ +3	(247) 97 Bk Berkelium $[Rn]\,5f^9 7s^2$ +3	(251) 98 Cf Californium $[Rn]\,5f^{10} 7s^2$ +3	(252) 99 Es Einsteinium $[Rn]\,5f^{11} 7s^2$ +3	(257) 100 Fm Fermium $[Rn]\,5f^{12} 7s^2$ +3	(258) 101 Md Mendelevium $[Rn]\,5f^{13} 7s^2$ +3	(259) 102 No Nobelium $[Rn]\,5f^{14} 7s^2$ +3

radioactive elements have masses in parenthesis

Millimoles

The mole (symbol: mol) is the System International (SI) base unit that measures an amount of substance. One mole contains Avogadro's number (approximately 6.022×10^{23}) entities. A mole is much like "a dozen" in that both are absolute numbers (having no units) and can describe any type of elementary object, although the mole's use is usually limited to measurement of subatomic, atomic, and molecular structures. The goal is to acquire a sufficient quantity of a material to be able to measure it, and since individual atoms and molecules are so small, having an entire mole of a substance makes it easier to work with. Once you acquire a mole of a particular substance, you may reference its atomic mass to know how much its mass in grams is. Let's look at carbon as an example:

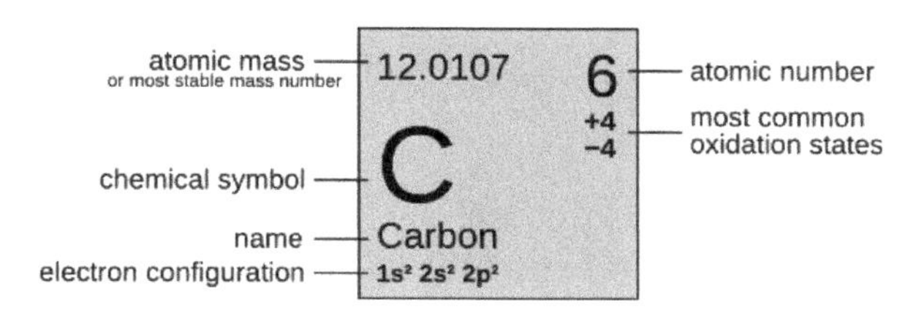

A mole of carbon-12 (which has an atomic mass of 12) would have a mass of 12 g.

In medicine we tend to look at items in slightly smaller quantities, hence the need to measure some substances in millimoles (symbol: mMol). A mMol is one-one thousandth (1/1000) of a mole and mass would be measured in milligrams instead; therefore **1 mol = 1,000 mMol** and a mMol of carbon 12 would weigh 12 mg.

Let's look at an example problem with dibasic sodium phosphate (Na_2HPO_4).

Example

A patient needs an infusion with 15 mMol of dibasic sodium phosphate (Na_2HPO_4) in 150 mL of NS. How many milligrams of dibasic sodium phosphate do you need to make this infusion?

Na = 23×2 = 46
H = 1
P = 31
O = 16×4 = 64

$46 + 1 + 31 + 64 = 142$ atomic mass units in a molecule of Na_2HPO_4.

Therefore, a mMol of Na_2HPO_4 would have a mass of 142 mg. With that in mind, the problem becomes pretty straightforward to solve from there.

$$\frac{15\,\text{mMol}}{1} \times \frac{142\,\text{mg}}{1\,\text{mMol}} = \mathbf{2130\ mg\ of\ Na_2HPO_4}$$

Attempt the practice problem on the next page to verify your comprehension of this concept.

Practice Problem

A patient needs an infusion with 12 mMol of monobasic potassium phosphate (KH_2PO_4) in 150 mL of NS. How many milligrams of monobasic potassium phosphate do you need to make this infusion?

1,632 mg of monobasic potassium phosphate

MilliEquivalents

The equivalent (symbol: Eq or eq) is a measurement unit used in chemistry and the biological sciences such as pharmacy. It is a measure of a substance's ability to combine with other substances, and is a comparison of an ion's charge in comparison to a mole of the product. As an example, if you look at potassium on a periodic table, you will find that it typically has an ionic charge of +1 when it binds to other ions. Therefore, 1 mole of potassium would equal 1 equivalent of potassium. Likewise, if we look at oxygen, it typically has a charge of -2 when it combines with other elements, so 1 mole of oxygen equals 2 equivalents of oxygen.

Much like pharmacy typically looks at items in slightly smaller quantities than moles, you'll find that it also looks at quantities slightly smaller than Eq. Just like 1,000 mMol are equal to 1 mole, 1,000 milliequivalents (symbol: mEq or meq) are equal to 1 Eq. Therefore, 1 mMol of potassium equals 1 mEq of potassium, and 1 mMol of oxygen equals 2 mEq of oxygen. Before expanding on this concept, try a couple of practice problems.

Practice Problems

Determine how many mEq are in a millimole of each of the following elements.

1) 1 mMol H 2) 1 mMol Ca 3) 1 mMol Cl

1) 1 mEq H 2) 2 mEq Ca 3) 1 mEq Cl

Now, let's add some steps to this concept using volume of a solution, mass of a compound, millimoles of that compound, and milliequivalents of a particular ion. Proceed to the next page for examples of using these concepts.

Example

1) If a patient required 4.2 g of sodium bicarbonate ($NaHCO_3$), how many mEq of sodium (Na) is the patient receiving?

   Na = 23
   H = 1
   C = 12
   O = 16×3 = 48

   23+ 1+ 12+ 48=84 is the atomic mass of $NaHCO_3$, therefore 1 mMol = 84 mg.

   Also, a mMol of $NaHCO_3$ contains 1 mMol of Na, and 1 mMol of Na = 1 mEq, so you can solve the problem as follows:

   $$\frac{4.2\text{ g}}{1} \times \frac{1000\text{ mg}}{1\text{ g}} \times \frac{1\text{ mMol}}{84\text{ mg}} \times \frac{1\text{ mEq}}{1\text{ mMol}} = \textbf{50 mEq of sodium}$$

2) How many mEq of sodium and how many mEq of chloride are in one liter of normal saline (0.9% w/v NaCl)?

   Na = 23
   Cl = 35.5

   23+35.5=58.5 is the atomic mass of NaCl, therefore 1 mMol = 58.5 mg.

   1 mMol Na = 1 mEq
   1 mMol Cl = 1 mEq

   $$\frac{1\text{ L}}{1} \times \frac{1000\text{ mL}}{1\text{ L}} \times \frac{0.9\text{ g}}{100\text{ mL}} \times \frac{1000\text{ mg}}{1\text{ g}} \times \frac{1\text{ mMol}}{58.5\text{ mg}} \times \frac{1\text{ mEq}}{1\text{mMol}} = 154\text{ mEq}$$

   154 mEq sodium
   154 mEq chloride

Now that you've seen a couple of example problems, attempt the following practice problems.

Practice Problems

1) How many mEq of calcium are in 111 mg of $Ca(OH)_2$.

2) How many mEq of sodium and how many mEq of chloride are in a 50 mL minibag of half-normal saline (0.45% w/v NaCl)?

1) 3 mEq Ca 2) 3.85 mEq Na & 3.85 mEq Cl

Millicurie

A curie (Ci) is a unit measuring the activity of radioactive isotopes and is named after French-born scientists Pierre and Marie Curie whom did a significant quantity of early research on radioactive isotopes.

Nuclear pharmacy involves the preparation of radioactive materials that will be used to diagnose and treat specific diseases. Nuclear pharmacies typically work with weak radioactive isotopes and will commonly measure the activity level of the substances they are working with a smaller unit known as a millicurie (symbol: mc or mCi).

In nuclear pharmacy, radioactive isotopes with relatively short half-lives are typically used in order to limit a patient's exposure. This also creates an added challenge for those actually preparing the kits, as they need to know not just what the activity level is when they are preparing it, but that it will have the correct activity level when given to the patient. While software exists to help pharmacy personnel perform these calculations, it is a good idea to understand the formula being used. The following is the exponential decay formula:

$$a_t = a_0 e^{-\lambda t}$$

a_t = quantity at end of time(t)
a_0 = quantity at beginning of time(t)
e = the value of base e from a natural logarithm (2.71828...)
λ = the exponential decay constant for a particular substance
t = time

Some common values for lambda (λ) when time (t) is provided in hours would include:

> Technetium-99m (^{99m}Tc) = 0.11514
> Iodine-131 (^{131}I) = 0.00359
> Thalium-201 (^{201}Tl) = 0.00950

Example

A common request for a stress test is to take a drug called sestamibi and reconstitute it with either technetium-99m or thalium-201, as the sestamibi is drawn to the heart and then the isotope helps it to show up for imaging. If a physician requests a dose of sestamibi with 10 millicuries of ^{99m}Tc for 0900, how many mCi would you need if you were drawing this dose up at 0100?

> a_t = 10 mCi
> a_0 = quantity at beginning of time, this is what you are solving for.
> e = 2.71828...
> λ = 0.11514
> t = 8 hours

$$10\,\text{mCi} = a_0 e^{-0.11514 \times 8}$$

$$\frac{10\,\text{mCi}}{e^{-0.11514 \times 8}} = a_0$$

$$a_0 = \mathbf{25.1\,mCi}$$

Now that you know you need 25.1 mCi of ^{99m}Tc, you find that your stock bottle has an activity level of 57.9 mCi/mL. When you are ready to draw it up, how many mL would you add to the sestamibi vial?

$$\frac{25.1\,\text{mCi}}{1} \times \frac{\text{mL}}{57.9\,\text{mCi}} = \mathbf{0.43\,mL}$$

Practice Problem

A physician requests a dose of tetrofosmin with 30 millicuries of ^{99m}Tc for 1100. How many mCi would you need if you were also drawing this dose up at 0100? If your stock bottle has an activity level of 57.9 mCi/mL when you are ready to draw it up, how many mL would you add to the tetrofosmin?

94.9 mCi and 1.64 mL

International Units

In pharmacology, the International unit (IU, alternatively abbreviated UI, from French unit internationale) is a unit of measurement for the amount of a substance, based on measured biological activity (or effect). It is used for vitamins, hormones, some drugs, vaccines, blood products and similar biologically active substances. Despite its name, the IU is not part of the International System of Units used in physics and chemistry.

The precise definition of one IU differs from substance to substance and is established by international agreement. To define an IU of a substance, the Committee on Biological Standardization of the World Health Organization provides a reference preparation of the substance, (arbitrarily) sets the number of IUs contained in that preparation, and specifies a biological procedure to compare other preparations to the reference preparation. The goal here is that different preparations with the same biological effect will contain the same number of IUs.

For some substances, the equivalent mass of one IU is later established, and the IU is then officially abandoned for that substance. However, the unit often remains in use nevertheless, because it is convenient. For example, Vitamin E exists in a number of different forms, all having different biological activities. Rather than specifying the precise type and mass of vitamin E in a preparation, for the purposes of pharmacology, it is sufficient to simply specify the number of IUs of vitamin E.

Below are example mass equivalents of 1 IU for selected substances:

1 IU Insulin: the biological equivalent of about 45.5 mcg pure crystalline insulin (1/22 mg exactly)
1 IU Vitamin A: the biological equivalent of 0.3 mcg retinol, or of 0.6 μg beta-carotene
1 IU Vitamin C: 50 mcg L-ascorbic acid
1 IU Vitamin D: the biological equivalent of 0.025 mcg cholecalciferol/ergocalciferol
1 IU Vitamin E: the biological equivalent of about 0.667 mg d-alpha-tocopherol (2/3 mg exactly), or of 1 mg of dl-alpha-tocopherol acetate

Example

A vial of insulin with 1000 units of insulin contains how many actual mg of insulin?

$$\frac{1000\text{ units}}{1} \times \frac{45.5\text{ mcg}}{1\text{ unit}} \times \frac{1\text{ mg}}{1000\text{ mcg}} = \mathbf{45.5\ mg}$$

Practice Problem

A vitamin E softgel capsule contains 500 IU of d-alpha-tocopherol and 500 IU of dl-alpha-tocopherol acetate. How many milligrams of each are in the capsule?

333 mg of d-alpha-tocopherol and 500 mg of dl-alpha-tocopherol acetate

Worksheet 17-1

Name:

Date:

Solve the following problems.

1) Determine how many of each type of atom are present in the following molecules:

 a) Calcium gluconate - $C_{12}H_{22}CaO_{14}$

 b) Monobasic sodium phosphate - NaH_2PO_4

 c) Potassium acetate - CH_3CO_2K

 d) Ethanol - C_2H_5OH

2) A patient needs an infusion with 7.5 mMol of monobasic potassium phosphate (KH_2PO_4) and 7.5 mMol of dibasic potassium phosphate (K_2HPO_4) in 150 mL of NS.

 a) How many mg of monobasic potassium phosphate (KH_2PO_4) are in the infusion?

 b) How many mg of dibasic potassium phosphate (K_2HPO_4) are in the infusion?

3) If a gram of calcium gluconate ($C_{12}H_{22}CaO_{14}$) was added to an IV bag:

 a) How many mMol of calcium gluconate ($C_{12}H_{22}CaO_{14}$) were added to the IV bag?

 b) How many mEq of calcium were added to the IV bag? (Calcium has a charge of +2)

4) A 10 mL potassium chloride (KCl) vial has a concentration of 2 mEq of potassium/mL. How many mg of KCl are in the vial?

5) A physician requests a dose of tetrofosmin with 30 millicuries of ^{201}Tl for 1100.

   a) How many mCi would you need if you were drawing this dose up at 0100?

   b) If your stock bottle has an activity level of 59.7 mCi/mL when you are ready to draw it up, how many mL would you add to the tetrofosmin vial?

6) If a patient with a thyroid carcinoma required a capsule of ^{131}I with an activity level of 150 mCi 72 hours from now, what should the activity level be when preparing the capsule?

7) A vitamin D-400 tablet contains 400 IU of cholecalciferol. How many mcg of cholecalciferol are in the tablet?

8) A particular vitamin C (ascorbic acid) injection has a concentration of 500 mg/mL. How many IU are in 1 mL of this ascorbic acid preparation?

Chapter 18

Powder Volume Calculations

An oddity of understanding this chapter is that you'll realize that Bill Cosby lied when he said, "There's always room for JELL-O."

--Sean Parsons

As we start this chapter, we should take a moment and consider why some medications come in lyophilized (freeze-dried) powders. The reasons are fairly straight forward, such as improved stability, longer expiration, and easier storage with concern to temperature requirements. There are some concepts we need to thoroughly address in this chapter such as:

- powder volume,
- diluent volume,
- total volume,
- final concentration, and
- how to obtain different concentrations.

As this chapter is obviously primarily focused on powder volume we should explore this concept. The volume or space that the powdered drug occupies after it is reconstituted is called powder volume. For some drugs, the powder volume is so small that it is considered negligible. Other drugs have substantial powder volume which needs to be taken into consideration when reconstituting.

A common comparison is that most people do not leave significant, if any, room for the sugar they add to their morning coffee. In this scenario we can say that powder volume is negligible. But an interesting experiment to do is to take two containers that can hold three liters of volume. In one container, add enough gelatin mix to make three liters of gelatin. In the other container, fill it with three liters of fluid. Then using a funnel, add all the water from the one container to the other one with the gelatin. If you pour out all the fluid, a significant amount of fluid will spill everywhere. In this case, we can say that the gelatin mix has a significant amount of powder volume that we should have accounted for.

From this basic idea, we can ascertain the following formula:

$$\text{Powder Volume} + \text{Diluent Added} = \text{Total Volume}$$

We should consider several different ways that we can find powder volume. Sometimes the package insert or vials will tell you. Other times you will be told how much to reconstitute it with and be given either the concentration or the total volume. From there you can find the powder volume by taking the total volume and subtracting the amount of diluent added.

Without wanting to over explain a simple concept, let's go to the next page and look at an example problem.

Example

If 95 mL of sterile water for injection (SWFI) is added to a 10 g bulk powdered drug pharmacy container, the concentration obtained is 100 mg/mL. What is the powder volume of the drug?

QUESTION
What is the powder volume of the drug?

DATA
95 mL SWFI added (this is the diluent added)
10 g powdered drug (this is the weight of drug)
100 mg/mL (this is our final concentration)

FORMULA/METHOD

$$\begin{array}{r} \text{Powder Volume} \\ \underline{+\ \text{Diluent Added}} \\ \text{Total Volume} \end{array}$$

MATH
We know the diluent added is 95 mL, and since we have the weight of the drug in the vial and the final concentration after reconstitution, we can find the total volume as follows:

$$\frac{\text{mL}}{100\,\text{mg}} \times \frac{1000\,\text{mg}}{1\,\text{g}} \times \frac{10\,\text{g}}{1} = 100\,\text{mL}$$ total volume, often referred to as final volume.

$$\begin{array}{rl} \text{Powder Volume} = & ??? \\ \underline{+\ \text{Diluent Added} =} & \underline{95\ \text{mL}} \\ \text{Total Volume} = & 100\ \text{mL} \end{array}$$

$100\,\text{mL} - 95\,\text{mL} = $ **5 mL Powder Volume**

DOES THE ANSWER MAKE SENSE?
Yes

It is noteworthy focusing on how we obtained the total volume. The final concentration and the quantity of drug in the vial are both reflective of what will be in the vial immediately after reconstitution is complete, and therefore are appropriate to use when determining the total volume in the vial after reconstitution.

Now, you should attempt some problems to ensure that the concepts are making sense thus far.

Practice Problems

1) You want to prepare a 500 mg dose of ceftriaxone for IM injection. If you reconstitute it with 1.8 ml of 2% lidocaine, you attain a concentration of 250 mg/ml. What is the powder volume of the drug?

2) A 20 million unit (MU) vial of penicillin g potassium has 8 mL of powder volume. *(Notice that this problem has already given you the powder volume)*

   a) How many mL will you reconstitute it with to attain a concentration of 500,000 unit/mL? *(Hint: You can use the concentration to determine the total volume, which will be necessary to determine the volume of diluent needed.)*

   b) How many mL of the reconstituted solution will you draw up for a 12 MU dose? *(Hint: This is similar to what you did in Chapter 15, so all you will need is the dose and the concentration to solve this.)*

1) 0.2 mL powder volume   2-a) 32 mL of diluent   2-b) 24 mL

One more concept to cover is the idea of changing the quantity of diluent used to reconstitute a preparation in order to obtain a different concentration. Let's look at an example problem to explain this scenario.

Example

Ordinarily IVIG is reconstituted to a concentration of 50 mg/mL. To obtain this concentration, a 5 g vial is reconstituted with 93 mL of diluent. What is the powder volume for a 5 g vial of IVIG? How much would you reconstitute it with if the patient was fluid restricted and the physician requested a concentration of 100 mg/mL?

You can find the powder volume by doing the same steps as we have already done on previous problems. The first step is to determine what the total volume would be, and conveniently, the problem already gave us a volume for diluent added if reconstituted to a concentration of 50 mg/mL.

$$\frac{\text{mL}}{50\,\text{mg}} \times \frac{1000\,\text{mg}}{1\,\text{g}} \times \frac{5\,\text{g}}{1} = 100\,\text{mL total volume}$$

Powder Volume = ???
\+ Diluent Added = 93 mL
Total Volume = 100 mL

100 mL – 93 mL = **7 mL powder volume**

Now that we've answered the first question about powder volume, we can determine the amount of diluent to add to obtain the desired concentration of 100 mg/mL. We will once again need to determine the total volume, but this time for our new concentration.

$$\frac{\text{mL}}{100\ \text{mg}} \times \frac{1000\ \text{mg}}{1\ \text{g}} \times \frac{5\ \text{g}}{1} = 50\ \text{mL total volume}$$

Powder Volume = 7 mL
+ Diluent Added = ???
Total Volume = 50 mL

50 mL – 7 mL = **43 mL of diluent** to obtain the requested concentration.

Now, using the example problem as a template, solve the practice problem below.

Practice Problem

A 1 g vial of vancomycin ordinarily is reconstituted with 19.5 mL of sterile water for injection (SWFI) to obtain a concentration of 50 mg/mL. What is the powder volume and how much should you reconstitute it with if you needed a concentration of 100 mg/mL?

The vial has 0.5 mL of powder volume and 9.5 mL of SWFI should be used to obtain a concentration of 100 mg/mL.

Worksheet 18-1

Name:

Date:

Solve the following powder volume problems.

1) If 192 mL of sterile water for injection (SWFI) is used to reconstitute a 20 g vial of cefazolin Na, the concentration obtained is 100 mg/mL. What is the powder volume of this drug?

2) Using another 20 g vial of cefazolin Na, how many mL of SWFI should be added to obtain a concentration of 200 mg/mL instead? *(Hint: The powder volume from the previous question is required to solve this.)*

3) A vial contains a combination drug of ampicillin/sulbactam. There is 1 g of ampicillin and 0.5 g of sulbactam in the vial. When the vial is reconstituted with 3.2 mL of sterile water, you end up with a total volume of 4.0 mL.

   a) How many mL of powder volume are there?

   b) What is the concentration of each drug?

4) The label on a bottle of oral amoxicillin suspension states you are to add 39 mL of purified or distilled water to the bottle to obtain a suspension with a concentration of 150 mg/tsp. The total amount of active ingredient in the bottle is 2 g. What is the powder volume?

5) The directions for a vial containing 1 g of lyophilized ceftriaxone states that the addition of 3.6 mL of SWFI will yield a solution with a concentration of 250 mg/mL. What is the powder volume of the drug?

6) If you dissolve 5 MU of penicillin with 8 mL of SWFI, and you know that this vial contains 2 mL of powder volume, what is the concentration of the drug in the solution?

7) A 4 g vial of a powdered drug is reconstituted with 4.9 mL of SWFI to obtain a concentration of 800 mg/mL.

   a) What is the powder volume of the drug?

   b) If 2.6 mL of the reconstituted solution is added to a 500 mL bag of D5W, how much medication is in the IV bag?

8) If a 20 MU vial of penicillin g potassium has a powder volume of 8 mL, what would be concentration obtained if each of the possible volumes of SWFI listed below were used to reconstitute the vial?

   a) 32 mL

   b) 42 mL

   c) 72 mL

   d) 92 mL

9) You add 4.3 mL of diluent to a 1 g vial and have a final volume of 5 mL.

   a) What is the powder volume?

b) What is the final concentration in mg/mL?

c) How many mL would you need to add to a 50 mL bag of NS if a dose of 250 mg were required?

10) A pharmacists asks you to prepare cefazolin eye drops. You will need to add 9.8 mL of NSS to a 500 mg vial of cefazolin which has 0.2 mL of powder volume according to the package insert. You will draw up 1 mL of the reconstituted solution and filter through a 0.5 micron filter into a sterile eye dropper. Then you will add 9 mL of NSS to the eye dropper (this will also need filtered). What is the final concentration of cefazolin in the eye dropper? *(Hint: You will first need to figure out the concentration of the solution in the vial before you can calculate the concentration in the eye dropper.)*

Worksheet 18-2

Name:

Date:

Solve the following powder volume problems.

1) If 95 mL of sterile water for injection (SWFI) is used to reconstitute a 10 g vial of vancomycin, the concentration obtained is 100 mg/mL. What is the powder volume of this drug?

2) Using another 10 g vial of vancomycin, how many mL of SWFI should be added to obtain a concentration of 200 mg/mL instead?

3) A bulk pharmacy vial contains 40.5 g of a combination drug of Zosyn (piperacillin/tazobactam). There are 36 g of piperacillin and 4.5 g of tazobactam in the vial. When the vial is reconstituted with 152 mL of sterile water you end up with a concentration of 225 mg/mL of piperacillin and tazobactam combined.

   a) What is the powder volume in the vial?

   b) What is the concentration (in mg/mL) of each active ingredient separately?

   c) If a physician ordered 4.5 g of Zosyn (piperacillin/tazobactam) in 100 mL of 0.45% sodium chloride, how many mL would you need to transfer from the bulk vial after reconstitution to the 100 mL minibag?

4) The package insert for a 1 g streptomycin vial instructs you to add 4.2 mL SWFI to obtain a concentration of 200 mg/mL. What is the powder volume for a 1 g vial of sreptomycin powder?

5) If you add 10 mL of SWFI to a 1 g vial of powdered drug that has a powder volume of 0.5 mL, what is the resulting concentration going to be in mg/mL?

6) Review the amoxicillin/clavulanate label below carefully:

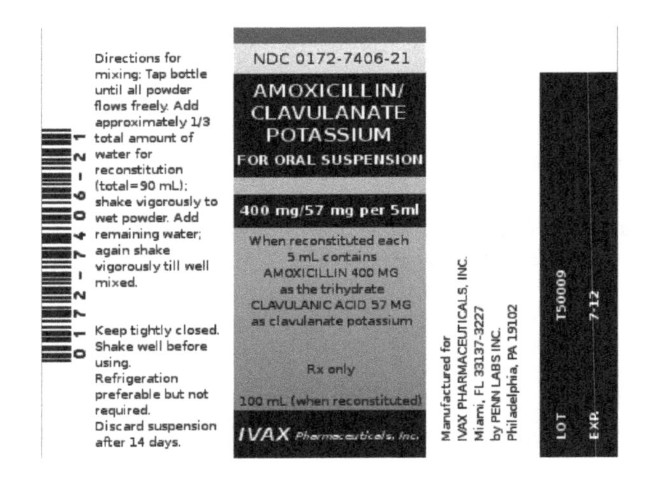

a) What is the powder volume?

b) How many grams of amoxicillin are in the bottle?

c) How many grams of clavulanate are in the bottle?

7) The directions for preparing a vial containing 1 g of lyophilized cefepime HCl for IM injection suggests the addition of 2.4 mL of 1% lidocaine HCl which will yield a solution with a concentration of 280 mg/mL.

a) What is the powder volume of the drug?

b) If a patient required only 500 mg to be given IM, how many mL would be needed to administer the desired dose?

8) An 8 g vial of a powdered drug is reconstituted with 19.6 mL of SWFI to obtain a concentration of 400 mg/mL.

a) What is the powder volume of the drug?

b) If a physician requests that 12 g of the reconstituted solution is to be added to a 500 mL bag of D5W, how many mL will you need to add to the IV bag?

9) If a 10 MU vial of medication has a powder volume of 4 mL, what would be the concentration obtained if each of the possible volumes of SWFI listed below were used to reconstitute the vial?

a) 16 mL

b) 21 mL

c) 36 mL

d) 46 mL

10) A 6 g bulk vial of ceftazidime has a final volume of 30 mL when reconstituted with 26 mL.

a) What is the powder volume?

b) What is the concentration in mg/mL?

c) How many mL of the reconstituted solution would be required to fill an order for 2 g of ceftazidime?

11) A 3.2 g vial of Timentin (ticarcillin/clavulanic acid) has 2.2 mL of powder volume.

a) If you reconstitute it with 17.8 mL of SWFI, what is the cocnentration in mg/mL?

b) How many mL of the reconstituted solution will you need for a dose of 1.6 g of Timentin?

12) Ordinarily, IVIG is reconstituted to a 5% (w/v) concentration and is available in vials of various quantities. For standard reconstitution, a 2.5 g vial is reconstituted with 46.5 mL of diluent, a 5 g vial is reconstituted with 93 mL of diluent, and a 10 g vial is reconstituted with 186 mL of diluent.

a) What does 5% (w/v) mean?

b) How much powder volume is in each size of vial?

c) If a patient needed 27.5 g of IVIG, how many of each size vial would be needed to fill this order? (I'm looking for the combination with the fewest possible vials that won't waste any drug.)

d) If the patient is fluid restricted and needs a 10% concentration of the drug instead, how many mL of diluent would you reconstitute each size vial with?

e) What would be the final volume of a 27.5 g solution at a 10% concentration?

Worksheet 18-3

Name:

Date:

Solve the following powder volume problems.

1) A 10 g bulk vial of vancomycin is supposed to be reconstituted with 95 mL of SWFI to achieve a concentration of 100 mg/mL. A technician receives an order for 1.5 g of vancomycin in 250 mL of 0.9% sodium chloride. The technician accidentally reconstitutes the vial with 100 mL of SWFI, instead of 95 mL, but does not want to waste the drug vial; therefore, we will need to do some calculations to figure out how to still use this solution.

   a) First, you will need to know the powder volume in the vial. This is easiest to calculate by using the suggested reconstitution information.

   b) Second, you will need to determine the total volume actually in the vial once it was incorrectly reconstituted.

   c) Third, you will want to know the concentration of the drug in the vial after it was reconstituted incorrectly.

   d) Fourth, you can now determine the quantity of solution that should be added to the bag from the incorrectly reconstituted vial.

   e) Often, you will want to know whether an IV can be infused peripherally or if it requires a central line. Vancomycin when given IV requires a central line if the IV bag has a concentration >5 mg/mL. Can this bag be infused peripherally or does it require a central line? Why or why not?

2) A vial contains 200 mg of gemcitabine, and after being reconstituted with 5 mL of NS, it has a concentration of 38 mg/mL. What is the powder volume of the gemcitabine in a 200 mg vial?

3) Another vial contains 1000 mg of gemcitabine, and after being reconstituted with 25 mL of NS it has a concentration of 38 mg/mL. What is the powder volume of the gemcitabine in a 1000 mg vial?

4) The pharmacy must add water to an oral suspension before it can be dispensed to the patient. The dose is to be 750 mg/Tbs, and the dry powder is 5g with a powder volume of 8.6 mL. How much water must you add?

5) The directions for a bulk pharmacy vial containing 10 g of oxacillin states that the addition of 93 mL of SWFI will provide a final concentration of 100 mg/mL.

   a) What is the powder volume?

   b) How many mL would be needed to fulfill a physician order for 1 g of oxacillin in 100 mL of half normal saline?

6) Using another 10 g vial of oxacillin, you receive a request to prepare a bottle with a final concentration of 200mg/mL.

   a) How many mL of SWFI will you need to reconstitute it with?

   b) How many mL would be needed to fulfill a physician order for 1 g of oxacillin in 100 mL of half normal saline using this new concentration?

7) A 3 g vial of powdered drug is reconstituted with 4.8 mL of SWFI to obtain a concentration of 600 mg/mL.

   a) What is the powder volume of the drug?

   b) If 3 mL of the reconstituted solution is added to a 150 mL bag of D5W, how much medication is in the IV bag?

8) You dissolve 5 MU of penicillin with 18 mL of SWFI, and according to the package the vial contains 2 mL of powder volume.

    a) What is the concentration of penicillin (in units/mL) in the vial after reconstitution?

    b) How many mL would be required to fill an order for 3.5 MU?

9) You add 8.6 mL of diluent to a 2 g vial and have a final volume of 10 mL.

    a) What is the powder volume?

    b) What is the final concentration in mg/mL?

    c) How many many mL would you add to a 50 mL bag of NS if a dose of 1.5 g were required?

10) A bulk pharmacy vial contains 40.5 g of a combination drug of Zosyn (piperacillin/tazobactam). There are 36 g of piperacillin and 4.5 g of tazobactam in the vial. When the vial is reconstituted with 62 mL of sterile water, you end up with a concentration of 450 mg/mL of piperacillin and tazobactam combined.

    a) What is the powder volume in the vial?

    b) What is the concentration (in mg/mL) of each active ingredient separately?

    c) If a physician ordered 3.375 g of Zosyn (piperacillin/tazobactam) in 100 mL of 0.45% sodium chloride, how many mL would you need to transfer from the bulk vial after reconstitution to the 100 mL minibag?

Chapter 19

Percentage Strength

Famous last words, "This chapter should be easy."
--Sean Parsons

Back in chapter 9, you learned that there are three kinds of percentage strength that you will frequently use when doing dosage calculations:

weight/weight (w/w) – examples include ointments, creams, etc.
volume/volume (v/v) – a common example is an alcohol preparation or an oil in water preparation.
weight/volume (w/v) – this is the most common group and includes items such as solutions, suspensions, etc.

Another kind of percentage strength that you should acquire knowledge about is called:

milligram percent (mg%) - a measurement frequently used when looking at lab values for patients.

When these concepts are well understood you can:

- calculate the percentage strength of a mixture when you know know both the quantity of the active ingredient and the total mixture,
- determine how much of a mixture can be prepared from a given weight of ingredient when trying to prepare a compound with a specific percentage strength,
- determine the amount of an ingredient that is present in a mixture if you already know its percentage strength,
- determine whether a patient's lab values are within the appropriate range.

We should start off reviewing the first three kinds of percentage strength. You probably remember that:

w/w% is *typically* measured as g/100 g so a 1% hydrocortisone cream would mean that there is 1 gram of hydrocortisone in every 100 g of cream
v/v% is *typically* measured as mL/100 mL so a 70% isopropyl alcohol solution would mean that there are 70 milliliters of isopropyl alcohol in every 100 milliliters of solution.
w/v% is *always* measured as g/100 mL so a 0.9% sodium chloride solution would mean that there 0.9 grams of sodium chloride for every 100 milliliters of solution.

Weight/Weight (w/w)

A weight/weight percentage strength expresses the number of parts in 100 parts of a preparation. As mentioned above, it is usually expressed as number of grams per 100 grams, but it could be expressed in any unit of weight (grams, grains, ounces, pounds, etc.) as long as the units on the top and bottom match. Let's look at an example to help this make more sense.

Example

What would be the weight, expressed in grams, of zinc oxide (zinc oxide is the active ingredient) in 120 grams of a 20% zinc oxide ointment (the ointment is the total mixture)?

QUESTION

How many grams of zinc oxide are in the ointment?

DATA

$120\ g\ of\ ointment \qquad 20\% = \frac{20\ g\ zinc\ oxide}{100\ g\ ointment}$

MATHEMATICAL METHOD / FORMULA

Ratio Proportion (This can be done other ways, but this is the easiest method to explain thoroughly.)

DO THE MATH

$$\frac{N}{120\ g\ ointment} = \frac{20\ g\ zinc\ oxide}{100\ g\ ointment}$$

$$N = \mathbf{24\ g\ zinc\ oxide}$$

DOES THE ANSWER MAKE SENSE?

Yes

By using the ratio-proportion method for solving the problem, it largely lines itself up because the active ingredient goes on top and the total mixture belongs on the bottom. Also, remember that a percent is always out of 100.

Let's look at a couple of practice problems based on w/w percentage strength.

Practice Problems

1) What would be the percentage strength of a zinc oxide ointment if you prepared 90 grams of an ointment that contained 4.5 grams of zinc oxide? *(Hint: In this problem you have the weight of the active ingredient and the weight of the mixture. You need to find the percentage strength which would be out of 100 grams of mixture.)*

2) If you had 2.5 grams of hydrocortisone powder on hand, how many grams of 0.5% hydrocortisone cream could you prepare?

1) 5% zinc oxide ointment 2) 500 grams of hydrocortisone cream

Volume/Volume (v/v)

Volume/volume percentage strength problems are worked out in a similar manner to w/w percentage strength problems, except now the ingredients are liquid form. Typically, it is expressed as number of milliliters per 100 milliliters, but it could be expressed in any unit of volume (liters, pints, fluid ounces, etc.) as long as the units on the top and bottom match. Look at the example problem below, and then complete the practice problem.

Example

How mL of isopropyl alcohol are in a 1 pint bottle of 70% isopropyl alcohol? Also, if the isopropyl alcohol solution is in sterile water for irrigation, how much water is in the solution?

QUESTION

How many mL of isopropyl alcohol and how many mL of water are in the bottle?

DATA

$1\,pint = 480\ mL\ of\ mixture$ $\quad 70\%\,isopropyl\ alcohol = \frac{70\ mL\ isopropyl\ alcohol}{100\ mL\ mixture}$

MATHEMATICAL METHOD / FORMULA

Ratio Proportion (This can be done other ways, but this is the easiest method to explain thoroughly.)

DO THE MATH

$$\frac{N}{480\ mL\ mixture} = \frac{70\ mL\ isopropyl\ alcohol}{100\ mL\ mixture}$$

$$N = \mathbf{336\ mL\ isopropyl\ alcohol}$$

$480\ mL - 336\ mL = \mathbf{144\ mL\ SWFI}$

DOES THE ANSWER MAKE SENSE?

Yes

Practice Problems

1) How many liters of 2% lysol solution could be made with 80 mL of lysol, and how many mL of SWFI would be required to make this preparation?

2) A pharmacy elixir recipe calls for a fluid ounce of 90% ethanol. How many milliliters of pure ethanol are going to be in this preparation?

1) 4 liters of 2% lysol solution; 3,920 mL of SWFI 2) 27 mL of ethanol

Weight/Volume (w/v)

Weight/volume (w/v) percentage strengths are the most common percentages worked with in institutional pharmacy. The units in this type of problem are **always** grams of drug in 100 milliliters of solution (or suspension). Let's look at a practice problem.

Example

How many grams of sodium chloride are in a 500 mL bag of half-normal saline (0.45% sodium chloride)?

QUESTION

How many g of sodium chloride are in the bag?

DATA

$500\ mL\ of\ mixture$ $\quad 0.45\%\, sodium\ chloride = \frac{0.45\ g\ sodium\ chloride}{100\ mL\ mixture}$

MATHEMATICAL METHOD / FORMULA

Ratio Proportion

DO THE MATH

$$\frac{N}{500\ mL\ mixture} = \frac{0.45\ g\ sodium\ chloride}{100\ mL\ mixture}$$

$$\mathbf{N = 2.25\ g\ sodium\ chloride}$$

DOES THE ANSWER MAKE SENSE?

Yes

As w/v percentage strength problems are the ones we will be working with most frequently, it is worth pointing out that our ratio proportions for w/v percentage strength problems always have grams on top and milliliters on the bottom. Let's look at some practice problems.

Practice Problems

1) A physician orders a TPN with 20 grams of dextrose. The pharmacy has a stock solution of 70% dextrose in water (D70W). How many milliliters of D70W are required to make the requested TPN?

2) An order for 1 liter of a 0.05% (w/v) solution is received in the pharmacy. In stock is 525 mg of the powdered drug. Is there enough drug to be able to fill this order?

1) 28.57 mL of D70W 2) Yes, there is enough drug on hand to fill the order.

Worksheet 19-1

Name:

Date:

Solve the following problems.

1) You need to prepare 120 g of a 2% zinc oxide ointment. How many grams of zinc oxide will be needed to prepare this ointment?

2) How many liters of a 0.02% solution of cinnamon oil in alcohol could be made from 5 mL of cinnamon oil?

3) A patient is to be given 1 g of magnesium sulfate IM. The label on a 10 mL vial reads "50% solution of Mag Sulf." How many mL will you need?

4) You have 20 mL of 2% lidocaine on the shelf. The order calls for 100 mg. How many mL do you need to use?

5) You have in stock a 250 mL stock bottle of 23.4% concentrated sodium chloride injection.

   a) How many grams of sodium chloride are in the bottle?

   b) How many mg are in 1 mL?

6) Calcium gluconate is available as a 10% solution. You receive an order for 1.25 g of calcium gluconate. How many mL of the solution should be used?

7) You have 50 mg of chloramphenicol powder in the pharmacy. How many mL of a 1% ophthalmic solution will you be able to prepare?

8) Hydrocortisone topical cream is available as a 0.2% concentration in a 1 pound jar. How many milligrams of hydrocortisone are in the product?

9) How many mL of ethanol are in 35 ml of a 50% ethanol solution?

10) How many grams of amino acid are in a 500 mL bottle of 5.2% amino acid?

11) How many mg of potassium chloride are in 2 tablespoonfuls of 10% KCl solution?

12) How many 150 mg capsules of clindamycin are needed to prepare 2 fluid ounces of 1% clindamycin solution?

13) Devonex is a topical ointment available in a 0.005% concentration. How many milligrams of active ingredient would be in 45 grams of ointment?

14) An order for 500 mL of a 1.5% (w/v) solution is received in the pharmacy. In stock is 7 grams of the powdered drug. Is there enough drug to be able to fill this order?

15) A powdered drug comes in a vial containing 1.8 g. If the total volume after the diluent is injected is 6 mL, what is the percentage strength of this solution?

16) A patient order calls for a liter bottle of acetic acid irrigation solution 0.25% (v/v).

a) How many milliliters of pure acetic acid is in a liter of 0.25% acetic acid?

b) How many milliliters of pure acetic acid is in a 30 mL dose of 0.25% acetic acid?

17) If you dissolve 12.5 g of a drug in 500 mL of sterile water, what will be the percentage strength of the final solution (assuming that powder volume is negligible)?

18) Sulfatrim suspension is a combination product that contains 200 mg of sulfamethoxazole and 40 mg of trimethoprim in every teaspoon. What are the percentage strengths of each drug?

19) Sulfatrim suspension uses ethanol as an excipient. The suspension is approximately 0.5% ethanol. How many milliliters of alcohol would be present in a teaspoonful of this medication?

20) You receive an order to prepare 2 fluid ounces of a 0.025% solution. How many mg of powder will you need to prepare this solution?

Milligram Percent (mg%)

If you are working with lab values, you will sometimes see another type of percentage strength referred to as a milligram percent (mg%). Milligram percents are not routinely used for dosing, but they are frequently used in reporting clinical laboratory test results. They are intended for reporting various chemicals that are present in the body in small quantities. Several examples are cholesterol, glucose, creatinine, and even some drugs.

You are already familiar with weight-in-volume percents, which were measured as g/100 mL. The only difference with milligram percent problems is that the numerator is in milligrams instead of grams. Also, instead of seeing the 100 mL in the denominator, you will see the letters dL, which stands for a "deciliter" and is the same as 100 mL. Sometimes, instead of being written as mg% you will see it written as mg/dL.

Let's look at an example problem below involving mg%.

Example

A patient in the ER was in a car accident. You receive the urine analysis and find that he had a blood alcohol content of 90 mg%. While individual states may set more stringent standards, the United States considers an individual legally intoxicated when their blood alcohol content is above 0.08%. Is the patient legally intoxicated based on federal standards?

$$\frac{90\,\text{mg}}{1} \times \frac{1\,\text{g}}{1000\,\text{mg}} = 0.09\,\text{g}$$

$$90\,\text{mg}\,\% = \frac{90\,\text{mg}}{100\,\text{mL}} = \frac{0.09\,\text{g}}{100\,\text{mL}} = \mathbf{0.09\,\%}$$

The patient **is** legally intoxicated.

Now, attempt a couple of practice problems prior to starting the next worksheet.

Practice Problems

Based on a patient with a reported serum glucose of 165 mg/dL, answer the following questions.

1) What would her serum glucose be if it were recorded as a mg%?

2) What would her serum glucose be if it were recorded as a traditional weight/volume percentage strength?

1) 165 mg% 2) 0.165%

Worksheet 19-2

Name:

Date:

Solve the following problems.

1) A patient has a blood glucose of 130 mg/dL. What would the patients blood glucose be if you recorded it as a mg%?

2) A patient has a serum creatinine of 0.15% (w/v). Express this as a mg%.

3) An order calls for a pint of 2.5% suspension. How many 200 mg tablets will be required to compound this suspension?

4) A physician orders 500 mcg of a medication. The drug is available in a concentration of 0.025%. How many milliliters of this 0.025% solution will be required to fill this order?

5) A 0.1% Protopic (tacrolimus) cream is temporarily unavailable from the manufacturer. How many 5 mg capsules of tacrolimus will you need to prepare 30 grams of this cream?

6) Pharmacies often use 70% isopropyl alcohol for wiping down items that need to maintain sterility. How many milliliters of isopropyl alcohol are in a pint of 70% isopropyl alcohol?

7) If a patient has a serum cholesterol level of 95 mg%, how many micrograms of cholesterol would be in 1 milliliter of serum?

8) If you dissolved 180 g of a drug in enough diluent to make a liter, what would be the resulting percentage strength?

9) The pharmacy has in stock a 50 mL vial of 25% mannitol injection. If an order request 4 g of mannitol, how many mL of solution will you need to withdraw from the vial?

10) How many mg of zinc sulfate are found in 8 oz. of a 0.02% solution?

11) How much of each ingredient must be weighed out to prepare the following ointment?

> Rx testosterone 2% and
> menthol 4.33% in hydrophilic petrolatum
> Disp: 90 g
> Sig: apply lightly q.i.d.

12) A patient is admitted to the hospital for a health condition unrelated to alcohol consumption. The physician decides to allow the patient to enjoy a 12 oz beer every evening. If the beer has an alcohol concentration of 5.9%, how much alcohol is this patient receiving every evening?

13) An order for 250 mL of a 2% (w/v) solution is received in the pharmacy. In stock is 5.5 grams of the powdered drug. Is there enough drug to be able to fill this order?

14) A child got into their parents' acetaminophen, and they aren't sure how much the child ingested. You checked a chart and saw that the blood serum for acetaminophen is not considered toxic till it reaches 15 mg%. If the child's acetaminophen level is high enough to be considered toxic, the physician will want to start an acetylcysteine infusion to try and prevent permanent damage to the child's liver, but the drug has the potential of damaging his kidneys in the process. The lab results show his serum concentration of acetaminophen to be 100 mcg/mL. Has the child reached a toxic level?

15) Someone signs out 100 mg of cocaine for you to prepare a 0.4% ophthalmic solution. How many mL of this ophthalmic solution will you be able to prepare?

16) You have 250 mg of chloramphenicol powder in the pharmacy. How many mL of a 1% ophthalmic solution will you be able to prepare?

17) A 50 mL vial of sulfamethoxazole and trimethoprim injection contains a concentration of 80 mg of sulfamethoxazole and 16 mg of trimethoprim per mL. What are the percentage strengths of each of the drugs?

18) A vial of lyophilized powder with 2 g of ampicillin and 1 g of sulbactam is reconstituted to have a volume of 8 mL. What are the percentage strengths of each of the drugs?

19) An insulin resistant patient is admitted and given the following insulin sliding scale to follow:

150-199 mg%: 2 unit bolus regular insulin
200-249 mg%: 4 units bolus regular insulin
250-299 mg%: 7 units bolus regular insulin
300-349 mg%: 10 units bolus regular insulin
Over 350 mg%: 12 units bolus regular insulin

If the patient has a blood glucose reading of 260 mg/dL, how many units of insulin should that patient receive?

20) What would be the percentage strength of a zinc oxide ointment if you prepared 60 grams of an ointment that contained 4.5 grams of zinc oxide?

Chapter 20

Ratio Strength

When it comes to the ratio of technicians to pharmacists, some might find it interesting that there are 16 states that have no limits on this ratio, 6 states allow a 4 to 1 ratio, and all the others allow either 3 to 1 or 2 to 1 technicians to pharmacist ratios.
--National Association of Boards of Pharmacy

Ratios are sometimes used to express the concentrations of particular drugs and are primarily used with very dilute solutions (although thi may be used with creams and other mixtures as well). Similar to the three major types of percentage strength, we also have three types of ratio strength in pharmacy:

- **weight : weight** (w:w) ~ This is typically expressed as grams of active ingredient to grams of mixture. This is sometimes referred to as *solids in solids.*
- **volume : volume** (v:v) ~ This is typically expressed as milliliters of active ingredient to milliliters of solution. This is sometimes referred to as *liquids in liquids.*
- **weight : volume** (w:v) ~ This is always expressed as grams of active ingredient to milliliters of solution. This is sometimes referred to as *solids in liquids.*

You have already learned about various other ways to measure the concentration of various preparations. In the previous chapter, we looked at percentage strength which expressed concentration per 100. Ratio strength is merely another way of expressing concentration. A 5% concentration could be expressed as a ratio by saying 5:100; although, you will typically see ratios expressed as one to something. Therefore, 5:100 would be reduced down to 1:20. Let's look at some examples of concentrations in various preparations and look at the corresponding ratio strength.

Examples

1) Determine the ratio strength of 20% mannitol.

$$\frac{20\,\text{g}}{100\,\text{mL}} = \frac{1\,\text{g}}{\text{N}}$$

N = 5 mL

The ratio strength is **1:5**.

2) A cefazolin vial has a concentration of 500 mg/10 mL. What is the ratio strength of the cefazolin solution?

First we need to convert our milligrams into grams:

$$\frac{500\,\text{mg}}{1} \times \frac{1\,\text{g}}{1000\,\text{mg}} = 0.5\,\text{g}$$

Now we can set-up our ratio:

$$\frac{0.5\,\text{g}}{10\,\text{mL}} = \frac{1\,\text{g}}{N}$$

N=20 mL

The ratio strength is **1:20**.

3) Epinephrine is available as a 1:1000 w:v solution. If the patient dose is 0.5 mg, how many mL are needed?

First we need to convert our milligrams into grams:

$$\frac{0.5\,\text{mg}}{1} \times \frac{1\,\text{g}}{1000\,\text{mg}} = 0.0005\,\text{g}$$

Now we can set-up our ratio:

$$\frac{0.0005\,\text{g}}{N} = \frac{1\,\text{g}}{1000\,\text{mL}}$$

N=0.5 mL

The volume that needs withdrawn to fill this order is **0.5 mL**.

Now that you have seen some examples, you should attempt the following practice problems.

Practice Problems

1) You need to prepare 1.5 liters of a 1:1000 neomycin bladder irrigation. How many grams of neomycin are required?

2) If you added 500 mcg of octreotide to a minibag with a final volume of 50 mL what would be both the resulting percentage strength and ratio strength? (*Hint: You are looking for two different answers.*)

3) If 150 mg of strychnine sulfate is intimately mixed with 7.35 g of lactose, what is the ratio strength of the strychnine sulfate compared to the total mixture?

1) 1.5 g of neomycin 2) 0.001%; 1:100,000 3) 1:50

Worksheet 20-1

Name:

Date:

Solve the following problems.

1) A physician orders a 1 liter polymixin-b sulfate bladder irrigation with a concentration of 1:1000. How many grams of polymyxin-b sulfate will be needed for the irrigation?

2) A technician is to prepare 200 mL of a 1:10,000 w/v solution. How many mg of the required drug will be needed to prepare the solution?

3) A drug solution is labeled 1:40 w/v. What is the percentage strength of the solution?

4) A medication is labeled with a 0.05% concentration.

   a) What is the ratio strength of this solution?

   b) What is the concentration of this solution in mcg/mL?

5) An order for 500 mL of a 1:1000 w:v solution is received in the pharmacy. There are 0.55 g of this medication available in the pharmacy. Will the pharmacy be able to prepare this order?

6) An order for 1 liter of a 1:2000 w:v solution is received in the pharmacy. In stock is 480 mg of the powdered drug. Will the technician be able to fill this order?

7) An order is received in the pharmacy for 5 mL of a 1:100,000 w:v solution. How many mcg are needed to fill this order?

8) An order is received in the pharmacy for 22.5 mL of a 1:50,000 solution. How many mcg of active ingredient are required to fulfill this request?

9) A physician uses the hospital's CPOE software to prescribe Racepinephrine inhale 0.5 mg per neb q3h prn asthma attack. If Racepinephrine has a concentration of 1:1000, how many milliliters will be required for a dose?

10) A patient is ordered 16 mg of norepinephrine in 250 mL of D5W. There are 4 mL ampules of norepinephrine with a concentration of 1:1000 available.

a) How many mL of norepinephrine will you need to add to the D5W bag?

b) How many ampules of norepinephrine will you need to prepare this solution?

11) The OR needs 0.5 mg of neostigmine to reverse the rocuronium that was used during surgery. The neostigmine you have on hand has a concentration of 1:2000 and comes in 10 mL vials. How many milliliters of neostigmine will you need?

12) You need to add 1.3 mg of a medication to a 50 mL bag of half-normal saline (0.45% NaCl). If the 2 mL ampule of this particular medication on your shelf has a ratio strength of 1:750, how many mL will you need to add to the half-normal saline bag?

13) If you have a 1:50 dilution of hydrocortisone lotion, what is the percentage strength?

14) If you have a 1:60 dilution of atropine in a petrolatum ointment, what is its percentage strength?

15) If you have 100 mg of drug dissolved in 100 mL of solution, what is its ratio strength?

16) If a particular vaccine contains 50 mcg of thimerosal in 0.5 mL of solution, answer the following.

a) What is the percentage strength?

b) What is its ratio strength?

17) Four hundred milligrams of drug is mixed with 4600 milligrams of sterile ophthalmic ointment base.

a) How many grams is the final mixture?

b) What is the resulting ratio strength?

18) If 140 mg of strychnine sulfate is intimately mixed with 1.26 g of lactose, what is the w/w ratio strength of the strychnine sulfate compared to the total mixture?

19) How many liters of a 1:5000 solution of cinnamon oil in alcohol can be made from 5 mL of cinnamon oil?

20) What is the ratio strength of a liquid in liquid solution if 130 mL of solution contains 0.65 mcL of active ingredient?

Worksheet 20-2

Name:

Date:

Solve the following problems.

1) A physician orders a 1 liter irrigation with 50 mg of amphotericin b. What is the ratio strength of this irrigation?

2) A technician is to prepare 250 mL of a 1:100,000 w/v solution. How many mg of the required drug will be needed to prepare the solution?

3) A drug solution is labeled 1:250 w/v. What is the percentage strength of the solution?

4) A medication is labeled with a 0.025% concentration.

   a) What is the ratio strength of this solution?

   b) What is the concentration of this solution in mcg/mL?

5) An order for 150 mL of a 1:1000 w:v solution is received in the pharmacy. There are 150 mg of this medication available in the pharmacy. Will the pharmacy be able to prepare this order?

6) An order for two bags, each bag being 1 liter in size, with a concentration of a 1:2000 w:v solution is received in the pharmacy. In stock is 960 mg of the powdered drug. Will the pharmacy be able to fill this order?

7) An order is received in the pharmacy for 7.5 mL of a 1:100,000 w:v solution. How many mcg are needed to fill this order?

8) An order is received in the pharmacy for 15 mL of a 1:50,000 solution. How many mcg of active ingredient are required to fulfill this request?

9) A physician uses the hospital's CPOE software to prescribe Racepinephrine inhale 1 mg per neb q3h prn asthma attack. If Racepinephrine has a concentration of 1:1000, how many milliliters will be required for a dose?

10) A patient is ordered 16 mg of epinephrine in NS with a final volume of 250 mL. There are 30 mL vials of epinephrine with a concentration of 1:1000 available.

a) How many mL of epinephrine will you need to add to the NS bag?

b) How many vials of epinephrine will you need to prepare this solution?

c) What is the ratio strength of epinephrine in the final NS bag?

11) The OR needs 1.5 mg of neostigmine to reverse the vecuronium that was used during surgery. The neostigmine you have on hand has a concentration of 1:2000 and comes in 10 mL vials. How many milliliters of neostigmine will you need?

12) You need to add 2.5 mg of a medication to a 100 mL bag of half-normal saline (0.45% NaCl). If the 2 mL ampule of this particular medication on your shelf has a ratio strength of 1:750, how many mL will you need to add to the half-normal saline bag?

13) If you have a 1:40 dilution of hydrocortisone lotion, what is the percentage strength?

14) If you have a 1:80 dilution of atropine in a petrolatum ointment, what is its percentage strength?

15) If you have 200 mg of drug dissolved in 100 mL of solution, what is its ratio strength?

16) If a particular vaccine contains 5 mcg of thimerosal in 0.5 mL of solution, answer the following.

a) What is the percentage strength?

b) What is its ratio strength?

17) Two hundred fifty milligrams of drug is mixed with 4750 milligrams of sterile ophthalmic ointment base.

a) How many grams is the final mixture?

b) What is the resulting ratio strength?

18) If 1400 mg of strychnine sulfate is intimately mixed with 12.6 g of lactose, what is the w/w ratio strength of the strychnine sulfate compared to the total mixture?

19) How many liters of a 1:5000 solution of cinnamon oil in alcohol can be made from 2.5 mL of cinnamon oil?

20) What is the ratio strength of a liquid in liquid solution if 250 mL of solution contains 0.5 mcL of active ingredient?

21) The OR wants you to make 25 phenylephrine syringes (10 mL/syringe) with a concentration of 1:12,500. You are to mix the phenylephrine (which is available in 10mg/mL 1 mL vials) with normal saline.

a) How many vials of phenylephrine will you need to make these syringes?

b) How many mL of normal saline will you need to make these syringes?

c) What is the final concentration of the phenylephrine in the syringes in mcg/mL?

Chapter 21

PPM, and Reducing & Enlarging Formulas

Someone once tried to tell me that this book was one in a million. Statistically speaking, it would be less significant than that as there are approximately 130 million books in the world.
--Number of books in the world according to Google

As a follow-up to the last chapter on ratio strength, we will look at some very closely related concepts including:

- parts per million (ppm) and parts per billion (ppb),
- reducing and enlarging formulas, and
- working with parts not associated with a specific unit.

Parts Per Million and Parts Per Billion

In the last chapter, we already learned that ratio strength is often used for quantifying the concentration of dilute solutions, but sometimes with *very* dilute solutions it may make more sense to refer to them in terms of parts per million (ppm) or parts per billion (ppb). These terms, ppm and ppb, are providing a ratio of the number of parts of a particular item within the mixture when compared to 1 million or 1 billion parts of the whole mixture.

Example

An example of ppm would be fluoridated drinking water which is typically listed as the fluoride being 1 ppm when compared to the water's total volume. If we know that this is a w:v ratio, how many mcg of fluoride are in 480 mL of fluoridated drinking water?

$$\frac{1\text{ g fluoride}}{1{,}000{,}000\text{ mL fluoridated drinking water}} = \frac{N}{480\text{ mL fluoridated drinking water}}$$

$$N = 0.00048\text{ g fluoride}$$

$$\frac{0.00048\text{ g}}{1} \times \frac{1000\text{ mg}}{1\text{g}} \times \frac{1000\text{ mcg}}{1\text{ mg}} = \mathbf{480\ mcg\ fluoride}$$

Practice Problem

To jump right into this concept, express 5 ppm of iron in water as both a percentage strength and a traditional ratio strength.

0.0005% is the percentage strength and 1:200,000 is the ratio strength

Reducing & Enlarging Formulas

Reducing and enlarging formulas is not an entirely new concept. You've already performed this in Chapter 12 when reducing and enlarging various compounding formulas. You may continue to use the methodology (ratio-proportions) employed in that chapter, but we will also introduce another means of performing this operation involving factors. Factors are based on the quantity of the formula desired divided by the quantity of the formula given.

$$\frac{\text{Quantity of formula desired}}{\text{Quantity of formula given}} = \text{Factor}$$

If the factor is greater than 1, it is often referred to as a growth factor, as you'll be using it to enlarge your recipe. If the factor is less than 1, you'll be using it to pair down your recipe. To get the final quantity of each ingredient, you can then multiply each ingredient in the original recipe by the factor. Let's look at an example of this concept.

Example

A prescription is written for a mouthwash containing 170 mL diphenhydramine elixir, 50 mL lidocaine viscous, 200 mL nystatin suspension, 52 mL erythromycin ethyl succinate suspension, and 28 mL of cherry syrup to make 500 mL of mouthwash. How much of each ingredient would be needed if you only wanted to prepare 240 mL of the mouthwash?

$$\frac{\text{Quantity of formula desired}}{\text{Quantity of formula given}} = \text{Factor}$$

$$\frac{240 \text{ mL}}{500 \text{ mL}} = 0.48$$

So, we have a factor of 0.48 that we should multiply all of our original ingredients by.

$$170 \text{ mL diphenhydramine elixir} \times 0.48 = \mathbf{81.6\ mL\ diphenhydramine\ elixir}$$

$$50 \text{ mL lidocaine viscous} \times 0.48 = \mathbf{24\ mL\ lidocaine\ viscous}$$

$$200 \text{ mL nystatin suspension} \times 0.48 = \mathbf{96\ mL\ nystatin\ suspension}$$

$$52 \text{ mL E.E.S. suspension} \times 0.48 = \mathbf{24.96\ mL\ E.E.S.\ suspension}$$

$$28 \text{ mL cherry syrup} \times 0.48 = \mathbf{13.44\ mL\ cherry\ syrup}$$

While this problem could have been solved using ratio-proportions, using a factor made this a much quicker problem. You should attempt the practice problem on the next page to verify your understanding of this concept.

Practice Problem

A prescription is written for a G.I. Cocktail containing 120 mL Donnatal elixir, 120 mL of lidocaine viscous solution, and 480 mL of Mylanta to make a total of 720 mL of G.I. Cocktail. What is the growth factor and how much of each ingredient would be needed if you only wanted to prepare 1 L of G.I. Cocktail?

Growth factor is 1.389
167 mL Donnatal elixir
167 mL lidocaine viscous solution
667 mL Mylanta

Parts

Occasionally, you may encounter a formula that indicates the ingredients in *parts* rather than in specific units such as grams or milliliters. The parts indicate the proportion of each of the ingredients in the recipe in comparison to the total number of parts. Sometimes the parts will be listed with each ingredient and sometimes they will be listed in a series separated with colons. If they are listed in a series, you may assume that the order of each number corresponds to the order of the list of each ingredient. Generally, in a recipe that involves a solid or semisolid (i.e., ointments, creams, powders, etc.) the parts will be based on weight (usually g); whereas a formula involving a liquid preparation (i.e., solutions, lotions, etc.) the parts are based on volume (usually mL).

Let's look at an example problem using parts.

Example

Consider the following recipe for an antacid powder. Determine how much of each ingredient you would need to make 300 g of this antacid powder.

Rx	calcium carbonate	5 parts
	magnesium oxide	1 part
	sodium bicarbonate	4 parts
	bismuth subcarbonate	3 parts

Mix i tsp c 8 oz of water and drink prn indigestion.

First, we will total all the parts up to be able to make a comparison between a part of something and the whole.

5+1+4+3=13 total parts

Now we are prepared to make a comparison between the parts and the whole in order to determine how much of each ingredient is needed to prepare 300 g of antacid powder.

$$\frac{5 \text{ parts calcium carbonate}}{13 \text{ parts antacid powder}} = \frac{N}{300 \text{ g antacid powder}}$$ N = **115 g calcium carbonate**

$$\frac{1 \text{ part magnesium oxide}}{13 \text{ parts antacid powder}} = \frac{N}{300 \text{ g antacid powder}}$$ N = **23 g magnesium oxide**

$$\frac{4 \text{ parts sodium bicarbonate}}{13 \text{ parts antacid powder}} = \frac{N}{300 \text{ g antacid powder}}$$ N = **92 g sodium bicarbonate**

$$\frac{3 \text{ parts bismuth subcarbonate}}{13 \text{ parts antacid powder}} = \frac{N}{300 \text{ g antacid powder}}$$ N = **69 g bismuth subcarbonate**

Now attempt the practice problem below in order to verify that the concept makes sense.

Practice Problem

Using the coal tar ointment recipe below, determine the quantity of each ingredient required to make 1 kg of coal tar ointment.

coal tar
zinc oxide 5:10:50
hydrophylic ointment

77 g coal tar
154 g zinc oxide
769 g hydrophilic ointment

Worksheet 21-1

Name:

Date:

Solve the following problems.

1) From the following formula, calculate the number of grams of each ingredient required to prepare 60 grams of the ointment:

Precipitated sulfur	10 g
Salicylic acid	2 g
Hydrophilic ointment	88 g

2) Based on the recipe in problem 1, how many grams of each ingredient would you need to prepare a kilogram of that ointment?

3) Here is the formula for zinc oxide paste written by parts:

zinc oxide	
starch	1:1:2
white petrolatum	

How much of each ingredient is required to make 60 g of zinc oxide paste?

4) A foot powder has this formula:

Talc:	15 parts
Benzoic acid:	1 part
Bentonite:	3 parts

Rewrite the formula so that it yield 60 g of foot powder.

5) Purified water contains not more than 10 ppm of total solids. Express this concentration as a percentage.

6) The formula for an analgesic powder is:

Aspirin
Phenacetin 6:3:1
Caffeine

All quantities are based on weight.

How many grams of each should be used to prepare 1.25 kg of the powder?

7) The formula for yellow ointment may be written:

Yellow wax	1 part
Petrolatum	19 parts

How many kilograms of yellow ointment can be prepared from 660 g of yellow wax? (*Hint: Read the question carefully before solving the problem.*)

8) The formula for camphorated parachlorophenol is

parachlorophenol	7 parts
camphor	13 parts

What quantities should be used to make 150 g of camphorated parachlorophenol?

9) Here is the formula for placebo tablets:

Lactose	100 g
Sucrose	150 g
Starch, direct compressing formula	250 g
Magnesium sulfate	0.5 g
Yield	1000 tablets

Rewrite the formula to make 50,000 tablets.

10) From the following calamine lotion formula, calculate the quantity of each ingredient required to make a pint of calamine lotion.

calamine	80 g
zinc oxide	80 g
glycerin	20 g
bentonite magma	250 mL
calcium hydroxide topical solution	q.s. 1000 mL

11) Trihalomethanes have a legal limit of 80 ppb in public drinking water. If a particular public drinking reservoir can hold 10,000 gallons, what is the maximum volume of trihalomethanes that could be legally dissolved in it? (1 gal = 3,785 mL)

12) If a particular insulin preparation contains 1 ppm of proinsulin, how many mcg would be present in a 10 mL vial?

Chapter 22

Dosage Calculations Based on Body Weight

If we follow the logic that 454 graham crackers equal a pound cake, does that mean that 2.2 pound cakes would make a kilo cake?
--A really bad attempt at humor from this book's author.

As we initially introduced the concept of medication doses based on body weight back in chapter 10, we will review a little before diving straight into this concept. Many drugs need to be calculated based on body weight. Some of the drugs where you will see this most often include chemotherapy, steroids, antibiotics, heparinoids, and drugs for pediatric and geriatric patients.

The specific ideas included in this chapter are:

- medication dosages in weight of drug,
- medication dosages in volume of drug, and
- verifying the appropriateness of a dosage when a drug dosage is based on body weight.

If a medication dose is to be based on a body weight, it will usually be requested in mg/kg of body weight. Sometimes a time element will be included as well. Typically, the time element will be kept with the body weight when performing dimensional analysis. A useful idea to keep in mind is always consider what labels you want to end up with in order to help you set up your factor label problems correctly. Let's look at some example problems.

Examples

1) Gentamicin is ordered 5 mg/kg/day divided into 3 equal doses for a patient with a serious infection who weighs 231 pounds. How many mg of gentamicin should the patient receive for each dose?

$$\frac{5\text{ mg}}{\text{kg day}} \times \frac{1\text{ day}}{3\text{ doses}} \times \frac{1\text{ kg}}{2.2\text{ pounds}} \times \frac{231\text{ pounds}}{1} = \mathbf{175\,mg/dose}$$

So the patient will need 175 mg of gentamicin in each dose. Some things to notice include kg and day being paired together in the dimensional analysis. It is also noteworthy that the question itself was about how many mg of gentamicin were needed in each dose, allowing you to know that your final answer should yield mg per dose.

2) A physician orders a 120 mcg/kg dose of vinblastine for a patient with testicular cancer. The drug is available in a 10 mg vial that you reconstitute with 10 mL of bacteriostatic saline to achieve a concentration of 1 mg/mL. If the patient weighs 176 pounds, how many mL are required for a dose?

$$\frac{\text{mL}}{1\text{ mg}} \times \frac{1\text{ mg}}{1000\text{ mcg}} \times \frac{120\text{ mcg}}{\text{kg}} \times \frac{1\text{ kg}}{2.2\text{ pounds}} \times \frac{176\text{ pounds}}{1} = \mathbf{9.6\ mL\ of\ vinblastine}$$

Something to be careful about when reading a problem is the potential for excess information. The fact that the vial contains 10 mg and that it was reconstituted with 10 mL did not matter for doing the math.

3) A 3-day old neonate weighing 7 pounds 11 ounces is ordered tobramycin 10 mg IM q12h. The recommended dosage of tobramycin for an infant is 2 mg/kg/dose IM q12h. Is 10 mg the appropriate dose for the neonate?

   First, to simplify things, we will want to convert the infant's weigh all into pounds.

   $$\frac{11\text{ oz}}{1} \times \frac{1\text{ pound}}{16\text{ oz}} = 0.6875\text{ pounds}$$

   $$7\text{ pounds} + 0.6875\text{ pounds} = 7.6875\text{ pounds}$$

   Now we can determine the normal dose.

   $$\frac{2\text{ mg}}{\text{kg dose}} \times \frac{1\text{ kg}}{2.2\text{ pounds}} \times \frac{7.6875\text{ pounds}}{1} = \mathbf{7\ mg/dose}$$

   So, the recommended dose would only be 7 mg per dose, which is less than what was originally written for by the physician.

Now that the example problems have allowed us to review some concepts, attempt the practice problems below.

Practice Problems

1) Ceftazidime is ordered for a 44 pound patient with meningitis at a dosage of 50 mg/kg every 8 hours. The drug when reconstituted has a concentration of 100 mg/mL. How many mL of the reconstituted solution will be required for each dose?

2) Sulfamethoxazole and trimethoprim is a combination antibiotic commonly used for enteritis, p. carinii pneumonia, and urinary tract infections. When treating enteritis, the pediatric dosage range is 8 to 10 mg/kg/day (based on trimethoprim) in 2 to 4 divided doses every 6, 8, or 12 hours for 5 days by IV infusion.

   a) If a physician orders 50 mg of this antibiotic every 12 hours for a patient weighing 25 pounds, is this within the suggested dosing guidelines?

b) If this antibiotic has a concentration of sulfamethoxazole 80 mg/mL and trimethoprim 16 mg/mL, how many milliliters of solution will you need for each dose?

3) A 115 pound patient is life flighted to your hospital after an accident resulting in a spinal cord injury. She received a bolus dose of methylprednisolone while en route to the hospital, but now requires a continuous infusion of methylprednisolone at a rate of 5.4 mg/kg/hr for 23 hours. How many grams of methylprednisolone will need to be placed in her continuous infusion?

1) 10 mL 2-a) Yes, this dose is within the acceptable range. 2-b) 3.125 mL 3) 6.5 g of methylprednisolone

Now that you've attempted the practice problems you are ready for the worksheets.

Worksheet 22-1

Name:

Date:

Solve the following problems.

1) A 66 pound pediatric patient is to receive cefuroxime 25 mg/kg q8h. How many mg per dose should this patient receive?

2) Ceftazidime is ordered for a 55 pound patient with meningitis at a dosage of 50 mg/kg every 8 hours. The drug when reconstituted has a concentration of 100 mg/mL. How many mL of the reconstituted solution will be required for each dose?

3) A physician orders cyclophosphamide 5 mg/kg in 50 mL of D5W to be administered twice a week for a 75 kg patient. If a 500 mg vial of cyclophosphamide is reconstituted with 25 mL of SWFI to obtain a concentration of 20 mg/mL, how many milliliters of the reconstituted solution will need to be added to the D5W bag?

4) A patient weighing 198 pounds is to receive 2.5 mg/kg/day of oral cyclophospamide. If cyclophosphamide is available in the pharmacy as 25 mg/tablet, how many tablets will he need?

5) A 4 pound 8 ounce neonate is to receive IV vancomycin 10 mg/kg q8h. How many milligrams of vancomycin should the neonate receive in each dose?

6) A 156 pound patient with an HIV infection is to receive IV zidovudine 1 mg/kg in 50 mL D5W minibags q4h atc. How many milligrams of zidovudine will the patient receive each day?

7) A physician orders a dose of vincristine, based on the patient's body weight, as 30 mcg/kg. The maximum dose of vincristine is capped at 2 mg, so if your calculated dose exceeds that it is automatically reduced to 2 mg. Look at the vial pictured below to obtain the concentration. If the patient weighs 143 pounds, how many mL are needed for a dose? Mark the syringe appropriately for the dose.

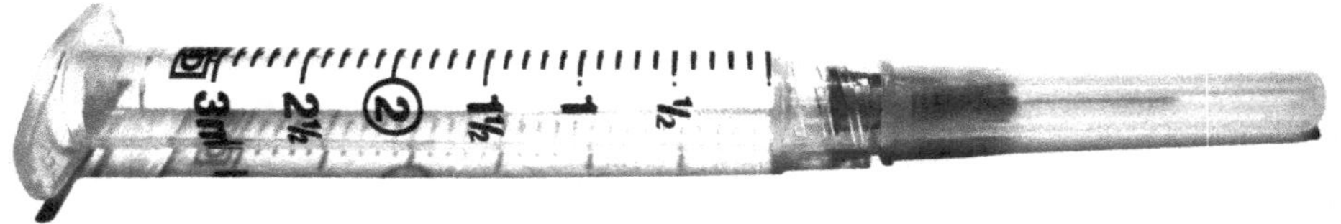

8) A pediatric patient with an upper respiratory tract infection weighing 44 pounds is receiving ampicillin 12.5 mg/kg in 50 mL of D5W q6h.

   a) How many milligrams will the patient receive for each dose?

   b) What is the total daily dose?

   c) If a 10 gram bulk vial is to be reconstituted with 35 mL of SWFI and contains 5 mL of powder volume, how many milliliters of this reconstituted solution will need to be added to each minibag?

9) Calculate both the loading dose and the maintenance dose of gentamicin for a 230 pound patient with pelvic inflammatory disease.

   a) If the loading dose is 2 mg/kg IV, how many mg of gentamicin will the patient need for their loading dose?

   b) If the maintenance dose is 1.5 mg/kg IV q8h, how many mg of gentamicin will the patient need for a single maintenance dose?

   c) If the gentamicin stocked in the pharmacy comes in a 30 mL MDV with a concentration of 40 mg/mL, how many milliliters of gentamicin will be required for both the loading dose and each maintenance dose?

10) Your pharmacy stocks a 10% immune globulin solution. The order for a 55 pound patient with primary immunodeficiency is 500 mg/kg every 4 weeks. How many mL of 10% immune globulin will need to be infused on this patient for each dose?

11) A medication order for a patient weighing 190 pounds requests amphotericin B 0.25 mg/kg in 1000 mL of D5W. Amphotericin B comes as a 50 mg lyophylized powder that gets reconstituted with 10 mL of SWFI and has negligible powder volume. How many milliliters of the reconstituted solution should be added to the 1 L bag?

12) A neonate weighing 4000 g is ordered tobramycin 10 mg IM q12h. The recommended dosage of tobramycin for an infant is 2.5 mg/kg/dose IM q12h.

a) Is 10 mg the appropriate dose for the neonate?

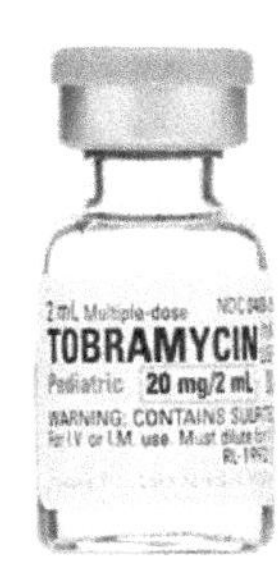

b) Looking at the vial, how many mL of solution are needed?

c) Appropriately mark the syringe for how much should be dispensed.

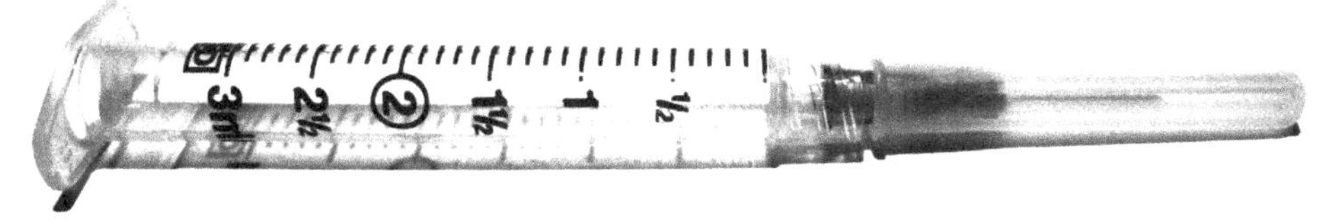

13) Sulfamethoxazole and trimethoprim is a combination antibiotic commonly used for enteritis, p. carinii pneumonia, and urinary tract infections. When treating p. carinii pneumonia the pediatric dosage range is 15 to 20 mg/kg/day (based on trimethoprim) in 3 or 4 divided doses every 6 or 8 hours for 14 days by IV infusion.

a) If a physician orders 75 mg of this antibiotic every 8 hours for a patient weighing 25 pounds, is this within the suggested dosing guidelines?

b) If this antibiotic has a concentration of sulfamethoxazole 80 mg/mL and trimethoprim 16 mg/mL, how many milliliters of solution will you need for each dose?

14) A 240 pound patient is admitted to the hospital due to herpes genitalis. The physician orders acyclovir 5 mg/kg in 100 mL of NS q8h. Acyclovir is available in 1 gram vials that get reconstituted with 20 mL of SWFI (powder volume is considered negligible). How many mL of the reconstituted acyclovir will you add to each 100 mL bag?

15) A pediatrician orders penicillin V potassium oral suspension 20,000 units/kg/dose q8h for 10 days for a patient weighing 66 pounds. Penicillin V potassium has a concentration of 250 mg/tsp and 1 mg of penicillin V potassium is equal to 1600 units of penicillin V potassium.

a) How many milliliters of this suspension will need to be used for each dose?

b) How much should be dispensed to cover all 10 days?

16) A 209 pound patient is rushed to the emergency room after a terrible car accident and requires the spinal cord protocol for two methylprednisolone infusions.

a) First, you need to make a bolus dose of 30 mg/kg in 50 mL of NS. How many mg of methylprednisolone will you need to make the bolus dose? (The bolus dose will be infused over 1 hour)

b) If methylprednisolone comes in a double chamber vial with a concentration of 1 g/8 mL, how many mL will you need to add to the 50 mL bag of NS?

c) After the bolus dose is infused, you need to prepare a continuous infusion of methylprednisolone at a rate of 5.4 mg/kg/hr for 23 hours. How many mg of methylprednisolone will you need to make the continuous infusion?

d) The continuous infusion is to be administered with normal saline as the diluent. If the final bag has to have an exact volume of 1000 mL, how many mL methylprednisolone will be needed since it comes in a double chamber vial with a concentration of 1 g/8 mL, and how many mL of NS will be needed?

Worksheet 22-2

Name:

Date:

Solve the following problems.

1) Enoxaparin 1 mg/kg SQ q12h is ordered for a 176 pound patient. How many mg of enoxaparin should the patient receive for each dose?

2) The usual dosage of dalteparin for patients with cancer and symptomatic venous thromboembolism for the first 30 days of treatment is 200 units/kg total body weight subcutaneously once daily. The maximum dose is 18,000 units subcutaneously once a day. How many units would be the recommended dose for each of the following body weights:

   a) 110 pounds

   b) 138 pounds

   c) 165 pounds

   d) 193 pounds

   e) 220 pounds

3) A 190 pound patient is receiving dopamine at 5 mcg/kg/min. If the physician does not titrate the dose up, how long should a 500 mL premixed bag of dopamine last if it has a concentration of 0.8 mg/mL?

4) A 231 pound patient is receiving dobutamine at 10 mcg/kg/min. If the dose remains the same,

how long should a 250 mL premixed bag of dobutamine last if it has a concentration of 1 mg/mL?

5) A five year old pediatric patient weighing 42 pounds with a severe infection is ordered cefotaxime 90 mg/kg/day by IV infusion in 6 equally divided doses every 4 hours in 50 mL of NS.

   a) How many mg of cefotaxime should this patient receive daily?

   b) How many mg will this patient receive for each dose?

   c) If a bulk 10 g vial of cefotaxime is reconstituted with 95 mL of SWFI and has 5 mL of powder volume, how many mL of the reconstituted solution will be shot into each minibag for the patient dose?

6) A pediatric patient weighing 20 kg with a severe skin infection that has probable penicillin resistance is ordered a combination of ampicillin and sulbactam 300 mg/kg/day (ampicillin 200 mg/sulbactam 100 mg) by IV infusion in 4 equally divided doses every 6 hours in 50 mL of NS.

   a) How many mg of ampicillin and sulbactam should this patient receive daily?

   b) How many mg will this patient receive for each dose?

   c) If a bulk 15 g vial of ampicillin and sulbactam (10 grams ampicillin and 5 grams sulbactam) is reconstituted with 32 mL of SWFI and has 8 mL of powder volume, how many mL of the reconstituted solution will be shot into each minibag for the patient dose?

7) The initial dosing guidelines for fluorouracil for a carcinoma of the stomach is 12 mg/kg once daily by IV bolus for 4 successive days. The daily dose should not exceed 800 mg. If no toxicity is observed, 6 mg/kg should be given on days 6, 8, 10, and 12 unless toxicity occurs. No therapy is given on days 5, 7, 9, and 11. Therapy is to be discontinued at the end of day 12, even if no toxicity has become apparent. If a 150 pound patient is receiving fluorouracil to treat a

stomach carcinoma, what would be the recommended doses for each of the first 12 days?

a) Day 1

b) Day 2

c) Day 3

d) Day 4

e) Day 5

f) Day 6

g) Day 7

h) Day 8

i) Day 9

j) Day 10

k) Day 11

l) Day 12

8) Based on the information in the previous problem and looking at the vial pictured to the right, how many milliliters of fluorouracil will the patient require each day?

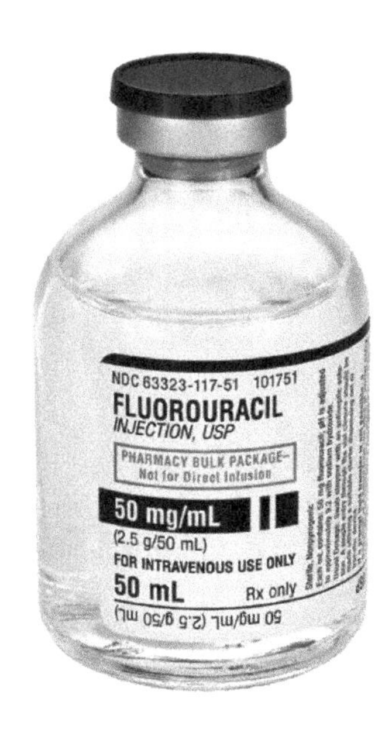

a) Day 1

b) Day 2

c) Day 3

d) Day 4

e) Day 5

f) Day 6

g) Day 7

h) Day 8

i) Day 9

j) Day 10

k) Day 11

l) Day 12

9) A 250 pound patient with Crohn's disease is ordered infliximab 5 mg/kg. How many mg of infliximab will the patient receive?

10) A one year old child that has been deemed as high risk for RSV is to receive palivizumab 15 mg/kg monthly throughout RSV season. If the child's body weight is 22 pounds when scheduled to receive his first dose, how many mg will the first dose be?

11) The initial oral dosage of tacrolimus after a heart transplant is 0.075 mg/kg/day administered in 2 divided doses (every 12 hours). A 180 pound patient is ordered oral tacrolimus 6 hours after completion of a heart transplant.

a) What would the recommended dose be for this patient?

b) Since tacrolimus is available in 0.5 mg, 1 mg, and 5 mg capsules, what combination of capsules makes the most sense to dispense for this patient for each dose?

12) A 154 pound patient with a systemic fungal infection is to receive amphotericin B lipid complex 5 mg/kg /day.

a) How many mg of amphotericin B lipid complex should this patient receive on a daily basis?

b) Amphotericin B lipid complex comes with a concentration of 5 mg/mL. How many mL of this stock solution is needed?

c) An IV bag of amphotericin B lipid complex should be diluted out with D5W till the amphotericin B lipid complex reaches a final concentration of 1 mg/mL. How many mL of D5W are needed for this IV?

13) A 4 pound 8 ounce neonate with a severe infection is ordered a continuous infusion of penicillin g potassium at a rate of 4,000 units/kg/hr. The nursing unit wants to hang a new bag every 12 hours. How many units of penicillin g potassium should be in each bag?

14) A physician orders potassium chloride 0.6 mEq/kg/day divided into 4 doses each in 100 mL of NS for an 198 pound patient suffering from hypokalemia. How many mL should be added to each IV bag based on the concentration listed on the vial? Mark the correct dose on the syringe below.

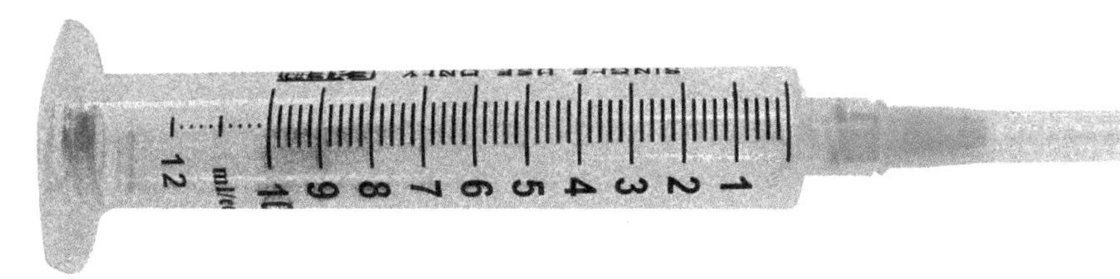

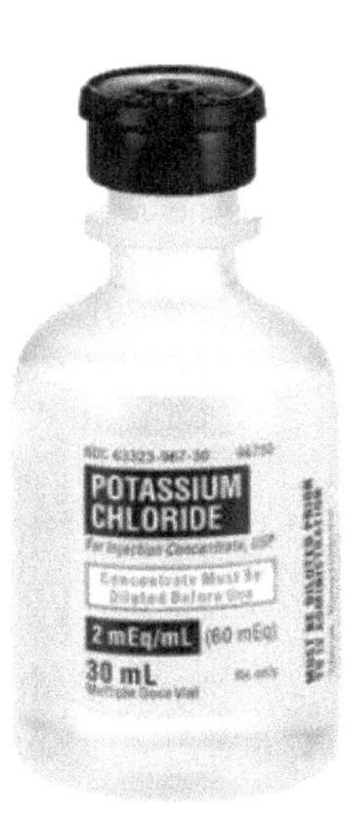

15) A 121 pound patient received 2 grams of a medication that should have been dosed at 50 mg/kg. Was the patient underdosed, correctly dosed or overdosed?

16) If a 147 pound patient were to receive an initial dose of 200 mg of fluconazole po followed by 3 mg/kg bid for 7 days, then how many 100 mg fluconazole tablets will the patient need to cover their course of therapy?

17) Phenytoin is a commonly used anticonvulsant. The dosage range for pediatric patients is 4 to 8 mg/kg/day in 2 to 3 divided doses when taken orally.

a) If a physician orders 25 mg of phenytoin suspension every 12 hours for a patient weighing 25 pounds, is this within the suggested dosing guidelines?

b) If phenytoin oral suspension has a concentration of 125 mg/5 mL, how many milliliters of phenytoin suspension will be needed for each dose?

18) The loading dose of indomethacin is 0.2 mg/kg IV for neonates with patent ductus arteriosus.

a) What would be the loading dose for a neonate weighing 5 pounds 8 ounces?

b) If a 1 mg vial of indomethacin is reconstituted to a final volume of 2 mL, how many mL will the loading dose be?

19) When treating hospital acquired pneumonia with vancomycin, the American Thoracic Society guidelines recommend 15 mg/kg every 12 hours in adults with healthy renal function. Assuming heathy renal function, what would be the appropriate dose for each of the following body weights?

a) 110 pounds

b) 165 pounds

c) 220 pounds

20) A 176 pound patient is rushed to the emergency room after a motorcycle accident and requires the spinal cord protocol for two methylprednisolone infusions.

a) First, you need to make a bolus dose of 30 mg/kg in 50 mL of NS. How many mg of methylprednisolone will you need to make the bolus dose? (The bolus dose will be infused over 1 hour)

b) If methylprednisolone comes in a double chamber vial with a concentration of 1 g/8 mL, how many mL will you need to add to the 50 mL bag of NS?

c) After the bolus dose is infused, you need to prepare a continuous infusion of methylprednisolone at a rate of 5.4 mg/kg/hr for 23 hours. How many mg of methylprednisolone will you need to make the continuous infusion?

d) The continuous infusion is to be administered with normal saline as the diluent. If the final bag has to have an exact volume of 1000 mL, how many mL methylprednisolone will be needed since it comes in a double chamber vial with a concentration of 1 g/8 mL, and how many mL of NS will be needed?

CHAPTER 23

DOSAGE CALCULATIONS BASED ON BSA

Average BSA is generally taken to be 1.73 m^2 for an adult (1.6 m^2 for women and 1.9 m^2 for men).
--Oxford Journals

The body surface area (BSA) is the measured or calculated surface of a human body, and it is measured in square meters (m^2). For many clinical purposes, BSA is a better indicator of metabolic mass than body weight because it is less affected by abnormal adipose mass.

Examples of uses for BSA include:

- renal clearance is usually divided by the BSA to gain an appreciation of the true required glomelular filtration rate (GFR),
- chemotherapy is often dosed according to the patient's BSA, and
- glucocorticoid dosing can also be expressed in terms of BSA for calculating maintenance doses or to compare high dose use with maintenance requirement.

There are number of ways to calculate BSA:

such as various formulas:
- Dubois & Dubois; BSA (m^2) = $0.007184 \times \text{weight in kg}^{0.425} \times \text{height in cm}^{0.725}$
- Mosteller; BSA (m^2) = $\sqrt{\frac{\text{weight in kg} \times \text{height in cm}}{3600}}$
- Haycock; BSA (m^2) = $0.024265 \times \text{weight in kg}^{0.5378} \times \text{height in cm}^{0.3964}$
- Gehan & George; BSA (m^2) = $0.0235 \times \text{weight in kg}^{0.51456} \times \text{height in cm}^{0.42246}$
- Fujimoto; BSA (m^2) = $0.008883 \times \text{weight in kg}^{0.444} \times \text{height in cm}^{0.663}$
- Takahira; BSA (m^2) = $0.007241 \times \text{weight in kg}^{0.425} \times \text{height in cm}^{0.725}$
- Boyd; BSA (m^2) = $0.0003207 \times \text{weight in g}^{(0.7285-0.0188 \log_{10}\text{weight in g})} \times \text{height in cm}^{0.3}$
- Schlick for women; BSA (m^2) = $0.000975482 \times \text{weight in kg}^{0.46} \times \text{height in cm}^{1.08}$
- Schlick for men; BSA (m^2) = $0.000579479 \times \text{weight in kg}^{0.38} \times \text{height in cm}^{1.24}$

using a nomogram (a chart method that can be based on any of the above formulas),
or some more peculiar ways, including geometry, thoroughly detailed 3D scans, even wrapping patients in aluminum foil.

As you look through the formulas, you will find numerous similarities as they are all trying to calculate the same thing in the end. Various formulas have been determined by looking at different segments of the population. Most hospitals and cancer clinics in the United States use either the Dubois & Dubois formula or the Mosteller formula for adults and most pediatricians utilize the Haycock method, the Mosteller formula, or the Dubois & Dubois formula. The nomograms used in this textbook are based on the Dubois & Dubois formula for calculating BSA. (*Note: the formula that is most often memorized is the Mosteller formula*)

A nomogram is a common method and is the way we will concentrate on. A nomogram typically has three columns:

- height based in both centimeters and inches,
- body surface area in m^2, and
- weight based in both kilograms and pounds.

To use a nomogram, the height and the weight of the patient are found on the nomogram and then a straight line is drawn connecting the two values. The BSA for that patient is found where the line intersects the BSA column. As an example, find 5' 3" (63 inches) and 110 pounds on the nomogram located on the next page. If you draw a line connecting the two values, you should get a BSA of 1.5 m^2.

Example

A reasonable exercise might be to try and plug 63 inches and 110 pounds into the Dubois and Dubois formula and see if you get the same answer. First though, we will need to turn those numbers into their metric equivalents in order to use them in the Dubois & Dubois formula.

$$\frac{63\,\text{inches}}{1} \times \frac{2.54\,\text{cm}}{1\,\text{inch}} = 160\,\text{cm}$$

$$\frac{110\,\text{pounds}}{1} \times \frac{1\,\text{kg}}{2.2\,\text{pounds}} = 50\,\text{kg}$$

$$\text{Dubois \& Dubois; BSA } (\text{m}^2) = 0.007184 \times \text{weight in kg}^{0.425} \times \text{height in cm}^{0.725}$$

$$0.007184 \times 50^{0.425} \times 160^{0.725} = \mathbf{1.50\ m^2}$$

Practice Problem

Attempt the process of determining the BSA of the author of this textbook using both a nomogram and the Dubois & Dubois formula. The height and weight of this book's author is 6'1" and 165 pounds.

Both answers should come out close to 1.98 m^2.

If your two answers were slightly different (within 5% of each other) that is considered an acceptable range when dealing with body surface area. This may result in classmates achieving slightly different answers in this chapter, but 5% is considered acceptable. That 5% also allows for the fact that not all oncologists may choose to use the same formulas. For example, if you calculated the BSA of a 6'1" 165 lb individual using the Mosteller formula instead, the BSA is 1.97 m^2, which is certainly within the 5% range.

Nomogram for Determination of Adult BSA from Height and Weight

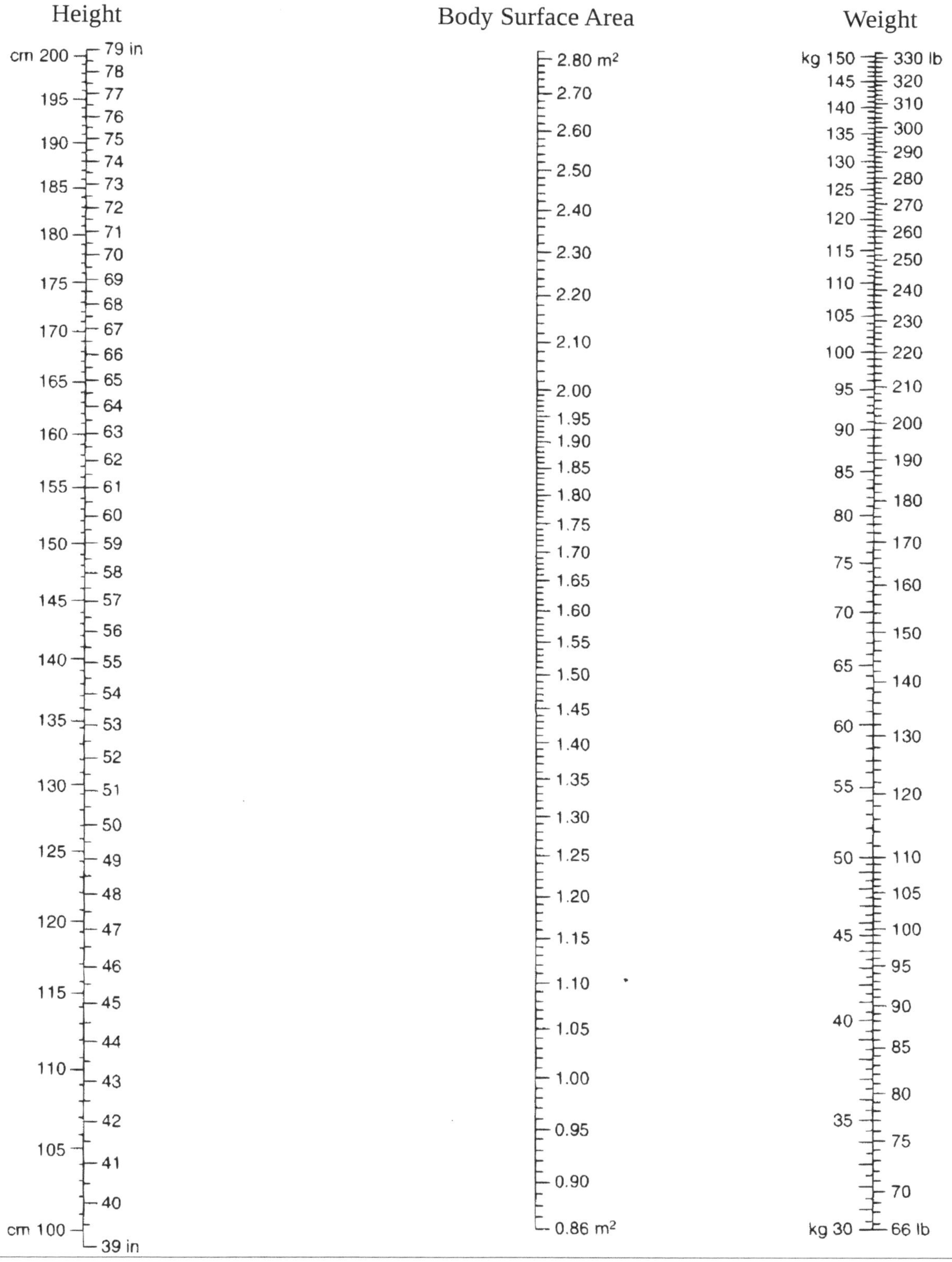

Height

cm 200 – 79 in
78
195 – 77
76
190 – 75
74
185 – 73
72
180 – 71
70
175 – 69
68
170 – 67
66
165 – 65
64
160 – 63
62
155 – 61
60
150 – 59
58
145 – 57
56
140 – 55
54
135 – 53
52
130 – 51
50
125 – 49
48
120 – 47
46
115 – 45
44
110 – 43
42
105 – 41
40
cm 100 – 39 in

Body Surface Area

2.80 m²
2.70
2.60
2.50
2.40
2.30
2.20
2.10
2.00
1.95
1.90
1.85
1.80
1.75
1.70
1.65
1.60
1.55
1.50
1.45
1.40
1.35
1.30
1.25
1.20
1.15
1.10
1.05
1.00
0.95
0.90
0.86 m²

Weight

kg 150 – 330 lb
145 – 320
140 – 310
135 – 300
290
130 – 280
125 – 270
120 – 260
115 – 250
110 – 240
105 – 230
100 – 220
95 – 210
90 – 200
85 – 190
80 – 180
170
75 – 160
70 – 150
65 – 140
60 – 130
55 – 120
50 – 110
105
45 – 100
95
90
40 – 85
80
35 – 75
70
kg 30 – 66 lb

Nomogram is based on the Dubois & Dubois formula.
BSA (m^2) = 0.007184 x weight in $kg^{0.425}$ x height in $cm^{0.725}$

Nomogram for Determination of Pediatric BSA from Height and Weight

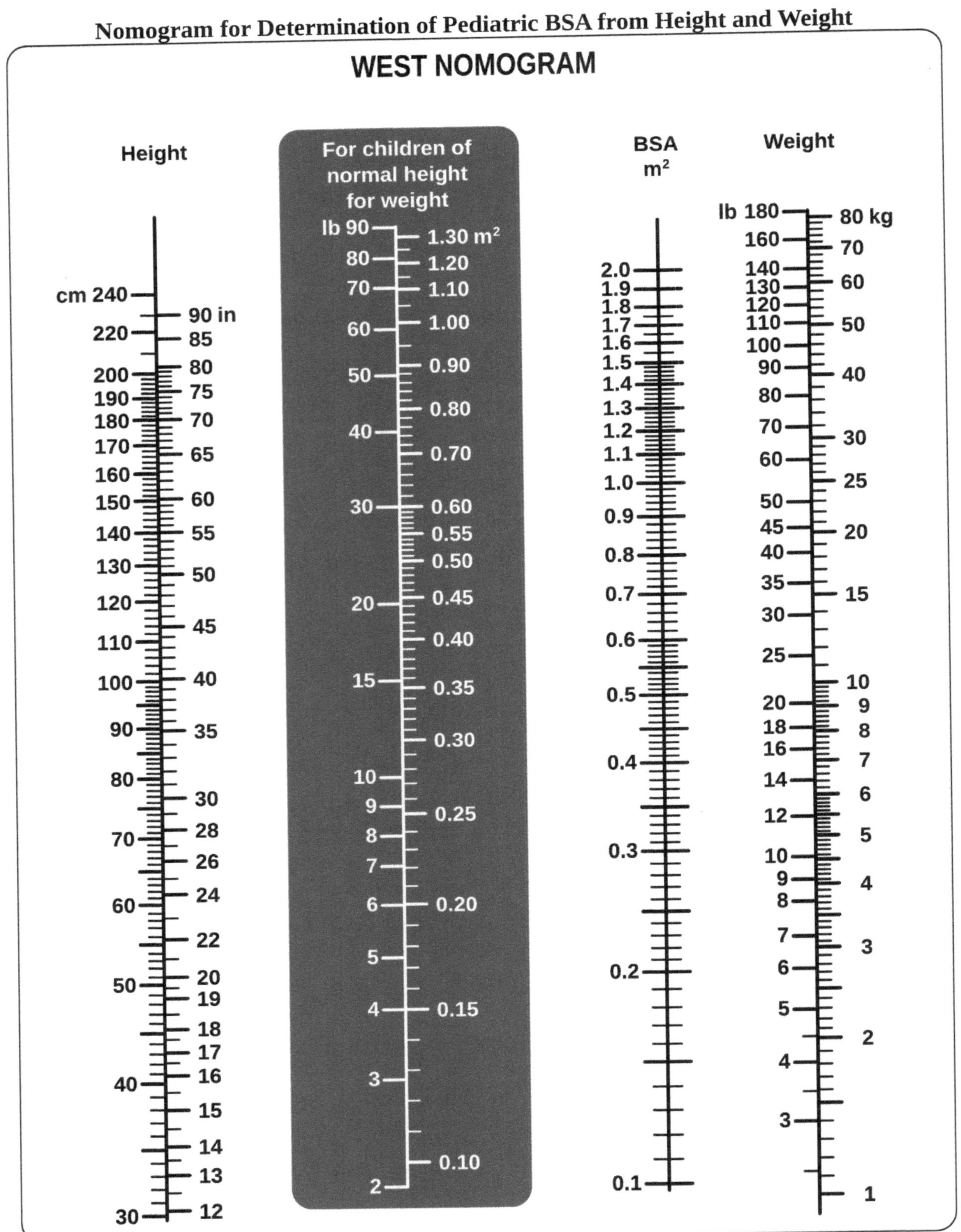

Notice that you can use the West Nomogram the same way as the adult nomogram, or if the child is a normal height for their weight you may use the darkened column instead as a shortcut.

At this point, we should look at some example problems requiring BSA.

Examples

1) A 5'3" woman weighing 120 pounds is to receive docetaxel 75 mg/m^2 every three weeks for breast cancer. How many mg should her dose be?

$$\frac{1.56\,\text{m}^2}{1} \times \frac{75\,\text{mg}}{\text{m}^2} = \mathbf{117\,mg}$$

2) A 4'7" 10 year old boy weighing 74 pounds is ordered etoposide IV 150 mg/m^2/day to treat acute nonlymphocytic leukemia. How many mg should his dose be?

$$\frac{1.16\,\text{m}^2}{1} \times \frac{150\,\text{mg}}{\text{m}^2} = \mathbf{174\,mg}$$

Practice Problems

1) A 5'6" woman weighing 320 pounds is to receive paclitaxel 260 mg/m^2 every three weeks for breast cancer. How many mg should her dose be?

2) A 4'2" 8 year old girl weighing 60 pounds is ordered methotrexate IV 3.3 mg/m^2 daily to treat leukemia. How many mg should her dose be?

1) 634.4 mg 2) 3.2 mg

While there are still more concepts to learn when dosing chemotherapy medications, this is a good point to reinforce the concepts already discussed with a worksheet.

Worksheet 23-1

Name:

Date:

Solve the following problems. You may need to refer to the nomograms in this book, or you may choose to use any of the appropriate formulas discussed in this chapter in order to attain the necessary BSA for these problems.

1) To get started, solve the BSA for the following adult patients based on their height and weight:

   a) Height 180 cm; Weight 80 kg

   b) Height 5' 11"; Weight 176 pounds

   c) Height 63"; Weight 143 lb

   d) Height 1.5 m; Weight 65 kg

2) Let's determine the BSA for several pediatric patients based on their height and weight:

   a) Height 39"; Weight 30 lb

   b) Height 112 cm; Weight 20 kg

   c) Height 4'; Weight 49 pounds

   d) Height 1.5 m; Weight 39 kg

3) An oncologist orders mitomycin 20 mg/m^2 for a 5' 9" patient weighing 165 lb. How many mg of mitomycin should the patient receive?

4) A patient with Hodgkin's disease is to receive a maintenance dose of bleomycin of 10 units/m^2. If the patient is 5' 7" and 160 lb, how many units of bleomycin should they receive?

5) An oncologist orders gemcitabine 1,250 mg/m^2 for a 5' 4" patient weighing 125 lb that has breast cancer.

   a) How many mg of gemcitabine should the patient receive?

   b) Gemcitabine is available in 200 mg vials and 1 g vials. How many of each vial size will you need to gather to make a dose of gemcitabine for this patient?

   c) The package insert instructs you to reconstitute 200 mg vials with 5 mL of diluent and 1 g vials with 25 mL of diluent providing a concentration of 38 mg/mL. How many mL of the reconstituted drug vials will you need to dispense this dose?

6) A pediatric patient is to receive doxorubicin 60 mg/m^2. Their height is 48 inches and weighs 54 pounds.

   a) How many mg of doxorubicin should this pediatric patient receive?

   b) Looking at the vials pictured, how many mL of doxorubicin should the patient receive?

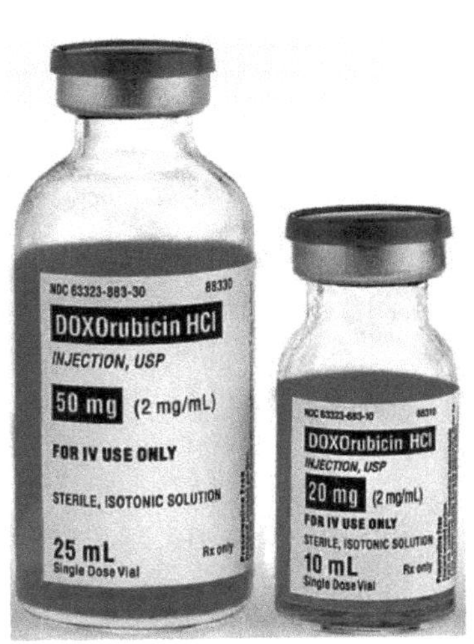

7) Adalimumab 24 mg/m^2 SQ is ordered for a juvenile. The child is 3' 11" and weighs 49 pounds. How many mg of adalimumab should this child receive?

8) An oncologist is initiating therapy with rituximab 375 mg/m^2 on a patient with chronic lymphocytic leukemia. The patient is 68" tall and weighs 140 lb. How many mg of rituximab should the patient receive?

9) A four agent dosage regimen, termed MOPP (which is related to the medications used), for the treatment of Hodgkin's lymphoma, includes the following drugs taken over a 28-day cycle:

mechlorethamine 6 mg/m^2, IV, days 1 & 8
vincristine 1.4 mg/m^2, IV, days 1 & 8
procarbazine 100 mg/m^2/day, po, days 1-14
prednisone 40 mg/m^2/day, po, days 1-14

If the patient is 5' 1" in height and weighs 116 lbs, calculate the following:

a) How many mL of mechlorethamine will you need for each dose if it comes in 10 mg vials that need to be reconstituted with either 10 mL of SWFI or NS (powder volume is considered negligible)?

b) If each dose of mechlorethamine is to have an exact volume of 100 mL (diluted in NS) how many mL of NS should be in the bag?

c) How many mL of vincristine will you need for each dose if it comes in a concentration of 1 mg/mL (vincristine dosing is capped at 2 mg/dose)?

d) How many 50 mg tablets of procarbazine should you dispense to cover this treatment cycle?

e) How many 20 mg tablets of prednisone should you dispense to cover this treatment cycle?

10) A 36 year old female patient is diagnosed with breast cancer. She weighs 106 kg and is 152 cm tall and she has elected to start a drug regimen known as AC therapy.

a) If the starting dose of doxorubicin is 50mg/m^2 and of cyclophosphamide is 25 mg/kg, what would be the appropriate dose of each drug for this patient?

b) This patient is later found to have signs of metastatic disease and decides to start Herceptin therapy. The initial dosing for Herceptin is 4mg/kg infused over 90 minutes. Subsequent dosing is 2mg/kg infused over 30 minutes. What dose of Herceptin should she receive on her first day of treatment? What dose should she receive weekly thereafter?

Carboplatin Dosing

Carboplatin is used for the treatment of ovarian carcinoma, small cell lung cancer and non-small cell lung cancer, squamous cell carcinoma of the head and neck, endometrial cancer, leukemia and for seminoma of testicular cancer. While carboplatin may be used either by itself or in combination with other chemotherapeutic agents for treating these diseases, its dosing is not usually based directly on body surface area or body weight. Carboplatin has its own dosing calculation referred to as the Calvert Formula (*it is noteworthy that the resulting dose is in mg and not mg/m²*).

Calvert Formula for carboplatin dosing

$$\text{carboplatin dose in mg} = \text{AUC} \times (\text{GFR} + 25)$$

AUC = area under the curve
GFR = glomerular filtration rate

The area under the curve is a measurement of how aggressive the treatment is going to be and will be determined by the oncologist. The AUC can range from 1 to 8, but is more commonly in the 4 to 6 range. GFR is a measurement of renal function and is particularly important for eliminating carboplatin from the body. Typically, creatinine clearance (CrCl) is substituted for GFR when using the Calvert Formula. A common way to calculate a patient's CrCl is with the Cockcroft and Gault equation (*it is noteworthy that if a patient's actual body weight is less than their ideal body weight you would use that instead*).

Cockcroft and Gault equation

for men

$$\text{CrCl} = \frac{(140 - \text{age}) \times \text{IBW}}{\text{Scr} \times 72}$$

for women

$$\text{CrCl} = \frac{(140 - \text{age}) \times \text{IBW}}{\text{Scr} \times 72} \times 0.85$$

IBW = ideal body weight
Scr = serum creatinine

Serum creatinine (Scr) is attainable through lab work, therefore the pharmacy will need the lab results in order to calculate this dosing. Ideal body weight can be calculated in a number of ways, but a common method to obtain an estimate is to use the Devine Formula (named for Dr. BJ Devine).

Devine Formula for an adult's ideal body weight in kg

for men

$$\text{IBW} = 50\,\text{kg} + [2.3\,\text{kg} \times (\text{height} - 60\,\text{inches})]$$

for women

$$\text{IBW} = 45.5\,\text{kg} + [2.3\,\text{kg} \times (\text{height} - 60\,\text{inches})]$$

Remember to calculate the height in inches minus 60 inches before multiplying it by 2.3 kg.

So in order to actually perform a carboplatin dosing calculation using the Calvert Formula, you will need to know the requested AUC, the lab values for the Scr, the patient's height, weight, and age. You would first calculate their IBW, then use either their IBW or actual body weight (whichever is less) to determine their CrCl. Then you will substitute GFR with CrCl to determine the actual carboplatin dose.

Let's perform an example problem with a patient to determine their carboplatin dosing using the Calvert Formula.

Example

A 5'7" 40 year old male weighing 176 pounds is ordered carboplatin with a requested AUC of 5. His lab value for serum creatinine (Scr) is 1.1 mg/dL. What will his carboplatin dosing be using the Calvert Formula?

First, we should look at his ideal body weight (IBW) compared to his actual body weight (ABW).

$$ABW = \frac{176\,\text{lb}}{1} \times \frac{1\,\text{kg}}{2.2\,\text{lb}} = 80\,\text{kg}$$

$$IBW = 50\,\text{kg} + [2.3\,\text{kg} \times (67 - 60)] = 66.1\,\text{kg}$$

Since his IBW is less we will use that to figure out his creatinine clearance (CrCl).

$$CrCl = \frac{(140 - 40) \times 66.1}{1.1 \times 72} = 83.5\,\text{mL/min}$$

Now, you'll substitute CrCl for glomerular filtration rate (GFR) in the Calvert Formula.

$$\text{carboplatin dose in mg} = 5 \times (83.5 + 25) = \mathbf{542.5\,mg\ of\ carboplatin}$$

Now attempt a practice problem using the Calvert Formula.

Practice Problem

A 5'3" 36 year old female weighing 120 pounds is ordered carboplatin with a requested AUC of 4. Her lab value for Scr is 0.8 mg/dL. What will her carboplatin dosing be using the Calvert Formula?

While the question only asked for the carboplatin dose, it may be useful to know the IBW and CrCl as well. Her IBW is 52.4 kg, her CrCl is 80.4 mL/min, and her carboplatin dose should be 422 mg.

Worksheet 23-2

Name:

Date:

Solve the following problems. You may need to refer to the nomograms in this book, or you may choose to use any of the appropriate formulas discussed in this chapter in order to attain the necessary BSA for these problems. Problems involving carboplatin may require you to look back at the Calvert Formula.

1) What is the BSA of a 6 foot tall patient that weighs 196 pounds?

2) A pediatric dose of a particular medication is 12.5 mg/m^2 bid for 7 days. Determine the following for a 4'1" tall 50 lb juvenile patient:

   a) How many milligrams of drug would be required for a single dose?

   b) How many milligrams of drug would be required for a full course of therapy?

3) An oncologist orders carboplatin with an AUC of 5 for 1 day and etoposide at 100 mg/m^2/day for 5 days for a 6'1" 35 year old male weighing 165 pounds that has small cell lung cancer. His lab value for Scr is 1.0 mg/dL.

   a) What is his actual body weight in kilograms?

   b) What is his estimated IBW using the Devine Formula?

   c) What is his CrCl using the Cockcroft and Gault equation?

   d) Substituting CrCl for GFR, what should his carboplatin dose be based on the Calvert Formula?

e) What is his BSA?

f) How many mg of etoposide should he receive for each dose?

g) How many mg of etoposide should he receive for the entire course of therapy?

4) Leukine (a drug used to increase neutrophils in patients receiving chemotherapy for Leukemia) is to be administered by IV at 250 mcg/m^2/day for 21 days. How many mcg will a patient receive each day if they are 5'10" and weigh 176 lbs?

5) An order is received in the pharmacy for 5-FU IV for a 5'4" patient weighing 125 lbs. The 5-FU solution is available in a 50 mg/ml concentration. The dosage schedule is as follows:

   Initial dose: 400 mg/m^2/day for 5 days IV push

   How many grams of 5-FU has the patient received after the first 5 days?

6) An order is received for sargramostim 250 mcg/m^2/day diluted to 50 mL with NS infused over 4 hours. The patient is a 5' 7" 150 lb woman.

   a) How many mcg of sargramostim should be administered each day?

   b) What is the final concentration of sargramostim in the IV bag in mcg/mL?

   c) Human albumin will need to be added to the IV bag prior to the sargramostim to keep the medication form being absorbed by the IV bag itself. If the human albumin in the IV bag needs to have a concentration of 0.1%, how many mg of human albumin will need to be added to the bag?

7) The pediatric dose of a particular drug is 300 mg/m^2. Calculate the dose for a 90 cm tall child weighing 20 kg.

8) An antineoplastic agent is available in a concentration of 5 mg/mL. A 4' 5" pediatric patient that weighs 59 pounds is ordered 59 mg/m^2 of this antineoplastic agent. How many mL of this drug will be required for this order?

9) A juvenile patient is to receive 240 mg/m^2 of a particular drug. The child is 142 cm in height and weighs 35 kg. If the drug is available in a concentration of 100 mg/mL, how many mL will you need to dispense for a dose?

10) A 5'2" 45 year old woman weighing 156 pounds is diagnosed with ovarian cancer. Her oncologist prescribes a 2 drug regimen of carboplatin in combination with cyclophosphamide. The physician orders carboplatin with an AUC of 4 and cyclophosphamide 600 mg/m^2 IV. The patient's Scr comes back with a lab value of 0.7 mg/dL.

   a) What is her BSA?

   b) How many mg of cyclophosphamide should she receive?

   c) What is her IBW?

   d) What is her CrCl?

   e) Based on the Calvert Formula, how many mg of carboplatin should she receive?

11) A medication order requests idarubicin 12 mg/m^2 daily for 3 days by slow (10 to 15 minute) IV injection. If the patient is 5 feet tall and weighs 120 pounds, how many mg of idarubicin should the patient receive daily?

12) A three drug regimen, known as VAD, can be used to treat multiple myeloma. It includes the following medications taken over a 28 day cycle:

vincristine 0.4 mg/day IV on days 1 through 4
doxorubicin 9 mg/m^2/day IV on days 1 through 4
dexamethasone 40 mg/day PO on days 1 through 4, days 9 through 12, and days 17 through 20

If a patient is 5' 7" and weighs 150 pounds, determine the total quantity of each drug given over the course of this treatment.

13) A patient with stage III non-Hodgkins lymphoma has elected to start a five drug regimen known as CHOP+R therapy every 3 weeks for 6 cycles. This patient is 178 cm tall and weighs 84.1 kg. On day 1 of his regimen, he arrives at the outpatient clinic where you are mixing IVs to receive chemotherapy as follows:

cyclophosphamide 750 mg/m^2 diluted in NS to have an exact volume of 500 mL
doxorubicin 50 mg/m^2 IVP
vincristine 1.4 mg/m^2 IVP (the maximum dose is capped at 2 mg)
prednisone 100 mg PO
rituximab 375mg/m^2 diluted in NS to have an exact volume of 500 mL

a) How many milligrams of cyclophosphamide will you prepare?

b) How many milligrams of doxorubicin will you prepare?

c) How many milligrams of vincristine will you prepare?

d) How many 50 mg prednisone tablets will you dispense?

e) How many milligrams of rituximab will you prepare?

14) A patient with metastatic colon cancer is given the following drug regimen:

irinotecan 125 mg/m^2 IV on days 1, 8, 15, and 22
fluorouracil 500 mg/m^2 IV on days 1, 8, 15, and 22
leucovorin 20 mg/m^2 IV on days 1, 8, 15, and 22

If the patient is 183 cm tall and weighs 78 kg, answer the following questions:

a) How many mg of irinotecan should the patient receive per dose?

b) If irinotecan has a concentration of 20 mg/mL, how many mL of irinotecan will be needed for a dose?

c) How many mg of fluorouracil should the patient receive per dose?

d) If fluorouracil has a concentration of 50 mg/mL, how many mL of fluorouracil will be needed for a dose?

e) How many mg of leucovorin should the patient receive per dose?

f) If the leucovorin the pharmacy has in stock has a concentration of 10 mg/mL, how many mL of leucovorin will be needed for a dose?

15) A 5'5" 60 year old woman weighing 145 pounds is diagnosed with non-small cell lung cancer. Her oncologist prescribes a 3 drug regimen of carboplatin in combination with paclitaxel and bevacizumab. The physician orders carboplatin with an AUC of 6, paclitaxel 200 mg/m^2 IV, and bevacizumab 15 mg/kg IV once on day 1 every 3 weeks for up to 6 cycles, then reevaluate progress. The patient's Scr comes back with a lab value of 0.9 mg/dL.

a) How many mg of bevacizumab should she receive for a single dose?

b) How many mg of paclitaxel should she receive for a single dose?

c) Based on the Calvert Formula, how many mg of carboplatin should she receive for a single dose?

Chapter 24

Drip Rates & Infusion Rates

This book has been infused with knowledge!
--This book's biased author.

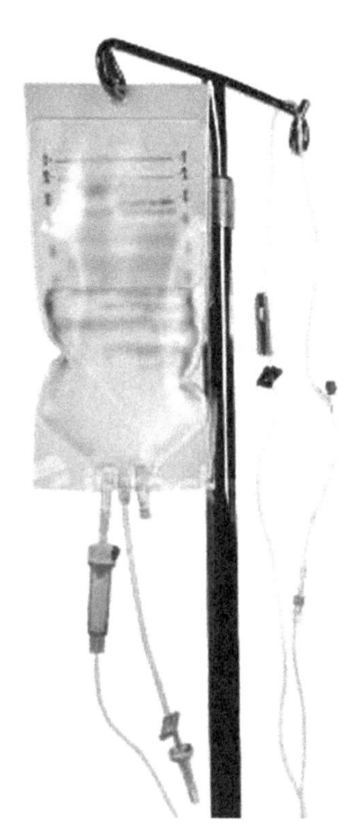

Physicians may order medications to be infused at a specific rate or over a set period of time. In other instances, institutions may have defined infusion rate protocols for a particular medication. There may even be recommendations from the drug manufacturer itself (listed in the drug literature) providing infusion rate information. Pharmacies often have the role of assisting with infusion rate calculations and verifying the appropriateness of various rates. The pharmacy will typically help determine items such as (but not limited to):

- milliliters per hour (mL/hr)
- milliliters per day (mL/day)
- drops per minute (gtt/min)
- grams per hour (g/hr)
- infusion time (minutes, hours, days)

When infusion rates were mentioned in an earlier chapter, some important concepts were mentioned that should be revisited, including a broad definition of infusion rates and the definition of the term drip rate. When discussing parenterals infusions, rates can be defined as a quantity of drug (mL, gtt, g, mcg, mEq, IU, etc.) provided over a quantity of time (minutes, hours, days, etc.).

$$\frac{\text{Quantity of Drug}}{\text{Time}} = \textbf{Infusion Rate}$$

Therefore, if you were told that a 1,000 mL bag were being infused over 8 hours, you could accurately state that the infusion rate is 125 milliliters per hour.

$$\frac{1000\,\text{mL}}{8\,\text{hr}} = \mathbf{125\,mL/hr}$$

Occasionally though, a parenteral may be timed with a venoclysis set (drip chamber) if an infusion pump is not readily available. A drip rate would then need to be calculated, and as drops are actually physically counted over a period of time, drip rates are done in drops per minute (gtt/min). Something certainly noteworthy is that a drip rate always needs to be calculated as a whole number of drops per minute, as a caregiver (most likely a nurse) will not be able to count a fraction of a drop when timing the drip. General rounding rules are applied to drip rates.

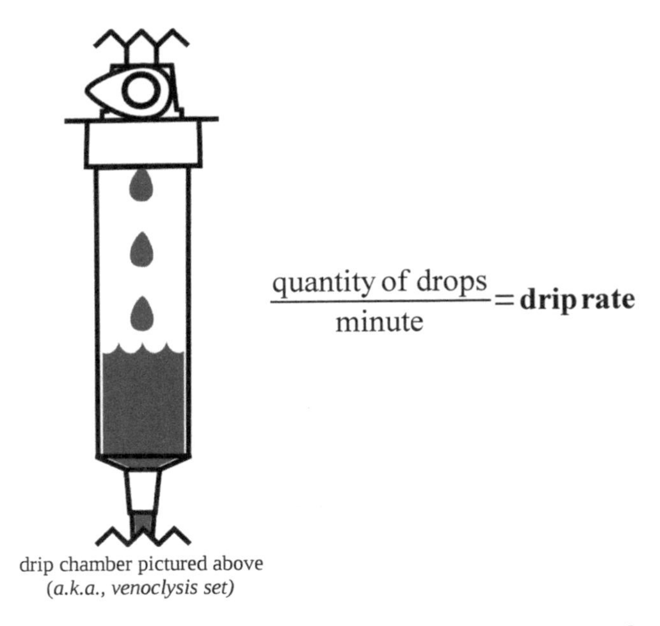

$$\frac{\text{quantity of drops}}{\text{minute}} = \textbf{drip rate}$$

drip chamber pictured above
(a.k.a., venoclysis set)

As you probably remember, you will need to know the drop factor on your tubing in order to accurately perform the necessary calculations. A drop factor is the number of drops required to add up to 1 mL. Common drop factors for standard tubing include 10, 15, and 20 drops per milliliter. Microdrip tubing sets have a drop factor of 60 drops per milliliter.

$$\frac{\#\text{ of drops}}{1\text{ mL}} = \textbf{drop factor}$$

Anything that has a drop factor can automatically be expressed as drops per milliliter (gtt/mL). Let's look at a couple of example problems involving infusion rates from various perspectives.

Examples

A 250 mL IV bag with 2 mg of isoproterenol is to be infused at a rate of 5 mcg/minute.

1) How long will it take to infuse?

$$\frac{\text{minute}}{5\text{ mcg}} \times \frac{1000\text{ mcg}}{1\text{ mg}} \times \frac{2\text{ mg}}{1} = 400\text{ minutes} = \textbf{6 hours 40 minutes}$$

2) What is its infusion rate in mL/hr?

$$\frac{250\text{ mL}}{6.67\text{ hrs}} = \textbf{37.5 mL/hr}$$

3) If the bag were infused using a venoclysis set, what would be the drip rate if the tubing had a drop factor of 20?

$$\frac{37.5\text{ mL}}{\text{hr}} \times \frac{\text{hr}}{60\text{ min}} \times \frac{20\text{ gtt}}{\text{mL}} = 12.5\text{ gtt/min} = \textbf{13 gtt/min}$$

Example problem 3 is a good reminder that you need to round your drip rate to a whole number of drops. Attempt the practice problems below.

Practice Problems

1) A medication order requests a 1 liter bag of NS to be infused over 8 hours. What is it infusion rate in milliliters per hour?

2) Ten mL of MVI, 1 mL of folic acid, 1 mL of trace elements, and 8 mL of 50% magnesium sulfate are being added to a 500 mL bag of D5W. If this bag were being infused over 5 hours via a venoclysis set with a drop factor of 15, what should the drip rate be?

3) A 1 liter IV solution was hung at 1300 on Tuesday and has an infusion rate of 75 mL/hr. When will the bag finish?

4) If a particular medication is to be infused at a rate of 50 mcg/kg/min, what would be the infusion rate for a 100 mL IV bag with 500 mg of drug if the patient receiving it weighed 180 lbs?

5) A 500 mL bag of 20% mannitol is ordered at a rate of 4 g/hr for a particular patient.

   a) What is the infusion rate in mL/hr?

   b) How long will the bag take to infuse?

1) 125 mL/hr 2) 26 gtt/min 3) 0220 on Wednesday 4) 49 mL/hr 5-a) 20 mL/hr 5-b) 25 hours

Worksheet 24-1

Name:

Date:

Solve the following problems.

1) A 50 mL IVPB with 1 g of nafcillin in NS is set to run over 30 minutes. What is its infusion rate in mL/hr?

2) A physician wants to slowly infuse a 2.5 liter TPN solution over 24 hours. What is the infusion rate in mL/hr?

3) The drip rate on a particular IV is 20 gtt/min. If the tubing set has a drop factor of 10, what is the infusion rate in mL/hr?

4) The recommended infusion rate for IV vancomycin when administered through a peripheral line is either 60 minutes or 10 mg/min, whichever is longer. What should the infusion rate be for a 200 mL premixed bag containing 1 g of vancomycin?

5) The OR intends to infuse a 6% hetastarch solution at a rate of 45 g/hr for the next two hours.

   a) What is the infusion rate in mL/hr?

   b) How many 500 mL bags of 6% hetastarch are they going to need for the full duration of the infusion?

6) A patient received a 1 liter infusion at a rate of 42 mL/hr.

   a) How long did it take to infuse?

b) If it were infused using a venoclysis set with microdrip tubing that has a drop factor of 60, what was its drip rate?

7) A 1 liter bag of D5W¼NS with 10 mEq of KCl is requested by a physician to infuse over 8 hours. What is the infusion rate?

8) An IV bag was infused on a patient for three and a half hours at an infusion rate of 50 mL/hr. How many milliliters were infused?

9) A physician orders a continuous infusion of D10W at a rate of 200 mL/hr on a patient for the next 24 hours.

a) How many 1 liter bags will you need to send to the floor for this patient?

b) If the first bag is hung at 0900 on Saturday, when will each of the other bags need to be scheduled?

10) A 500 mL solution is to be infused over 4 hours. If the administration set has a drop factor of 20, what should the drip rate be?

11) A 250 mL bag of D5W was administered at an infusion rate of 25 mL/hr. If the bag was hung at 0800 hours, when should it finish infusing?

12) A 176 pound patient with unstable angina is admitted to the hospital. The cardiologist orders a heparin bolus of 60 units/kg initially, followed by a maintenance infusion of 12 units/kg/hr. Answer the following questions if both the bolus dose and the infusion are administered from the same premixed heparin IV that has a concentration of 50 units/mL and is in a volume of 250 mL.

a) How many milliliters of heparin will be infused for the bolus?

b) What is the flow rate of the heparin bag for the maintenance dose?

c) How long will this specific bag last?(*Hint: remember to remove the appropriate volume from the bag for the bolus dose prior to determining how long the bag will last.*)

13) A physician orders a daily amphotericin B lipid complex 5 mg/kg IV for a 165 pound patient with a systemic infection. Amphotericin B lipid complex comes in 50 mg vials that have a concentration of 5 mg/mL, and when shot into an IV bag should be diluted out with D5W to a final concentration of 1 mg/mL. The recommended infusion practice is to infuse it at a rate of 2.5 mg/kg/hr.

a) How many mg of amphotericin B lipid complex should the patient receive daily?

b) How many 50 mg vials of amphotericin B lipid complex will you need for each dose?

c) How many mL of amphotericin B lipid complex and how many mL of D5W will you need to reach the final desired concentration?

d) What will the infusion rate of this bag be in mL/hr?

e) How long will it take to infuse this IV?

f) If this IV were infused with a venoclysis set that had a drip factor of 20, what would be the drip rate?

14) A medication cassette with 2 mg of epoprostenol in a total volume of 100 mL was just titrated up to a new dose of 14 ng/kg/min on a 75 kg patient.

a) How many mL/hour should the infusion pump be set for?

b) How many mL are being infused each day?

15) A nephrologist orders continuous venous to venous hemodialysis (CVVHD) on a patient at a rate of 1500 mL/hr. You should be able to get the first bag to the floor by noon. The dialysis bags that you are preparing are 5 liters each. Your shift that day is 0630 to 1500, and you are to make enough bags to get the patient four hours past the end of your shift.

a) How many bags should you send initially?

b) Then, at 1400 the nephrologist decides to increase the rate to 2000 mL/hr. How many bags should be left at 1400 from what you originally prepared?

c) How many additional bags will you need to prepare?

Worksheet 24-2

Name:

Date:

Solve the following problems.

1) A 50 mL IVPB is set to run over 20 minutes. What is its infusion rate in mL/hr?

2) A 50 mL minibag with 1.5 grams of an antibiotic is set to be infused over 15 minutes.

   a) What is its infusion rate in mL/hr?

   b) If your infusion set had a drop factor of 10, what would the drip rate be?

3) Some institutions have a policy of infusing hyperalimentations over 16 hours. In an institution with such a policy, what would the infusion rate be for a 2.5 liter hyperalimentation?

4) The drip rate on a particular IV is 30 gtt/min. If the tubing set has a drop factor of 20, what is the infusion rate in mL/hr?

5) The recommended infusion rate for IV vancomycin when administered through a peripheral line is either 60 minutes or 10 mg/min, whichever is longer. What should the infusion rate be for a 100 mL premixed bag containing 500 mg of vancomycin?

6) Using the infusion rate guidelines from the previous problem, what should the drip rate be for a 150 mL premixed bag containing 750 mg of vancomycin if you were infusing it using a venoclysis set with a drop factor of 10?

7) A patient received a 1 liter infusion at a rate of 120 mL/hr.

   a) How long did it take to infuse?

   b) If it were infused using a venoclysis set with a drop factor of 15, what should the drip rate have been?

8) A 1 liter bag of D5W¼NS with 20 mEq of KCl is requested by a physician to infuse over 8 hours. How many mEq of potassium chloride is the patient receiving each hour?

9) An IV bag was infused on a patient for three and a half hours at an infusion rate of 80 mL/hr. How many milliliters were infused?

10) A 500 mL bottle containing 18 g of tromethamine is to be infused on a 154 lb patient at the maximum allowable rate of 500 mg/kg/hr. What will the infusion rate be in mL/hr?

11) A patient is receiving a continuous infusion of NS at 125 mL/hr. If a patient receives a 1 gram dose of cefazolin in a 100 mL IVPB every 12 hours, how long will each dose take to infuse if it uses the Y-site on the continuous infusion's tubing? (*Hint: under these circumstances the IVPB will infuse at the same rate as what the continuous infusion is running.*)

12) A physician orders a 24 hour continuous infusion of LR solution running at 125 mL/hr starting at 1100 on day 1.

    a) How many bags will the patient need over a 24 hour period if the bags are 1 liter in size?

    b) What time should each bag be hung?

c) If the bags are being infused with tubing that has a drop factor of 10, what should the drip rate be?

13) A physician orders a 24 hour continuous infusion of D5W½NS running at 60 mL/hr starting at 0900 on day 1.

a) How many bags will the patient need over the 24 hour period if the bags are 1 liter in size?

b) What time should each bag be hung?

c) If the bags are being infused with tubing that has a drop factor of 20 gtt/mL, what should the drip rate be?

14) A solution containing 500,000 units of polymixin B is added to a 250 mL bag of D5W. The final volume is 260 mL and is to be infused over 2 hours.

a) What is the infusion rate in mL/hr?

b) If the infusion set is calibrated to have a drop factor of 10, what should the drip rate be?

15) If a patient received a 500 mL transfusion of whole blood starting at 2113 at a drip rate of 30 gtt/min via an administration set with a drop factor of 10, when should the transfusion finish?

16) You are working in the pharmacy when a nurse calls and asks how many mL/day she should set her CAD pump for to infuse her patient's troprostenil? The nurse says the physician ordered 104 ng/kg/min. You pull up the patient profile and see that the pharmacy sent a troprostenil cassette with 20 mg of troprostenil and 98 mL of NS for a total volume of 100 mL, and you confirm the patient's weight of 58 kg that you have on file is correct. What is the infusion rate in mL/day?

17) A 154 pound patient is to receive 2000 mg/kg/dose of immune globulin every 2 months. The immune globulin solution in the pharmacy has a 5% concentration.

a) How many grams of immune globulin should the patient receive for a dose?

b) How many milliliters of 5% solution will be required for each dose?

c) During the first 30 minutes of infusion, the IV should run at a rate of 30 mg/kg/min. If well tolerated, the rate may be increased to 60 mg/kg/min for the next 30 minutes. If the infusion is still well tolerated, the rate may then be increased to 90 mg/kg/min. After another 30 minutes, if the infusion is still well tolerated, the rate may be increased to a maximum of 200 mg/kg/min. What is the least amount of time that this immune globulin IV would take to infuse?

18) A nephrologist orders continuous venous to venous hemodialysis (CVVHD) on a patient at a rate of 2000 mL/hr. You should be able to get the first bag to the floor by 1000. The dialysis bags that you are preparing are 5 liters each. Your shift that day is 0630 to 1500, and you are to make enough bags to get the patient four hours past the end of your shift.

a) How many bags should you send initially?

b) Then, at 1230 the nephrologist decides to increase the rate to 2500 mL/hr. How many bags should be left at 1230 from what you originally prepared?

c) How many additional bags will you need to prepare?

Chapter 25

Dilutions & Alligations

"Dilution is the solution to pollution."
--A misguided statement from the EPA in the 70s.

In a previous chapter, the concepts of dilutions, alligations, and various mixtures were introduced along with problem solving methods with simple names like the dilution formula, the alligation method and ratio-proportions. Previously you were either given problems that could have been solved with any of the aforementioned ways, or if they could not, you were instructed as to which method to use. This chapter is going to review those concepts, but also help you identify which ways each problem could be solved. So, we will be reviewing and/or learning the following concepts in this chapter:

- the dilution formula,
- the alligation method,
- a series of ratio-proportions, and
- determining which method(s) to use.

Let's begin with some simple definitions for dilution, alligation, and mixture with respect to healthcare.

dilution – Mixing **a** substance (such as a medication) with **a** diluent/solvent to decrease its concentration is considered a dilution. As this method is intended only for when mixing 2 items, one with a particular ingredient, and the other without the same ingredient, this can always be solved with either the dilution formula or as a series of ratio-proportions. If the starting concentration of the ingredient is known as well as its final (or desired) concentration, then you can use the alligation method also because the diluent/solvent has a concentration of 0%.

alligation – An alligation is when **two** items with different concentrations of the same substance are being mixed together, and the final (or desired) concentration (which must be in-between the other two concentrations) is known. The alligation method may be used for this kind of problem. If one of the starting ingredients is actually a diluent/solvent with a concentration of 0%, then you may also solve this with either the dilution formula or as a series of ratio-proportions.

mixture – A mixture involves mixing **2 or more** ingredients of varying concentrations together. If the mixture only contains one starting ingredient with active substance in it, it may be treated as a dilution. If it has two initial components containing active ingredient, and you know what the final (or desired) concentration is supposed to be, you may treat it as an alligation. If it has two or more components containing the measured ingredient, and you need to solve for the final concentration, then you would use a series of ratio-proportions.

Now, we should take a moment and look at several practice problems and initially determine which methods can be utilized to solve each one. After we have successfully identified problem solving methods, we will review the actual processes used to solve these problems.

Practice Problems

When looking at the problems below, only determine which method(s) (alligation method, dilution formula, and a series of ratio-proportions) would function for solving them. You do not need to solve them.......yet.

1) You have 50 mL of a 15% solution. You dilute the solution to 150 mL. What is the final percentage strength of this solution?

2) Available in stock is a 70% dextrose solution and a 40% dextrose solution. How many mL of each stock solution would be needed to prepare a liter of 45% dextrose?

3) On hand in the pharmacy are three different percentage strength solutions; 200 mL of 80% solution, 100 mL of 40% solution, and 50 mL of 20% solution. If all three solutions are mixed together, what would be the final volume and the final concentration?

4) Determine how many mL of a 14.6% sodium chloride solution and how many mL of SWFI need to be mixed together to make 500 mL of ¼NS (0.225% sodium chloride).

1) This problem could be solved with either the dilution formula or with a series of ratio proportions.
2) This problem could be solved with the alligation method.
3) This problem could be solved with a series of ratio-proportions.
4) This problem could be solved with the alligation method, the dilution formula, or with a series of ratio proportions.

Identifying problem solving methods is something that improves with proper practice. Now let's solve each of these problems using the various methodologies discussed. This will provide a good opportunity to reacquaint yourself with the actual problem solving methods themselves.

Examples

1) You have 50 mL of a 15% solution. You dilute the solution to 150 mL. What is the final percentage strength of this solution?

   First, let's demonstrate this using a series of ratio-proportions:

$$\frac{15\,g}{100\,mL}=\frac{N}{50\,mL}$$
$$N=7.5\,g$$

$$\frac{7.5\,g}{150\,mL}=\frac{N}{100\,mL}$$
$$N=\mathbf{5\,\%}$$

Now let's solve this with the dilution formula:

The dilution formula is $C_1Q_1 = C_2Q_2$

C_1 = stock concentration is 15%
Q_1 = stock quantity is 50 mL
C_2 = final concentration is ???
Q_2 = final quantity is 150 mL

$$(15\%)(50\,mL)=(C_2)(150\,mL)$$

$$\frac{(15\%)(50\,mL)}{150\,mL}=\frac{(C_2)(\cancel{150\,mL})}{\cancel{150\,mL}}$$

$$C_2=\mathbf{5\,\%}$$

2) Available in stock is a 70% dextrose solution and a 40% dextrose solution. How many mL of each stock solution would be needed to prepare a liter of 45% dextrose?

We will solve this problem using the alligation method.

high concentration = 70%
desired concentration = 45%
low concentration = 40%
final quantity = 1000 mL

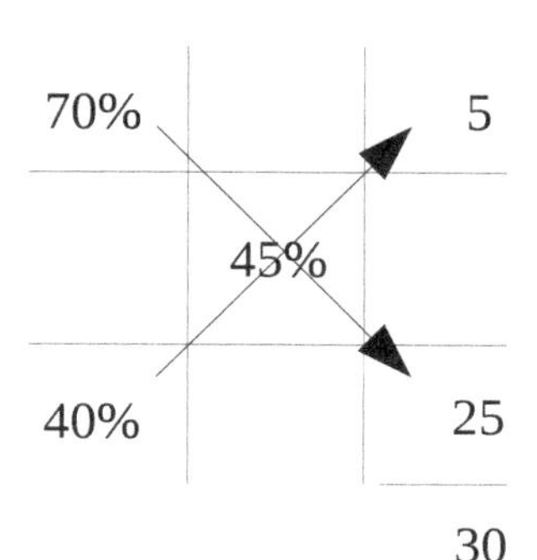

$$\frac{5}{30}\times 1000\,mL=\mathbf{167\,mL\ of\ 70\,\%\ solution.}$$

$$\frac{25}{30}\times 1000\,mL=\mathbf{833\,mL\ of\ 40\,\%\ solution.}$$

3) On hand in the pharmacy are three different percentage strength solutions; 200 mL of 80% solution, 100 mL of 40% solution, and 50 mL of 20% solution. If all three solutions are mixed together, what would be the final volume and the final concentration?

We will solve this problem using a series of ratio-proportions.

$$\frac{A}{200\,mL}=\frac{80\,g}{100\,mL} \qquad \frac{B}{100\,mL}=\frac{40\,g}{100\,mL} \qquad \frac{C}{50\,mL}=\frac{20\,g}{100\,mL}$$
$$A=160\,g \qquad B=40\,g \qquad C=10\,g$$

$$160\,g + 40\,g + 10\,g = 210\,g$$

$$200\,mL + 100\,mL + 50\,mL = 350\,mL$$

$$\frac{210\,g}{350\,mL} = \frac{N}{100\,mL}$$

$$N = \mathbf{60\,\%}$$

Practice Problem

The following problem we've already determined can be solved using all three methods. Attempt to solve it using the alligation method, the dilution formula, and as a series of ratio-proportions.

Determine how many mL of a 14.6% sodium chloride solution and how many mL of SWFI need to be mixed together to make 500 mL of ¼NS (0.225% sodium chloride).

7.7 mL of 14.6% sodium chloride solution, 492.3 mL of SWFI

Worksheet 25-1

Name:

Date:

Determine which method(s) (alligation method, dilution formula, and a series of ratio-proportions) would function for solving each scenario and then choose one of those methods to answer each question.

1) If 250 mL of a 12% solution is diluted to 1 L, what will be the percentage strength?

2) If 15 g of a 5% topical ointment is mixed with 135 g of ointment base, what will be the final concentration? (*Hint: Don't forget to add the weights of both ointments together when determining the final concentration.*)

3) If 500 mL of a 20% solution is diluted to 2 L, what will be the final percentage strength?

4) If 100 mL of a 1:1500 methylbenzethonium chloride lotion is diluted with an equal volume of water, what will be the ratio strength of this dilution? (*Hint: Although you could convert your ratio strengths to percentages, it is not necessary for solving this problem.*)

5) The pharmacy has on hand a 100 mL container of 23.4% concentrated sodium chloride solution. How many mL of half-normal saline could be made and how many mL of diluent would be required? (*Hint: Remember that normal saline is 0.9% sodium chloride solution, so half-normal saline would be 0.45% sodium chloride solution.*)

6) A request is made to the pharmacy for a pint of 30% isopropyl alcohol solution to soak sponges in. How many mL of 70% isopropyl alcohol and how many mL of water will be required to prepare this dilution?

7) A physician prescribes an otic suspension to contain 100 mg of cortisone acetate in a final volume of 8 mL. On hand in the pharmacy is a 2.5% cortisone acetate suspension. How many mL of the stock solution are required to make this and how many mL of NS will you use as a diluent? (*Hint: Normal saline does not contain any of the active ingredient in it.*)

8) What is the percentage strength of alcohol in a mixture containing 150 mL of witch hazel (14% alcohol), 200 mL glycerin, and 500 mL of 50% alcohol? (*Hint: Glycerin does not contain any ethyl alcohol.*)

9) A technician is to compound a hyperalimentation with 500 mL of 5.2% amino acid, 250 mL of 70% dextrose, 72.4 mL of various micronutrients, and 177.6 mL of sterile water. What will the final concentration of the amino acid and the dextrose be? (*Hint: The amino acids and the dextrose are different active ingredients.*)

10) You are to prepare 1 L of a 1:50 solution from 1:10 and 1:1000 stock solutions. How many mL of each of the stock solutions do you need? (Hint: You may want to convert the ratio strengths into percentages to simplify your work.)

11) Available in the pharmacy is 946 mL of a 10% povidone-iodine wash concentrate. The physician orders a diluted wash. If a pint of 1% povidone-iodine wash is ordered, how many mL of the 10% solution will be needed?

12) If the pharmacy technician is to prepare 2 liters of a 6% betadine solution, what volume of a 10% solution would be used if it is mixed with distilled water?

13) On hand in the pharmacy are three different percentage strength solutions; 60 mL of 75% solution, 26 mL of 15% solution, and 14 mL of 12.5% solution. If all three solutions are mixed together, what would be the final volume and the final concentration?

14) An order for a parenteral nutrition requests 60 mL of micronutrients, a final concentration of 4% amino acids, a final concentration of 5% dextrose, qs with sterile water to 1000 mL. If your stock solutions are 8.5% amino acid and 70% dextrose, how many mL of each ingredient (micronutrients, amino acids stock solution, dextrose stock solution, and sterile water) will you need?

15) What is the percentage strength of alcohol in a preparation if it contains 520 mL of a 95% solution, 1 L of a 70 % solution, and a pint of a 50% solution?

Worksheet 25-2

Name:

Date:

Answer the following problems appropriately.

1) Define the term dilution with respect to pharmacy math.

2) Define the term alligation with respect to pharmacy math.

3) Define the term mixture with respect to pharmacy math.

4) A 500 mL bag of 3% sodium chloride is requested.

    a) How many milliliters of a 23.4% stock solution of sodium chloride would be required to prepare this?

    b) How many mL of sterile water for injection (SWFI) will be required to make this bag?

5) How many mL of a 2% stock solution should be used to prepare 500 mL of a 0.025% solution?

6) How many mL of 95% ethyl alcohol and how many mL of NS will be needed to prepare 3 L of 50% ethyl alcohol?

7) The OR requests 3.9 L of 0.1% quaternary ammonium compound to cleanse their surgical instruments. If you have on hand a 6.5% concentrated quaternary ammonium compound, how many mL of this and how many mL of sterile water will you need to prepare the requested solution?

8) On hand in the pharmacy are several different percentage strength solutions; 50 mL of 75% solution, 115 mL of a 50% solution, 25 mL of 15% solution, and 10 mL of 12.5% solution.

   a) If all of these solutions are mixed together what would be the final volume?

   b) What will be the final concentration of this mixture?

9) An old pharmacy recipe is supposed to make a pint of a 117 alcohol proof spirit (2 proof = 1%). If 95% alcohol is used to prepare this, how many mL of 95% alcohol will be required?

10) An order is for 60 g of a 0.5% ointment. On hand is 454 g of a 10% ointment. The 10% ointment must be diluted with petrolatum. How many grams of each ingredient must be used?

11) One mL of a 1:5000 isoproterenol HCl solution is diluted to 10 mL with NS.

   a) What is the percentage strength of the final product?

   b) What is its ratio strength?

   c) What is its concentration in mcg/mL?

12) A compounded medication requires that two syrups be mixed together in equal parts dissolving a medication that has negligible powder volume. The final suspension contains 2 ounces of a syrup that has an 85% w/v sucrose content, and 2 ounces of another syrup that has a 60% w/v sucrose content.

a) What is the final percentage strength of sucrose in this suspension?

b) How many grams of sucrose would be in each tablespoon of this suspension?

13) Respiratory requests 3 mL of a 3% hypertonic sodium chloride solution intended for inhalation. If you decide to use a 14.6% sodium chloride stock solution and normal saline, how many mL of each will you need to prepare the requested solution?

14) A physician orders 250 mL of a 1:200 solution of potassium permanganate ($KmnO_4$) to be used as a foot wash on a patient. In stock the pharmacy has a 5% $KMnO_4$ solution. How many mL of stock solution and how many mL of diluent (water) will be needed to prepare the requested order?

15) You want to make 120 mL of 50% IPA solution, and you have on hand 70% IPA and purified water. How many mL of each do you need?

16) You receive a request for 6 mL of a 0.2% tobramycin ophthalmic solution. The pharmacist asks you to prepare it using 0.3% tobramycin ophthalmic solution and NS. How many mL of each do you need?

17) You receive another request for a tobramycin ophthalmic solution, but this time it needs to be fortified. The order wants 6 mL of a 1.5% tobramycin ophthalmic solution. The pharmacist asks you to use a 0.3% tobramycin ophthalmic solution and 80 mg/2 mL tobramycin injection to make this preparation. How many mL of each do you need?

18) A physician orders a vial that will be used for IM injections with 10 mL of 200 mg/mL edetate and 3.3 mL of 1% tetracaine.

a) What is the final concentration of edetate in mg/mL?

b) What is the final percentage strength of tetracaine?

19) A nurse wants half-strength Dakin's solution (0.25% sodium hypochlorite) for dressing changes on a patient. The pharmacy agrees to send a liter of it to the unit for the nurse to use. On hand in the pharmacy is a 3% sodium hypochlorite solution and sterile water for irrigation. How many mL of each will you need to prepare this?

20) A large volume parenteral ended up leaking in the refrigerator. To try and prevent excessive microbial growth, you are asked to prepare 250 mL of a 0.3% sodium hypochlorite solution that the inside of the refrigerator can be wiped down with. On hand you have a 3% sodium hypochlorite solution and sterile water for irrigation. How many mL of each will you need to prepare this?

21) A two liter bag of a 3-in-1 parenteral nutrition is ordered for a patient.

a) If the stock solution of amino acid has a concentration of 8.5%, and the final concentration is to be 4%, how many mL of stock solution is required?

b) If the stock solution of dextrose has a concentration of 70%, and the final concentration is to be 19%, how many mL of stock solution is required?

c) If this PN includes 250 mL of 20% lipid emulsion, what is the final concentration of lipids in this preparation?

d) If there are 127 mL of micronutrients being added to this bag and all the rest is sterile water, how many mL of sterile water will be needed?

e) If this bag is being infused over 24 hours what is the infusion rate in mL/hr?

f) If this is being infused through standard tubing with a drop factor of 20, what is the drip rate?

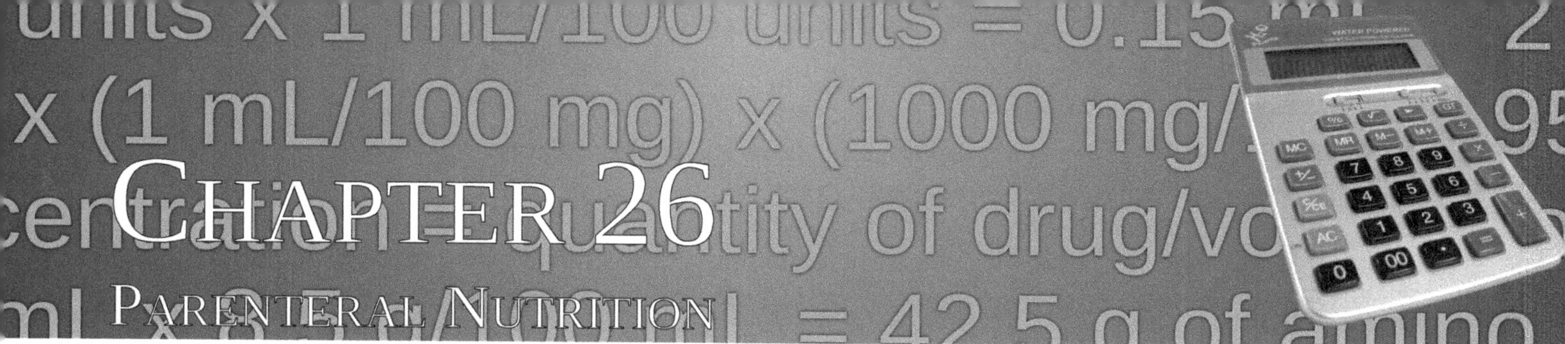

Chapter 26

Parenteral Nutrition

Perhaps we should say, "You are what's infused into you!" for this chapter.
--A horrible attempt at humor from this book's author.

A simple definition of parenteral nutrition (PN) would be a method of acquiring nutrition in a route other than through the digestive tract. It is intended for patients that are unable to obtain an adequate quantity of nutrients through their digestive tract. Indications for parenteral nutrition may include diseases and conditions associated with a nonfunctional gastrointestinal tract (bowel obstruction, severe pancreatitis, severe malabsorption), cancer therapy (radiation therapy, antineoplastic drugs, bone marrow transplantation), organ failure, hyperemesis during pregnancy, severe eating disorders (anorexia nervosa) when the patient cannot tolerate enteral nutrition, or failure when a patient attempted a trial of enteral nutrition.

This chapter will present the following concepts related to parenteral nutrition:

- various terms used to refer to parenteral nutrition,
- macronutrients and micronutrients commonly used,
- quantities and concentrations of nutrients in PN,
- precipitation concerns,
- osmolarity and implications involving infusion options,
- determining appropriate PN volume for a patient, and
- calculating calorie requirements for patients.

Commonly used terms for parenteral nutrition

Parenteral nutrition is referred to by many different names including:

- **total parenteral nutrition** (TPN) – As the name implies, it involves a patient receiving all of their nutritional needs parenterally
- **partial parenteral nutrition** (PPN) – This is when a patient is receiving part of their dietary needs through their digestive tract, but it is insufficient for all their needs, so they also receive a portion of their dietary needs parenterally
- **hyperalimentation** (HAL) or **intravenous hyperalimentation** (IVH) – This term may be defined in the same broad way as parenteral nutrition. The name itself is referring to the idea that the patient is receiving something outside of their alimentary canal.
- **total nutrient admixture** (TNA) – This is a term that could be used interchangeably with total parenteral nutrition.
- **3-in-1 admixture** or **all-in-one admixture** – This name is intended to inform the medical staff and caregivers that all three major bases (dextrose, amino acids, and lipids) along with micronutrients are included directly in the parenteral nutrition, as sometimes the lipids are infused separately from the rest of the PN.
- **centrally infused parenteral nutrition** (CPN) – Parenteral nutrition is always infused either centrally or peripherally. This name ensures that everyone knows the proper route of

administration, which is important when looking at the osmolarity.

- **peripherally infused parenteral nutrition** (PPN) – Parenteral nutrition is always infused either centrally or peripherally. This name ensures that everyone knows the proper route of administration, which is important when looking at the osmolarity. It is important to know whether the term PPN is being used to refer to a partial parenteral nutrition or a peripherally infused parenteral nutrition.

Macronutrients and micronutrients

Many people are used to looking at the nutrition labels on the foods they buy. You will notice that there are usually major items like carbohydrates, protein, and fat followed by other items like vitamins and minerals. Parenteral nutrition still functions on the same concepts but often uses different terms for the nutrients. The table below provides a comparison of these items between what we traditionally call them when dealing with a typical diet compared to the terms we'll use to prepare PN therapy.

enteral nutrition	***parenteral nutrition***	***function/purpose***
Macronutrients		
water	water	Water is the most necessary substance for life as it forms the solution base for all metabolic processes and makes up around 50 – 60% of body weight.
protein	amino acids	Amino acids form proteins and provide the major structural building blocks of the body and are needed for the daily activities of living.
carbohydrates	dextrose	Carbohydrates are the primary source of cellular energy. Dextrose, a simple sugar, provides this role in PN.
fat	lipids	Fatty acids perform important physiological functions.
Micronutrients		
vitamin	multivitamin injection (MVI)	Vitamins are important for biochemical processes within the human body. MVI typically includes fat soluble (A,D,E, and K) and water soluble (thiamine, riboflavin, niacin, pantothenic acid, pyridoxine, ascorbic acid, folic acid, B12, and biotin) vitamins.
trace elements	trace elements	These are elements that you only require small quantities of. MTE7 includes Zinc (Zn), copper (Cu), manganese (Mn), chromium (Cr), selenium (Se), iodine (I), and molybdenum (Mo). Other trace elements include Cobalt (Co), Vanadium (V), Nickel (Ni), and Flouride (F).
electrolytes	electrolytes	Electrolytes are necessary for optimal physiological fuction. Common electrolytes include sodium (na), potassium (K), chloride (Cl), magnesium (Mg), phosphate (PO_4), calcium (Ca), etc. There is a potential concern of the calcium and phosphate electrolytes forming a precipitate (discussed further later).
N/A	drug additives	While not technically a nutrient, PN will often include drug additives such as regular insulin, H2 antagonists, heparin, etc.

Quantities and concentrations of nutrients in parenteral nutrition

Now that we have introduced the most common ingredients of parenteral nutrition, let's take a moment and look at a PN that has been ordered and determine the quantities and concentrations of its nutrients.

Example

The pharmacy receives a request for a partial parenteral nutrition to be infused through a central line over 16 hours. On the left are the requested quantities/concentration and on the right are the available components.

Requested PN	*Source components*
amino acids 2.125%	8.5% amino acid solution
dextrose 20%	50% dextrose solution
sodium chloride 15 mEq	14.6% sodium chloride (2.5mEq/mL, 146 mg/mL)
potassium phosphate 15 mMol	potassium phosphate 3 mMol/mL
calcium gluconate 2.5 mEq	10% calcium gluconate (4.65 mEq/10 mL)
MVI 10 mL	MVI 10 mL vial
trace elements 1 mL	trace elements 1 mL vial
regular insulin 15 units	Humulin R U-100 (100 units/mL)
SWFI qs 1000 mL	Sterile Water for Injection

amino acids

$$\frac{2.125\,\text{g}}{100\,\text{mL}} = \frac{N}{1000\,\text{mL}}$$

$$N = 21.25\,\text{g}$$

$$\frac{21.25\,\text{g}}{N} = \frac{8.5\,\text{g}}{100\,\text{mL}}$$

$$N = \mathbf{250\,mL\ of\ 8.5\,\%\ amino\ acid}$$

dextrose

$$\frac{20\,\text{g}}{100\,\text{mL}} = \frac{N}{1000\,\text{mL}}$$

$$N = 200\,\text{g}$$

$$\frac{200\,\text{g}}{N} = \frac{50\,\text{g}}{100\,\text{mL}}$$

$$N = \mathbf{400\,mL\ of\ 50\,\%\ dextrose}$$

sodium chloride

$$\frac{15\,\text{mEq}}{1} \times \frac{\text{mL}}{2.5\,\text{mEq}} = \mathbf{6\,mL\ sodium\ chloride}$$

potassium phosphate

$$\frac{15\,\text{mM}}{1} \times \frac{\text{mL}}{3\,\text{mM}} = \mathbf{5\,mL\ potassium\ phosphate}$$

calcium gluconate

$$\frac{2.5\text{ mEq}}{1}\times\frac{10\text{ mL}}{4.65\text{ mEq}}=\mathbf{5.4\ mL\ calcium\ gluconate}$$

MVI

10 mL MVI

trace elements

1 mL trace elements

regular insulin

$$\frac{15\text{ units}}{1}\times\frac{\text{mL}}{100\text{ units}}=\mathbf{0.15\ mL\ regular\ insulin}$$

sterile water

$$\begin{array}{rl} 1000 & \text{mL} \\ -250 & \text{mL} \\ -400 & \text{mL} \\ -6 & \text{mL} \\ -5 & \text{mL} \\ -5.4 & \text{mL} \\ -10 & \text{mL} \\ -1 & \text{mL} \\ -0.15 & \text{mL} \\ \hline \mathbf{322.45} & \mathbf{mL\ sterile\ water} \end{array}$$

Calcium and phosphate solubility

Another thing to look at briefly is the solubility of phosphate and calcium in a parenteral nutrition admixture. These two ions have a strong affinity for each other and they may form a solid that precipitates out of solution. The following factors affect it including: order of mixing, pH, dextrose concentration, calcium salt form, storage temperature and time, amino acid profile, and drug additives. There is a formula that many institutions employ to estimate whether or not a solution is likely to form a precipitate.

calcium-phosphate solubility estimate

$$\frac{\left(\frac{2\text{ mEq}}{1\text{ mMol}}\times\text{phosphate mMol}\right)+\text{calcium mEq}}{\text{volume in liters}} < \mathbf{46\ mEq/liter}$$

As various factors in PN may affect the nature of the phosphates (monobasic vs dibasic) you automatically double the millimoles of phosphates when converting them to milliequivalents, treating them as all being in the dibasic form. While this formula does not claim to be perfect, it does provide a good estimate. Typically if the phosphate values and the calcium values total to less than 46 mEq per liter, you expect your PN to be stable for its intended purpose. It is also noteworthy that there are other variations on this formula along with visual charts. Familiarize yourself with the tools used wherever you practice. Let's go back and look at the PPN in the previous example and make sure that the calcium and phosphates should be stable.

Example

Using the numbers from the previous example problem, determine if the PPN should precipitate.

> There were 15 mMol of potassium phosphate and 2.5 mEq of calcium gluconate in a total volume of 1 liter.
>
> $$\frac{\left(\frac{2\text{ mEq}}{1\text{ mMol}} \times 15\text{ mMol of phosphate}\right) + 2.5\text{ mEq of calcium}}{1\text{ liter}} = \mathbf{32.5\,mEq/liter}$$
>
> It should not precipitate, as 32.5 mEq/liter is less than 46 mEq/liter.

A lot of material has already been covered in this chapter. Attempt a practice problem to help ensure that all the concepts make sense so far.

Practice Problem

A patient is to receive a 2000 mL all-in-one TPN admixture infused through a central line infused over 24 hours. On the left are the requested quantities/concentration and on the right are the available source components. Answer the following questions (the answers are on the next page):

a) What is the infusion rate in mL/hr?
b) Determine the quantity of each ingredient required.
c) Using the information here determine whether or not the calcium and phosphate is likely to precipitate.

Requested PN	*Source Components*
4% amino acid	8.5% amino acid solution
19% dextrose	70% dextrose solution
250 mL of 20% lipid emulsion	250 mL of 20% lipid emulsion
sodium chloride 100 mEq	14.6% sodium chloride (2.5mEq/mL, 146 mg/mL)
potassium acetate 80 mEq	potassium acetate 2 mEq/mL
calcium gluconate 9.4 mEq	10% calcium gluconate (4.65 mEq/10 mL)
magnesium sulfate 16 mEq	50% magnesium sulfate (4.06 mEq/mL)
sodium phosphate 30 mMol	sodium phosphate 3 mMol/mL
MVI 10 mL	MVI 10 mL vial
trace elements 1 mL	trace elements 1 mL
famotidine 40 mg	famotidine 20 mg/2 mL
SWFI qs 2000 mL	Sterile Water for Injection

a) 83.3 mL/hr b) 941 mL of 8.5% amino acid; 543 mL of 70% dextrose; 250 mL of 20% lipid emulsion; 40 mL of 14.6% sodium chloride; 40 mL of potassium acetate 2 mEq/mL; 20.2 mL of 10% calcium gluconate; 3.9 mL of 50% magnesium sulfate; 10 mL of sodium phosphate 3 mMol/mL; 10 mL of MVI; 1 mL of trace elements; 4 mL of famotidine 20 mg/2 mL; 137 mL of SWFI c) Since the calculations come out to 34.7 mEq/L, it is not likely to precipitate.

Worksheet 26-1

Name:

Date:

Solve the following problems involving parenteral nutrition using the source components listed on this page. Be aware that even though most institutional facilities offer various concentrations of amino acid bases and lipid emulsions, the orders on this worksheet will specify which ones to use.

Source Components

Macronutrients

- 10% amino acid solution
- 8.5% amino acid solution
- 5.2% amino acid solution (renal formula)
- 70% dextrose
- 10% lipid emulsion
- 20% lipid emulsion
- sterile water for injection

Micronutrients

- potassium chloride 2 mEq/mL
- sodium chloride 14.6% (2.5 mEq/mL)
- calcium gluconate 10% (4.65 mEq/10 mL)
- magnesium sulfate 50% (4.06 mEq/mL)
- sodium acetate 2 mEq/mL
- sodium phosphate 3 mMol/mL (4 mEq/mL)
- potassium acetate 19.6% (2 mEq/mL)
- potassium phosphate 3 mMol/mL (4.4 mEq/mL)
- multivitamin injection (MVI) – standard 10 mL
- trace elements – standard 1 mL
- vitamin C 250 mg/2 mL
- folic acid 5 mg/mL
- insulin regular U-100 (100 units/mL)
- famotidine 20 mg/2 mL

1) Answer the questions pertaining to the hyperalimentation ordered below:

Additive	*Ordered Quantity*	*Volume*
Macronutrients		
Amino Acid 10%	1000 mL	
Dextrose 70%	1000 mL	
Sterile Water for Injection	qs 2500 mL	
Micronutrients		
sodium chloride	50 mEq	
potassium chloride	60 mEq	
magnesium sulfate	24 mEq	
potassium phosphate	30 mMol	
calcium gluconate	46.5 mEq	
MVI	10 mL	
trace elements	1 mL	
vitamin C	250 mg	
regular insulin	55 units	
folic acid	5 mg	
sodium acetate	80 mEq	
potassium acetate	28 mEq	
sodium phosphate	6 mMol	

a) What is the appropriate volume of each ingredient? List them on the chart above.

b) If this is to be infused over a 24 hour period, what is the infusion rate?

c) What are the final concentrations of amino acid and dextrose?

d) Based on the calcium-phosphate solubility estimate, is this HAL likely to precipitate?

2) Answer the questions pertaining to the parenteral nutrition ordered below:

Additive	*Ordered Quantity*	*Volume*
Macronutrients		
Amino Acid 5.2%	800 mL	
Dextrose 70%	700 mL	
Sterile Water for Injection	qs 2000 mL	
Micronutrients		
sodium chloride	24 mEq	
potassium chloride	40 mEq	
magnesium sulfate	40 mEq	
potassium phosphate	17.6 mEq	
calcium gluconate	1.4 g	
MVI	10 mL	
trace elements	1 mL	
vitamin C	500 mg	
regular insulin	95 units	
folic acid	7.5 mg	
sodium acetate	70 mEq	
potassium acetate	40 mEq	
sodium phosphate	40 mEq	

a) What is the appropriate volume of each ingredient? List them on the chart above.

b) If this is to be infused over a 24 hour period, what is the infusion rate?

c) What are the final concentrations of amino acid and dextrose?

d) Based on the calcium-phosphate solubility estimate, is this PN likely to precipitate? (*Hint: Be careful to convert both of your phosphates into mMol prior to plugging them into the formula.*)

3) Answer the questions pertaining to the parenteral nutrition admixture ordered below:

Additive	*Ordered Quantity*	*Volume*
Macronutrients		
Amino Acid 8.5%	1000 mL	
Dextrose 70%	1000 mL	
Sterile Water for Injection	qs 3000 mL	
Micronutrients		
sodium chloride	60 mEq	
potassium chloride	40 mEq	
magnesium sulfate	4 g	
potassium phosphate	30 mMol	
calcium gluconate	5 g	
MVI	10 mL	
trace elements	1 mL	
vitamin C	250 mg	
regular insulin	40 units	
folic acid	2 mg	
sodium acetate	60 mEq	
potassium acetate	28 mEq	
sodium phosphate	15 mMol	

a) What is the appropriate volume of each ingredient? List them on the chart above.

b) If this is to be infused over a 24 hour period, what is the infusion rate?

c) What are the final concentrations of amino acid and dextrose?

d) Based on the calcium-phosphate solubility estimate, is this parenteral nutrition admixture likely to precipitate?

Osmolarity of parenteral nutrition

Osmolarity is the measure of osmoles of solute per liter of solution (osmol/L). Osmoles are only looking at the moles of chemical compound that contribute to the osmotic pressure of a solution. When looking at the constituents of parenteral nutrition, dextrose and amino acids are the primary contributors to osmolarity. Lipids do not contribute to the solution's osmolarity, but electrolytes do. You can usually use the concentrations of dextrose and amino acid to estimate the osmolarity of PN. When looking at IV infusions, we traditionally measure the osmolarity in milliosmoles per liter (mOsmol/L).

Extracellular fluid generally has an osmolality of approximately 285-295 mOsmol/kg[1]. A challenge is that parenteral nutrition has a much higher concentration making it hypertonic and therefore potentially irritating to the veins it is being infused through, this despite its slow infusion rate. While institutions may have varying policies, a common practice is if a PN has an osmolarity less than a 1000 mOsmol/L it can be infused peripherally, versus an osmolarity of 1000 mOsmol/L or more is infused through a central line (an IV line that feeds into the superior vena cava). You can quickly estimate the osmolarity of PN by multiplying the grams of dextrose per liter by 5 and adding it to the grams of amino acid per liter multiplied by 10. Let's look at a quick example problem.

Example

If a TNA has a final concentration of 10% dextrose and 2.75% amino acid, what is the estimated osmolarity and could it be infused peripherally?

dextrose

$$\frac{10\,\text{g}}{100\,\text{mL}} = \frac{N}{1000\,\text{mL}}$$
$$N = 100\,\text{g}$$
$$\frac{100\,\text{g}}{\text{L}} \times \frac{5\,\text{mOsmol}}{\text{g}} = 500\,\text{mOsmol/L}$$

amino acid

$$\frac{2.75\,\text{g}}{100\,\text{mL}} = \frac{N}{1000\,\text{mL}}$$
$$N = 27.5\,\text{g}$$
$$\frac{27.5\,\text{g}}{\text{L}} \times \frac{10\,\text{mOsmol}}{\text{g}} = 275\,\text{mOsmol/L}$$

$$500\,\text{mOsmol/L} + 275\,\text{mOsmol/L} = \mathbf{775\,mOsmol/L}$$

This TNA could be infused peripherally.

Now attempt a practice problem.

Practice Problem

If a TPN has a final concentration of 15% dextrose and 4% amino acid, what is the estimated osmolarity and could it be infused peripherally?

The estimated osmolarity is 1150 mOsmol/L, therefore this TPN could not be infused peripherally.

1 Extracellular fluids are generally measured in osmolality (mOsmol/kg) as opposed to IV fluids measuring osmolarity (mOsmol/L). The two numbers are typically very close, as a liter of water weighs 1 kilogram. The mass of the dissolved solutes makes a slight difference.

Fluid requirements

The fluid requirements for parenteral nutrition are usually calculated as either 30 mL/kg of body weight, 1500 mL per m^2 of BSA, or 1 mL/kcal of nutrition required. These are just guidelines and may be adjusted for various reasons such as dehydration or fluid restriction. Let's look at these guidelines with an example.

Example

Use all three methods for determining PN fluid volume for a 6'1" patient weighing 168 pounds (BSA of 2 m^2) if they are receiving 2400 kilocalories per day.

$$\frac{168\,\text{lb}}{1} \times \frac{1\,\text{kg}}{2.2\,\text{lb}} \times \frac{30\,\text{mL}}{\text{kg}} = \mathbf{2291\,mL}$$

$$\frac{2\,\text{m}^2}{1} \times \frac{1500\,\text{mL}}{\text{m}^2} = \mathbf{3000\,mL}$$

$$\frac{2400\,\text{kcal}}{1} \times \frac{1\,\text{mL}}{\text{kcal}} = \mathbf{2400\,mL}$$

You will notice a a range of answers using the above guidelines. It is a good practice to learn which guidelines are used at the institution where you practice.

Practice Problem

Use all three methods for determining PN fluid volume for a 5'4" patient weighing 125 pounds (BSA of 1.6 m^2) if they are receiving 1800 kilocalories per day.

1705 mL when based on weight; 2400 mL when based on BSA; 1800 mL when based on caloric intake.

Caloric requirements

The kilocalorie (kcal, or Calorie, C, or Cal) is the basic unit used for quantifying the amount of energy that is in a potential sustenance. A patient's energy requirements may vary based on their body type/size, medical condition (called stress factors), and activity level. While there are other ways to determine a patient's Calorie needs, the Harris-Benedict equation is a very common methodology taking the patient's basal energy expenditure (BEE) multiplied by an activity factor and multiplied again by a stress factor. On the next page we will look at this equation.

Harris-Benedict equation

$$BEE \times \text{activity factors} \times \text{stress factors} = \text{Total Daily Energy Expenditure}$$

Basal Energy Expenditure is determined the following way:

for males

$$66.67 + (13.75 \times \text{weight in kg}^{*}) + (5 \times \text{height in cm}) - (6.76 \times \text{age in years}) = BEE$$

for females

$$655.1 + (9.56 \times \text{weight in kg}^{*}) + (1.86 \times \text{height in cm}) - (4.68 \times \text{age in years}) = BEE$$

*If a patient is obese, their ideal body weight (IBW) will be used instead.

Activity factors:

- Confined to bed: 1.2
- Ambulatory: 1.3

Stress factors:

- Surgery 1.2
- Infection 1.4-1.6
- Trauma: 1.3-1.5
- Burns: 1.5-2.1

This provides the total daily energy expenditures for a patient, but does not account for the make-up of amino acids, dextrose, and lipids. Traditionally, you determine how many calories the patient needs from amino acids and lipids, and whatever is left is provided by the dextrose.

Amino acid requirements

Use the following guidelines to estimate the patient's amino acid needs (use IBW if obese):

- 0.8 g/kg in an unstressed patient
- 0.8 to 1 g/kg for a mildly stressed patient
- 1.2 g/kg for a renal dialysis patient
- 1.1 to 1.5 g/kg for a moderately stressed patient
- 1.5 to 2 g/kg for a severely stressed patient
- 3 g/kg for a severely burned patient

Amino acid provides 4 kcal/g, therefore 50 g of amino acid would provide 200 kilocalories.

Lipid requirements

The proportion of calories provided by lipids are in the 30 to 40% range of the patient's total caloric intake. Many people will simply split the difference and base it on 35%. Each gram of lipids provides 9 kilocalories, therefore a patient that requires 900 kilocalories of lipids would only need 100 g of lipids.

Dextrose quantity

After the kcal from both the amino acids and the lipids have been subtracted from the total daily expenditure, the rest is provided by dextrose. Each gram of dextrose given parenterally provides 3.4 kilocalories. Therefore, if a patient receiving 2800 kcal per day had 200 calories of amino acids and

900 kcal of lipids, they would need 1700 kcal of dextrose or 500 g of dextrose.

Micronutrients

A patient will ordinarily receive a standard quantity of multivitamins and trace elements every day. There are also standard quantities of electrolytes in each liter of PN (35 mEq of sodium, 30 mEq of potassium, 5 mEq of magnesium, 3 mEq of calcium, 15 mM of phosphorous, and a 1:1 ratio of acetate to chloride). The quantity of various electrolytes may be adjusted based on disease state and lab values. Physician's may choose to add additional vitamins, electrolytes, elements, and drug additives to the bag.

Let's do a practice problem to emphasize all these ideas. This problem will be broken into many steps so you can review each concept.

Practice Problem

The 6'1", 165 lb author of this text is severely burned (use a stress factor of 2) in a fire at the age of 35 and the medical staff at the burn center where he is hospitalized has confined him to bed and is initiating parenteral nutrition on him using an all-in-one admixture that is to be infused over 24 hours.

a) Determine his BEE.

1788.42 kcal

b) Determine his total daily energy expenditure.

4292 kcal

c) How many grams of amino acid does he require and how many kcal does that translate into?

225 g; 900 kcal

d) If the facility decides 35% of his kcal should come from lipids, how many kcal does he require and how many grams of lipids does that translate into?

166.9 g; 1502 kcal

e) How many kcal should he receive from dextrose and how many grams does that translate into?

1890 kcal; 555.9 g

f) If they decide to base the fluid volume of his PN on 1 mL/kcal of nutrition, how many mL will this entire admixture need to be?

4292 mL

g) How many mEq or mMol of each electrolyte will the patient require if his electrolytes are ordered as follows: sodium chloride 35 mEq/L , potassium acetate 20 mEq/L , calcium gluconate 4.5 mEq/L , potassium phosphate 15 mMol/L , and magnesium sulfate 5 mEq/L?

sodium chloride 150.2 mEq; potassium acetate 85.8 mEq; calcium gluconate 19.3 mEq; potassium phosphate 64.4 mMol; magnesium sulfate 21.5 mEq

h) Using the concentrations of the source components you've already calculated along with a vial of MVI and a vial of trace elements, fill in the ordered quantities and determine the appropriate volume of each ingredient to make this PN.

Source Components	*Fill in the Ordered Quantity*	*Volume Required*
Macronutrients		
Amino Acid 10%		
Dextrose 70%		
Lipid Emulsion 20%		
Sterile Water for Injection		
Micronutrients		
sodium chloride 23.4% (4 mEq/mL)		
potassium acetate 2 mEq/mL		
calcium gluconate 4.65 mEq/10 mL		
potassium phosphate 3 mMol/mL		
magnesium sulfate 50% (4.06 mEq/mL)		
MVI 10 mL/vial		
trace elements 1 mL/vial		

This is what the chart on the previous page should look like when completed.

Source Components	*Fill in the Ordered Quantity*	*Volume Required*
Macronutrients		
Amino Acid 10%	225 g	2250 mL
Dextrose 70%	555.9 g	794.1 mL
Lipid Emulsion 20%	166.9 g	834.5 mL
Sterile Water for Injection	q.s. 4292 mL	253.6 mL
Micronutrients		
sodium chloride 23.4% (4 mEq/mL)	150.2 mEq	37.6 mL
potassium acetate 2 mEq/mL	85.8 mEq	42.9 mL
calcium gluconate 4.65 mEq/10 mL	19.3 mEq	41.5 mL
potassium phosphate 3 mMol/mL	64.4 mMol	21.5 mL
magnesium sulfate 50% (4.06 mEq/mL)	21.5 mEq	5.3 mL
MVI 10 mL/vial	1 vial	10 mL
trace elements 1 mL/vial	1 vial	1 mL

i) What will the infusion rate be in mL/hr?

178.8 mL/hr

j) If the tubing being used has a drop factor of 20, what will the drip rate be?

60 gtt/min

k) Use the calcium-phosphate solubility formula to check whether or not this solution is likely to precipitate.

The formula works out to 34.5 mEq/L; therefore it is not likely to precipitate.

l) What are the final percentage strengths of both the dextrose and the amino acid of this admixture?

13% dextrose; 5.2% amino acid

m) Based on the estimated osmolarity of this bag, could it be infused peripherally?

The estimated osmolarity is 1170 mOsmol/L; therefore this bag must be infused centrally.

Worksheet 26-2

Name:

Date:

Review the patient case and determine the correct quantities for preparing parenteral nutrition for the patient. Your answers should be based on the Harris-Benedict Equation.

Pt Case: An obese patient with a GI obstruction due to complications from a gastric bypass has developed severe pneumonia, and her care team has decided she should be placed on parenteral nutrition using a 3-in-1 admixture as opposed to a starvation diet. The patient is a 5'6", 290 lbs, 40 year old female and is to continue to be ambulatory. She does not have an allergy to eggs, as an allergy to eggs would impact her ability to receive the lipid emulsion.

1) Since this patient is obese, you will need to use her ideal body weight (IBW) for determining her BEE and her amino acid requirements. Use the Devine formula for determining her IBW (it is on page 531 if you need to look it up).

2) Using her IBW, calculate her basal energy expenditure.

3) Determine her total daily energy expenditure now that she is ambulatory after her recent surgery, but has has developed pneumonia. With these factors in mind, treat her as having a stress factor of 1.6.

4) How many grams of amino acid does she require, and how many kcal does that translate into remembering to use her IBW? She is considered a moderately stressed patient, so her amino acids should be calculated as 1.3 g/kg.

5) If the facility decides 30% of her kcal should come from lipids, how many kcal does she require and how many grams of lipids does that translate into?

6) How many kcal should she receive from dextrose and how many grams does that translate into?

7) What is her BSA? Remember to use her actual body weight. You may need to refer to the nomogram on page 523. (Your answer may vary depending on method used, but for consistency, with the answer key, use 2.34 m^2.)

8) If they decide to base the fluid volume of her PN on 1500 mL per m^2 of BSA, how many mL will this entire admixture need to be?

9) How many mEq or mMol of each electrolyte will the patient require if her electrolytes are ordered as follows (according to her lab values her sodium was a little high and her potassium was a little low): sodium chloride 25 mEq/L , potassium chloride 10 mEq/L, potassium acetate 20 mEq/L , calcium gluconate 4.5 mEq/L , potassium phosphate 15 mMol/L , and magnesium sulfate 5 mEq/L?

10) Using the concentrations of the source components you've already calculated, along with 5 mg of folic acid, a vial of MVI, a vial of trace elements, and 40 mg of famotidine, fill in the ordered quantities and determine the appropriate volume of each ingredient on the chart on the next page.

Source Components	*Fill in the Ordered Quantity*	*Volume Required*
Macronutrients		
Amino Acid 8.5%		
Dextrose 70%		
Lipid Emulsion 20%		
Sterile Water for Injection		
Micronutrients		
sodium chloride 23.4% (4 mEq/mL)		
potassium chloride 2 mEq/mL		
potassium acetate 2 mEq/mL		
calcium gluconate 4.65 mEq/10 mL		
potassium phosphate 3 mMol/mL		
magnesium sulfate 50% (4.06 mEq/mL)		
folic acid 5 mg/mL		
MVI 10 mL/vial		
trace elements 1 mL/vial		
famotidine 20 mg/2 mL		

11) What will the infusion rate be in mL/hr if the facility wants to infuse it over 24 hours?

12) If the tubing being used has a drop factor of 20, what will the drip rate be?

13) Use the calcium-phosphate solubility formula to check whether or not this solution is likely to precipitate.

14) What are the final percentage strengths of both the dextrose and the amino acid of this admixture?

15) Based on the estimated osmolarity of this bag, could it be infused peripherally?

Worksheet 26-3

Name:

Date:

Answer/solve the following questions/problems.

1) What are the various names and corresponding acronyms for parenteral nutrition?

2) Are lipids always mixed directly into a parenteral nutrition admixture?

3) List several reasons for placing a patient on parenteral nutrition.

4) What are the macronutrients used in parenteral nutrition and what are the corresponding designations of a traditional enteral diet?

5) Which two ions commonly used in parenteral nutrition raise a concern about precipitation?

6) Is PN hypertonic, isotonic, or hypotonic?

7) What is the common cut off for when PN can be infused peripherally or centrally?

8) Pt. Case: A patient's estimated nutritional requirements have been assessed at approximately 95-105 grams protien/day and 1800-2100 nonprotien kcal/day. The patient has no history of hyperlipidemia or allergy to eggs and is not fluid restricted. The PN solution will be compounded as an individualized regimen using a single bag, 24-hour infusion of amino acid/dextrose/intravenous lipid emulsion combination. All lab values for vitamins and electrolytes are within normal ranges.

a) Determine the appropriate volume for each ingredient.

Source Components	*Ordered Quantity*	*Volume Required*
Macronutrients		
Amino Acid 8.5%	4% final conc.	
Dextrose 70%	19% final conc.	
Lipid Emulsion 10%	250 mL	
Sterile Water for Injection	q.s. 2000 mL	
Micronutrients		
sodium chloride 14.6% (2.5 mEq/mL)	50 mEq/L	
potassium acetate 2 mEq/mL	40 mEq/L	
calcium gluconate 4.65 mEq/10 mL	4.7 mEq/L	
sodium phosphate 3 mMol/mL	15 mMol/L	
magnesium sulfate 50% (4.06 mEq/mL)	8 mEq/L	
MVI 10 mL/vial	1 vial	
trace elements 1 mL/vial	1 vial	
famotidine 20 mg/2 mL	40 mg	

b) What is the appropriate infusion rate in mL/hr?

c) Are the calcium and phosphate ions likely to precipitate?

d) Could this bag be infused peripherally?

9) Pt Case: Patient's first attempt at enteral feeding has failed as patient is unable to attain a sufficient quantity of nutrition through their GI tract. Physician has decided to place patient on a PPN during the night.

a) Determine the appropriate volume for each ingredient.

Source Components	*Ordered Quantity*	*Volume Required*
Macronutrients		
Amino Acid 8.5%	4% final conc.	
Dextrose 70%	5% final conc.	
Sterile Water for Injection	q.s. 1000 mL	
Micronutrients		
sodium chloride 14.6% (2.5 mEq/mL)	50 mEq/L	
potassium acetate 2 mEq/mL	40 mEq/L	
calcium gluc. 10% (4.65 mEq/10 mL)	1 g/L	
sodium phosphate 3 mMol/mL	15 mMol/L	
magnesium sulfate 50% (4.06 mEq/mL)	2 g/L	
MVI 10 mL/vial	1 vial	
trace elements 1 mL/vial	1 vial	

b) What is the appropriate infusion rate in mL/hr if the physician requests this admixture to be infused over 8 hours?

c) Are the calcium and phosphate ions likely to precipitate?

d) Could this bag be infused peripherally?

10) Pt Case: A 58 year old female who is 5' 3" tall and weighs 140 lbs has been determined to have the following nutritional requirements for her TPN (the Harris-Benedict equation was employed to derive these numbers).

a) Determine the appropriate volume for each ingredient.

Source Components	*Ordered Quantity*	*Volume Required*
Macronutrients		
Amino Acid 8.5%	50.91 g	
Dextrose 70%	186.65 g	
Lipid Emulsion 10%	41.03 g	
Sterile Water for Injection	q.s. ad 30 mL/kg	
Micronutrients		
sodium chloride 14.6% (2.5 mEq/mL)	35 mEq/L	
potassium acetate 2 mEq/mL	35 mEq/L	
calcium gluconate 4.65 mEq/10 mL	9.6 mEq/L	
potassium phosphate 3 mMol/mL	6.7 mMol/L	
magnesium sulfate 50% (4.06 mEq/mL)	8 mEq/L	
folic acid 5 mg/mL	1.7 mg	
MVI 10 mL/vial	1 vial	
trace elements 1 mL/vial	1 vial	
regular insulin U-100	40 units	

b) What is the appropriate infusion rate in mL/hr if the physician requests this admixture to be infused over 24 hours?

c) Are the calcium and phosphate ions likely to precipitate?

d) Could this bag be infused peripherally?

CHAPTER 27

ALIQUOT

"Aliquot! Is that anything like a kumquat?"
--A bad joke that at least one student in every class makes.

There are times when you may need to measure a particularly small quantity of a particular medication or substance. This is a frequent scenario with pediatrics/neonates and desensitization therapies. An aliquot is a small portion of something that goes into a larger quantity a specific number of times, and therefore, the aliquot method is a method by which these small portions may be accurately obtained despite generally being below the minimum measurable quantity (MMQ) of your measuring device. The aliquot method may be used for both solids and liquids. In this chapter you will learn:

- terminology associated with aliquots,
- calculating minimum measurable quantity (MMQ),
- how to obtain quantities of liquids below your MMQ,
- how to obtain quantities of solids and semi-solids below your MMQ,
- how to calculate desensitization therapies with double-aliquots.

Terminology

Before getting started, it may be useful to learn some terminology commonly associated with aliquots.

aliquot signifying or relating to a divisor of a quantity or number. Three is an aliquot part of 12 as it can divide into 12 four times. In pharmacy math, an aliquot number is typically rounded up to the next whole number if there is a non zero value anywhere to the right of the decimal. An aliquot value of 4.001 would therefore be rounded up to 5.
aliquot method is a method of measuring ingredients below the sensitivity of a scale by proportional dilution with inactive ingredients.
sensitivity requirement is the quantity of material (weight or volume) required to move one point on an index plate on a balance or one marking on a syringe or graduated cylinder. On an electronic balance, it would be the smallest quantity that the balance responds to below which is not registered by the balance.
acceptable error is the error rate from which the measuring device may reasonably deviate according to accuracy of the device and accepted practice for what is being prepared. The acceptable error is commonly listed as a percentage (*i.e.*, 5%).
minimum measurable quantity (MMQ) is the smallest amount you can accurately measure with a given device and is determined by taking 100 percent of the sensitivity requirement and dividing it by the acceptable error. Some will also refer to this as minimum weighable quantity (MWQ) when dealing with prescription balances.
desensitization therapy is the concept of exposing the patient to exceedingly small quantities of a substance that they are believed to have an allergy to. The exposure is slowly increased in order to desensitize the patient to the offending substance and eventually allow them to have an appropriate

response to a normal quantity of the substance. The patient should be closely monitored with each dose to ensure there is no severe reaction.

Minimum measurable quantity

The minimum measurable quantity (MMQ) is often a given, where your pharmacy will know in advance what the smallest quantity of a substance is that you can accurately measure with a given device. If such information is not known in advance or it is not easily obtainable, it may be derived as follows:

$$\frac{100\,\% \times \text{Sensitivity Requirement}}{\text{Acceptable Error}} = \text{MMQ}$$

Let's pause and look at an example.

Example

A pharmacy has a prescription balance and you need to determine its MMQ. After the balance is leveled, it requires 6 mg to be added to one of the pans before the red marker moves one full spot on the index plate. This means the sensitivity requirement is 6 mg. A common standard in pharmacy is to use 5% as your acceptable error rate. With that in mind we can now calculate the MMQ.

$$\frac{100\% \times 6\text{ mg}}{5\%} = \mathbf{120\,mg\ is\ the\ MMQ}$$

Now, attempt to calculate the MMQ of a syringe.

Practice Problem

If a 1 mL syringe has a volume mark every 0.01 mL and your acceptable error rate is considered 5%, what is your MMQ with this syringe?

0.2 mL is the MMQ

Aliquot method

Now that we've discussed the concept of MMQ, we now need to determine a manner of how to accurately measure a substance that is below our MMQ. The aliquot method provides a means for achieving this. The steps are as follows:

1) Determine the minimum measurable quantity. This amount will vary because it depends on the sensitivity of the syringe, graduated cylinder, or balance.
2) Determine the growth factor by dividing the MMQ by the quantity of the active ingredient desired and rounding up to the next whole number (don't use a decimal point).
3) Determine the total quantity of the active ingredient required for the aliquot by multiplying the

desired quantity of the active ingredient by the same multiple (factor) found in Step 2.

4) Determine the total quantity of the mixture by multiplying the volume of the active ingredient by the same multiple (factor) found in Step 2.
5) Determine the amount of diluent needed by subtracting the amount of active ingredient calculated from the mixture total.
6) Determine the amount of mixture needed to provide the originally ordered amount by using the ratio-proportion method.

Simply reading this process can leave it sounding rather abstract. Let's look at an example problem with actual numbers to help make sense of this methodology.

Example

You receive a prescription for 1 mg of a drug that is available as 4 mg/mL. The minimum volume that can be accurately measured in your pharmacy is 1.0 ml. What amount of aliquot mixture will contain the ordered quantity of drug?

Step 1:

The **MMQ is 1.0 mL** as it was given in the question.

Step 2:

Now divide the MMQ by the desired volume, but to do that we need to calculate our desired volume first.

$$\frac{\text{mL}}{4\text{ mg}} \times \frac{1\text{ mg}}{1} = 0.25\text{ mL desired volume}$$

Now that we have our desired volume we can divide our MMQ by that to determine our growth factor.

$$\frac{1.0\text{ mL}}{0.25\text{ mL}} = \textbf{a growth factor of 4}$$

Step 3:

Now we can determine how many mL of our active ingredient we will need to use in order to make our aliquot by multiplying our desired volume by our growth factor.

$$0.25\text{ mL} \times 4 = \textbf{1 mL active ingredient}$$

Step 4:

Now we can determine our volume of mixture by taking our quantity of active ingredient and multiplying it by our growth factor.

$$1\text{ mL} \times 4 = \textbf{4 mL mixture}$$

Step 5:

We can determine the amount of diluent needed by subtracting our quantity of active ingredient from our quantity of total mixture.

$$4\text{ mL} - 1\text{ mL} = \textbf{3 mL diluent}$$

Step 6:

Now we can determine the amount of mixture needed to provide the originally ordered amount by using the ratio-proportion method.

$$\frac{1\text{ mL active ingredient}}{4\text{ mL mixture}} = \frac{0.25\text{ mL active ingredient}}{N}$$

$$N = \textbf{1 mL mixture}$$

The actual process would require you to use 1 mL of active ingredient with 3 mL of diluent in order to prepare a 4 mL mixture. Then, you would use 1 mL of this mixture to provide the appropriate dose for the patient.

Attempt a practice problem with a similar scenario, although it will involve weight measurements instead.

Practice Problem

You receive a request for 5 mg of a powdered drug which you can dilute out with methylcellulose powder . The minimum weighable quantity that can be accurately measured on your balance is 120 mg. What amount of powdered drug, and how much methylcellulose will be required to prepare this mixture? How much of the resulting mixture will contain the requested quantity of drug?

120 mg powdered drug; 2760 mg methylcellulose; and 120 mg of the mixture will be required to provide the requested quantity of the mixture

Desensitization therapy

As stated earlier in the chapter, desensitization therapy is the concept of exposing the patient to exceedingly small quantities of a substance that they are believed to have an allergy to. The exposure is slowly increased in order to desensitize the patient to the offending substance and eventually allow them to have an appropriate response to a normal quantity of the substance. The patient should be closely monitored with each dose to ensure there is no severe reaction. At the first sign of a sensitivity reaction to the allergen, desensitization therapy should be abandoned.

The process required to make the very small doses often required of such therapy is called a double aliquot although it may be more accurately called serial dilution. The reason for doing a serial dilution is when the initial dose is so small that it no longer makes sense, or can not be done with a single dilution.

Antibiotics are the most common use of drug desensitization therapy, although you may see it done with other medication classes as well, such as corticosteroids and salicylates. Occasionally, a physician may decide that the best course of therapy for a patient involves the use of a medication that a patient is allergic to; and therefore, the physician may decide that the best course of therapy is to perform drug desensitization therapy. Let's look at an example problem involving a serial dilution.

Example

A patient on one of the units in the hospital has an infection that the physician wants to treat with a combination drug of piperacillin and tazobactam, but the patient is listed as having a penicillin allergy. The physician orders a drug desensitization therapy on a patient using piperacillin and tazobactam. The doses are ordered as follows:

> Each of the following doses is based on just the piperacillin component of a combination product with both piperacillin and tazobactam.
>
> Dose 1 ~ 0.1 mg
> Dose 2 ~ 0.2 mg
> Dose 3 ~ 0.4 mg
> Dose 4 ~ 0.8 mg
> Dose 5 ~ 1.6 mg
> Dose 6 ~ 3 mg
> Dose 7 ~ 6 mg
> Dose 8 ~ 12 mg
> Dose 9 ~ 25 mg
> Dose 10 ~ 50 mg
> Dose 11 ~ 100 mg
> Dose 12 ~ 200 mg
> Dose 13 ~ 400 mg
> Dose 14 ~ 800 mg
>
> Each dose should be injected into a NS bag with a final volume of 50 mL. Run each bag over 15 minutes with 15 minutes between each dose. After 14th dose wait 1 hour and start 2.25 g piperacillin and tazobactam IV q8h, check BP before each dose.

The pharmacy stocks 50 mL premixed bags with 2 g of piperacillin and 250 mg of tazobactam and the piperacillin has a concentration of 40 mg/mL. The pharmacist asks you to prepare the following dilutions to make all of your doses:

- Initial stock bag has a piperacillin concentration of 40 mg/mL
- Dilution 1 is made using 20 mL from the stock bag along with 180 mL of NS providing a piperacillin concentration of 4 mg/mL
- Dilution 2 is made using 20 mL from Dilution 1 along with 180 mL of NS providing a piperacillin concentration of 0.4 mg/mL
- Dilution 3 is made using 20 mL from Dilution 2 along with 180 mL of NS providing a piperacillin concentration of 0.04 mg/mL

Also, because of the low acceptable error rate in these preparations, the pharmacist informs you that you have a MMQ of 1 mL. Based on all this information, how would you prepare the patient doses?

Dose 1 (using Dilution 3)

$$\frac{0.1\,\text{mg}}{1} \times \frac{\text{mL}}{0.04\,\text{mg}} = \mathbf{2.5\,mL\ of\ Dilution\ 3\ ,\ 47.5\ mL\ of\ NS}$$

Dose 2 (using Dilution 3)

$$\frac{0.2\,\text{mg}}{1} \times \frac{\text{mL}}{0.04\,\text{mg}} = \mathbf{5\,mL\ of\ Dilution\ 3\ ,\ 45\ mL\ of\ NS}$$

Dose 3 (using Dilution 3)

$$\frac{0.4\,\text{mg}}{1} \times \frac{\text{mL}}{0.04\,\text{mg}} = \mathbf{10\,mL\ of\ Dilution\ 3\ ,\ 40\ mL\ of\ NS}$$

Dose 4 (using Dilution 3)

$$\frac{0.8\,\text{mg}}{1} \times \frac{\text{mL}}{0.04\,\text{mg}} = \mathbf{20\,mL\ of\ Dilution\ 3\ ,\ 30\ mL\ of\ NS}$$

Dose 5 (using Dilution 3)

$$\frac{1.6\,\text{mg}}{1} \times \frac{\text{mL}}{0.04\,\text{mg}} = \mathbf{40\,mL\ of\ Dilution\ 3\ ,\ 10\ mL\ of\ NS}$$

Dose 6 (using Dilution 2)

$$\frac{3\,\text{mg}}{1} \times \frac{\text{mL}}{0.4\,\text{mg}} = \mathbf{7.5\,mL\ of\ Dilution\ 2\ ,\ 42.5\ mL\ of\ NS}$$

Dose 7 (using Dilution 2)

$$\frac{6\,\text{mg}}{1} \times \frac{\text{mL}}{0.4\,\text{mg}} = \mathbf{15\,mL\ of\ Dilution 2\ ,\ 35\ mL\ of\ NS}$$

Dose 8 (using Dilution 2)

$$\frac{12\,\text{mg}}{1} \times \frac{\text{mL}}{0.4\,\text{mg}} = \mathbf{30\,mL\ of\ Dilution\ 2\ ,\ 20\ mL\ of\ NS}$$

Dose 9 (using Dilution 1)

$$\frac{25\,\text{mg}}{1} \times \frac{\text{mL}}{4\,\text{mg}} = \mathbf{6.25\,mL\ of\ Dilution\ 1\ ,\ 43.75\ mL\ of\ NS}$$

Dose 10 (using Dilution 1)

$$\frac{50\,\text{mg}}{1} \times \frac{\text{mL}}{4\,\text{mg}} = \mathbf{12.5\,mL\ of\ Dilution\ 1\ ,\ 37.5\ mL\ of\ NS}$$

Dose 11 (using Dilution 1)

$$\frac{100\,\text{mg}}{1} \times \frac{\text{mL}}{4\,\text{mg}} = \mathbf{25\,mL\ of\ Dilution\ 1\ ,\ 25\ mL\ of\ NS}$$

Dose 12 (using Dilution 1)

$$\frac{200\,\text{mg}}{1} \times \frac{\text{mL}}{4\,\text{mg}} = \mathbf{50\ mL\ of\ Dilution\ 1\ ,\ 0\ mL\ of\ NS}$$

Dose 13 (using Initial Bag)

$$\frac{400\,\text{mg}}{1} \times \frac{\text{mL}}{40\,\text{mg}} = \mathbf{10\,mL\ of\ Initial\ Bag\ ,\ 40\ mL\ of\ NS}$$

Dose 14 (using Initial Bag)

$$\frac{800\,mg}{1}\times\frac{mL}{40\,mg}=\mathbf{20\,mL\ of\ Initial\ Bag\ ,\ 30\ mL\ of\ NS}$$

Something you may find interesting is that all 14 bags were prepared using only one premixed bag of piperacillin and tazobactam. Now that we've looked over the example problem, prepare a similar set of bags for a drug desensitization therapy below.

Practice Problem

A patient on one of the units in the hospital has an infection that the physician wants to treat with a combination drug of ampicillin and sulbactam, but the patient is listed as having a penicillin allergy. The physician orders a drug desensitization therapy on a patient using ampicillin and sulbactam. The doses are ordered as follows:

Each of the following doses are based on just the ampicillin component of a combination product with both ampicillin and sulbactam.

Dose 1 ~ 0.05 mg	Dose 8 ~ 6 mg
Dose 2 ~ 0.1 mg	Dose 9 ~ 12 mg
Dose 3 ~ 0.2 mg	Dose 10 ~ 25 mg
Dose 4 ~ 0.4 mg	Dose 11 ~ 50 mg
Dose 5 ~ 0.8 mg	Dose 12 ~ 100 mg
Dose 6 ~ 1.6 mg	Dose 13 ~ 200 mg
Dose 7 ~ 3 mg	Dose 14 ~ 400 mg

Each dose should be injected into a NS bag with a final volume of 50 mL. Run each bag over 15 minutes with 15 minutes between each dose. After 14th dose, wait 1 hour and start 1.5 g ampicillin and sulbactam IV q6h, check BP before each dose.

The pharmacy stocks vials with 1 g of ampicillin and 500 mg of sulbactam that get mixed with 50 mL of NS making the ampicillin have a concentration of 20 mg/mL. The pharmacist asks you to prepare the following dilutions to make all of your doses:

- Initial stock bag has an ampicillin concentration of 20 mg/mL
- Dilution 1 is made using 20 mL from the stock bag along with 180 mL of NS providing an ampicillin concentration of 2 mg/mL
- Dilution 2 is made using 20 mL from Dilution 1 along with 180 mL of NS providing an ampicillin concentration of 0.2 mg/mL
- Dilution 3 is made using 20 mL from Dilution 2 along with 180 mL of NS providing an ampicillin concentration of 0.02 mg/mL

Also, because of the low acceptable error rate in these preparations, the pharmacist informs you that you have a minimum measurable quantity of 1 mL. Based on all this information, how would you prepare the patient doses?

Dose 1 requires 2.5 mL of Dilution 3 and 47.5 mL of NS.
Dose 2 requires 5 mL of Dilution 3 and 45 mL of NS.
Dose 3 requires 10 mL of Dilution 3 and 40 mL of NS.
Dose 4 requires 20 mL of Dilution 3 and 30 mL of NS.
Dose 5 requires 40 mL of Dilution 3 and 10 mL of NS.
Dose 6 requires 8 mL of Dilution 2 and 42 mL of NS.
Dose 7 requires 15 mL of Dilution 2 and 35 mL of NS.
Dose 8 requires 30 mL of Dilution 2 and 20 mL of NS.
Dose 9 requires 6 mL of Dilution 1 and 44 mL of NS.
Dose 10 requires 12.5 mL of Dilution 1 and 37.5 mL of NS.
Dose 11 requires 25 mL of Dilution 1 and 25 mL of NS.
Dose 12 requires 50 mL of Dilution 1 and 0 mL of NS.
Dose 13 requires 10 mL of Initial Bag and 40 mL of NS.
Dose 14 requires 20 mL of Initial Bag and 30 mL of NS.

Worksheet 27-1

Name:

Date:

Solve problems 1-8 on pages 597-600.

1) A pharmacy has a prescription balance and you need to determine its MMQ. After the balance is leveled, it requires 6 mg to be added to one of the pans before the red marker moves one full spot on the index plate. This means the sensitivity requirement is 6 mg. Your pharmacy has chosen to set a standard of 4% as your acceptable error rate. What is the MMQ in this scenario?

2) If a syringe has a volume mark every 0.01 mL, and your acceptable error rate for something that needs to be very precise is considered 1%, what is your MMQ with this syringe?

3) A 1 mg dose of phytonadione is ordered and it comes in a 10mg/mL ampule. If 1.0 mL is the minimum volume that can be accurately measured, how many mL of phytonadione, and how much diluent will be required to prepare this mixture? How much of the resulting mixture will contain the requested quantity of drug?

4) The pharmacy receives a prescription requiring you to prepare capsules with 0.25 grain of codeine phosphate.

   a) If the balance has a MMQ of 120 mg, how would you prepare this aliquot using lactose as your diluent?

   b) Using the chart on page 323, what size capsule shell will you need?

5) A prescription order requests 100 mg of chlorpheniramine maleate powder. If the prescription balance has a sensitivity requirement of 6 mg, and you are using an acceptable error rate of 5% how would you prepare this dose?

6) A prescription requires 0.05 mL of a particular substance. The most accurate syringe in the pharmacy has a sensitivity of 0.01 mL and your acceptable error rate is 5%. If you use SWFI as your diluent, how would you prepare the requested dose?

7) A patient on one of the units in the hospital has an infection that the physician wants to treat with vancomycin, but the patient claims to have a vancomycin allergy. The physician orders a drug desensitization therapy on a patient using vancomycin. The doses are ordered as follows:

The following is a vancomycin desensitization therapy

Dose 1 ~ 0.1 mg	Dose 8 ~ 12 mg
Dose 2 ~ 0.2 mg	Dose 9 ~ 25 mg
Dose 3 ~ 0.4 mg	Dose 10 ~ 50 mg
Dose 4 ~ 0.8 mg	Dose 11 ~ 100 mg
Dose 5 ~ 1.6 mg	Dose 12 ~ 200 mg
Dose 6 ~ 3 mg	Dose 13 ~ 400 mg
Dose 7 ~ 6 mg	

Each dose should be injected into a D5W bag with a final volume of 100 mL. Infuse each bag over 30 minutes. After 13th dose, wait 1 hour and start 1000 mg vancomycin IV q12h, check BP before each dose.

The pharmacy stocks 200 mL premixed bags with 1 g of vancomycin. The pharmacist asks you to prepare the following dilutions to make all of your doses:

- Initial stock bag has a vancomycin concentration of 5 mg/mL
- Dilution 1 is made using 20 mL from the stock bag along with 180 mL of D5W providing a vancomycin concentration of 0.5 mg/mL
- Dilution 2 is made using 20 mL from Dilution 1 along with 180 mL of D5W providing a vancomycin concentration of 0.05 mg/mL

Also, because of the low acceptable error rate in these preparations, the pharmacist informs you that you have a MMQ of 1 mL. Based on all this information, how would you prepare the patient doses?

8) A patient with an aspirin hypersensitivity reaction is ordered drug desensitization therapy. The physician wants the patient to receive the following doses: 0.1 mg, 0.3 mg, 1 mg, 3 mg, 10 mg, 30 mg, 40 mg, 81 mg, 162 mg, and 325 mg of aspirin. Part of the challenge is that you will need to prepare the first 7 doses as powder packets to be dissolved in water immediately prior to consumption. Dose 8 can be an 81 mg low-dose aspirin tablet, dose 9 can be 2 low-dose 81 mg aspirin tablets, and dose 10 can be a standard 325 mg tablet. Your torsion prescription balance has a sensitivity of 0.004 g and you are allowed an error rate not greater than 5%. If you are using powdered sugar as your diluent with aspirin powder, how would you prepare the first 7 doses of this desensitization therapy? So that none of your powder packets are too hard to dissolve the pharmacist suggests that you keep each of your powder packets below 10 grams. (If you check your answer against the back of the book, know that there are a multitude of ways that this could be solved.)

Answer Key

Chapter 1 - Numeral Systems Used in Pharmacy

Worksheet 1-1

1) i
2) ii
3) iii
4) iv
5) v
6) vi
7) vii
8) viii
9) ix
10) x
11) xx
12) xl
13) xlv
14) c
15) cd
16) m
17) cm
18) lxix
19) xxiv
20) mcmxcix
21) 9
22) 18
23) 24
24) 36
25) 3
26) 240
27) 55
28) 555
29) 1111
30) 999
31) 4
32) 7
33) 12
34) 16
35) 22
36) 42
37) 31
38) 1337
39) 2008
40) 1783

Worksheet 1-2

1) 29
2) 23
3) 5
4) 24
5) 4
6) 12
7) xx
8) xxx
9) iv
10) xii
11) v
12) xxiv
13) ix
14) vii

	I	**II**	**III**	**IV**	**V**	**VI**	**VII**	**VIII**	**IX**	**X**
I	i	ii	iii	iv	v	vi	vii	viii	ix	x
II	ii	iv	vi	viii	x	xii	xiv	xvi	xviii	xx
III	iii	vi	ix	xii	xv	xviii	xxi	xxiv	xxvii	xxx
IV	iv	viii	xii	xvi	xx	xxiv	xxviii	xxxii	xxxvi	xl
V	v	x	xv	xx	xxv	xxx	xxxv	xl	xlv	l
VI	vi	xii	xviii	xxiv	xxx	xxxvi	xlii	xlviii	liv	lx
VII	vii	xiv	xxi	xxviii	xxxv	xlii	xlix	lvi	lxiii	lxx
VIII	viii	xvi	xxiv	xxxii	xl	xlviii	lvi	lxiv	lxxii	lxxx
IX	ix	xviii	xxvii	xxxvi	xlv	liv	lxiii	lxxii	lxxxi	xc
X	x	xx	xxx	xl	l	lx	lxx	lxxx	xc	c

Worksheet 1-3

1) hundreds place
2) hundred-thousands place
3) ten-millions place
4) tenths place
5) ten-thousandths place
6) ones place
7) 12,004,025
8) 31,337
9) 5,318,008
10) 0.7
11) 0.39
12) 15.7
13) 6.07
14) 705.0107

15) 0.7
16) 95.07
17) 0.57
18) 0.0019
19) 406.0214
20) 507.0112
21) $\frac{6}{10}$
22) $\frac{85}{100}$
23) $\frac{6}{1,000}$
24) $\frac{574}{10,000}$
25) $13 \frac{13}{1,000,000}$
26) $80 \frac{8,135}{100,000}$

Worksheet 1-4
1) 103
2) 1
3) 88
4) 100
5) 187
6) 29,002
7) 1.0
8) 99.7
9) 2.5
10) 29,001.5
11) 187.5
12) 13.0
13) 187.50
14) 12.99
15) 29,000.50
16) 1,234,567.89
17) 66.67
18) 5,454.55
19) 3
20) 4
21) 2
22) 3
23) 3
24) 1
25) 2
26) 2
27) 1

Worksheet 1-5
1) 55,886
2) 5.25
3) 43.79
4) 97.16
5) 35.75
6) 86.964
7) 73
8) 8238
9) 54.17
10) 88.10
11) 37.03
12) 34.342
13) 696
14) 4.91
15) 6.39
16) 96.98
17) 26.35
18) 83.436
19) 1664
20) 15.97
21) 88.06
22) 0.958
23) 0.45
24) 0.839
25) 0.086 kg
26) 37.4° C

Worksheet 1-6
1) 780,690,145
2) 1.015
3) 366.212
4) 4.1454
5) 370.962
6) 41.8481
7) 16,273,937
8) 669.837
9) 1.7616
10) 0.01164
11) 18.225
12) 0.106512
13) 81 mg
14) $210.72
15) 734.4 mg

Worksheet 1-7
1) 105
2) 16.24
3) 1.362
4) 1151.5
5) 6.774
6) 0.043
7) 0.015
8) 1.836
9) 440.4
10) 80.833
11) 0.9
12) 8.491
13) 32 bottles
14) 200 days
15) 120 capsules
16) 8 days' supply

Worksheet 1-8
1) xxxviii
2) dli
3) xxiv
4) 21
5) 400
6) 48
7) ten-thousands place
8) hundredths place
9) tenths place
10) 2,015,600
11) 0.4004

12) 9,876.5432
13) 32,349
14) 82.25
15) 23.7
16) 40.842
17) 111.95
18) 134.79
19) 7,385
20) 81.89
21) 22.3
22) 92.9
23) 102.65
24) 29.01
25) 240
26) 74.4
27) .1107
28) 10.4016
29) 0.1176
30) 20.99579
31) 61.5
32) 104.33
33) 1.36
34) 41.18
35) 0.2
36) 0.12
37) 454 grams
38) 2.3 kilograms
39) $29.88
40) 14 days

Chapter 2 - Fractions

Worksheet 2-1

1) 0.2
2) 0.5
3) 0.5
4) 0.1
5) 0.083
6) 0.01
7) 0.001
8) 0.667
9) 0.75
10) 0.8
11) 0.375
12) 0.417
13) 0.389
14) 0.364
15) 0.091
16) 0.143
17) 0.164
18) 0.206
19) 0.211
20) 0.277
21) $\frac{5}{1}$
22) $\frac{2}{1}$
23) $\frac{4}{2}$
24) $\frac{10}{1}$
25) $\frac{12}{1}$
26) $\frac{100}{1}$
27) $\frac{1000}{1}$
28) $\frac{3}{2}$
29) $\frac{4}{3}$
30) $\frac{5}{4}$
31) $\frac{8}{3}$
32) $\frac{12}{5}$
33) $\frac{18}{7}$
34) $\frac{11}{4}$
35) $\frac{22}{2}$
36) $\frac{35}{5}$
37) $\frac{55}{9}$
38) $\frac{63}{13}$
39) $\frac{71}{15}$
40) $\frac{83}{23}$

Worksheet 2-2

1) $\frac{1}{2}$
2) $\frac{1}{3}$
3) $\frac{1}{8}$
4) $\frac{1}{3}$
5) $\frac{1}{12}$
6) $\frac{1}{5}$
7) $\frac{11}{15}$
8) $\frac{4}{5}$
9) $\frac{7}{10}$
10) $\frac{1}{5}$
11) $\frac{7}{18}$
12) $\frac{7}{11}$
13) 0
14) $\frac{2}{3}$
15) $\frac{2}{9}$
16) $\frac{12}{35}$

17) $\frac{1}{4}$

18) $\frac{1}{10}$

19) $\frac{1}{9}$

20) $\frac{1}{2}$

21) $\frac{1}{3}$

22) $\frac{1}{6}$

23) $\frac{1}{15}$

24) $\frac{2}{9}$

25) $\frac{1}{5}$

26) $\frac{7}{15}$

27) $\frac{7}{12}$

28) $\frac{3}{7}$

29) $\frac{3}{5}$

30) $\frac{10}{33}$

Worksheet 2-3

1) $\frac{2}{3}$

2) $\frac{1}{4}$

3) $\frac{1}{2}$

4) $\frac{9}{25}$

5) $\frac{5}{16}$

6) $\frac{5}{8}$

7) $\frac{6}{7}$

8) $1\frac{1}{4}$

9) $\frac{6}{19}$

10) $\frac{1}{4}$

11) $\frac{5}{24}$

12) $\frac{13}{100}$

13) $5\frac{703}{1000}$

14) $\frac{343}{432}$

15) $1\frac{73}{100}$

16) $\frac{23}{36}$

17) $\frac{11}{18}$

18) $\frac{8}{9}$

19) $1\frac{1}{4}$

20) $1\frac{3}{38}$

21) $1\frac{1}{4}$

22) $1\frac{41}{60}$

23) $1\frac{1}{4}$

24) $\frac{34}{81}$

Worksheet 2-4

1) $\frac{1}{9}$

2) $\frac{3}{64}$

3) $\frac{2}{9}$

4) $\frac{15}{64}$

5) $\frac{3}{25}$

6) $\frac{21}{64}$

7) 0

8) $\frac{3}{10}$

9) $\frac{1}{18}$

10) $\frac{2}{35}$

11) $\frac{27}{40}$

12) $\frac{9}{44}$

13) $\frac{3}{40}$

14) $\frac{91}{120}$

15) $\frac{42}{125}$

16) $\frac{49}{200}$

17) 18

18) $\frac{15}{2}$ or $7\frac{1}{2}$

19) $\frac{4}{9}$

20) $\frac{1}{21}$

21) $\frac{28}{5}$ or $5\frac{3}{5}$

22) 20

23) $\frac{7}{3}$

24) 18

Worksheet 2-5

1) 1
2) $\frac{1}{3}$
3) $\frac{2}{3}$
4) $\frac{4}{5}$
5) $1\frac{1}{3}$
6) $\frac{2}{3}$
7) $\frac{1}{15}$
8) 1
9) $1\frac{24}{25}$
10) 32
11) $2\frac{14}{15}$
12) $\frac{1}{4}$
13) $3\frac{1}{16}$
14) $12\frac{1}{2}$
15) 80 capsules
16) 25 syringes

Worksheet 2-6

1) 0.467
2) 0.325
3) 0.625
4) 0.867
5) 0.567
6) 0.875
7) 0.333
8) 0.68
9) 0.351
10) $\frac{4}{3}$
11) $\frac{5}{1}$
12) $\frac{24}{15}$ or $\frac{8}{5}$
13) $\frac{8}{2}$ or $\frac{4}{1}$
14) $\frac{585}{65}$ or $\frac{9}{1}$
15) $\frac{1}{1}$
16) $\frac{3}{4}$
17) $\frac{1}{4}$
18) $\frac{5}{8}$
19) $\frac{1}{4}$
20) $\frac{1}{3}$
21) $\frac{3}{4}$
22) $\frac{3}{5}$
23) $\frac{17}{31}$
24) $\frac{13}{37}$
25) $\frac{2}{3}$
26) $\frac{11}{8}$ or $1\frac{3}{8}$
27) $\frac{7}{24}$
28) $\frac{1}{6}$
29) $\frac{19}{24}$
30) $\frac{3}{16}$
31) $\frac{31}{42}$
32) $\frac{23}{55}$
33) $\frac{2}{21}$
34) $3\frac{1}{2}$
35) $4\frac{1}{2}$
36) $\frac{1}{16}$
37) 20
38) $3\frac{1}{16}$
39) $\frac{1}{4}$
40) $1\frac{1}{4}$
41) $2\frac{3}{8}$ mL
42) $\frac{7}{8}$ L
43) 5 mg
44) 40 doses
45) $\frac{7}{12}$ ounces

Chapter 3 - Percentages

Worksheet 3-1

1) 24.44%
2) 30%
3) 50%
4) 12.5%
5) 75%
6) 2%
7) 9%
8) 10%
9) 80%
10) 36%
11) 52%
12) 40%
13) 65%
14) 2.5%

15) 3.5%
16) 5.5%
17) 0.4%
18) 110%
19) 175%
20) 200%
21) 0.33
22) 0.24
23) 0.333
24) 0.505
25) 0.2
26) 0.47
27) 0.93
28) 0.325
29) 0.75
30) 0.8332
31) 0.6666667
32) 0.185
33) 0.013
34) 0.0025
35) 0.00125
36) 80%
37) 20%
38) 87.5%
39) 75%
40) 85.7% (rounded)
41) 85%
42) 62.5%
43) 42.9% (rounded)
44) 30%
45) 60%
46) 87.5%
47) 80%
48) 55%
49) 88.9% (rounded)
50) 50%

Worksheet 3-2

1) $\frac{1}{2}$
2) $\frac{3}{4}$
3) $\frac{3}{5}$
4) $\frac{7}{20}$
5) $\frac{3}{10}$
6) 1
7) $\frac{17}{25}$
8) $\frac{41}{100}$
9) $\frac{93}{100}$
10) $\frac{3}{250}$
11) $\frac{1}{8}$
12) $\frac{67}{100}$
13) $\frac{333}{500}$
14) $\frac{9}{25}$
15) $\frac{11}{10}$ or $1\frac{1}{10}$
16) 150
17) 6
18) 3
19) 37.5
20) 3
21) 32
22) 90
23) 1.7
24) 5.5
25) 29.7
26) 2.5
27) 36
28) 20
29) 90
30) 280
31) 10
32) 8.4
33) 4.8
34) 7.5
35) 12
36) 12
37) 50
38) 12.5
39) 27
40) 50

Worksheet 3-3

Percent	***Decimal***	***Reduced Fraction***
20%	0.2	1/5
6%	0.06	3/50
87.5%	0.875	7/8
15%	0.15	3/20
11%	0.11	11/100
80%	0.8	4/5
96%	0.96	24/25

Percent	*Decimal*	*Reduced Fraction*
55%	0.55	11/20
12%	0.12	3/25
53%	0.53	53/100
100%	1	1
7.5%	0.075	3/40

1) 2
2) 16%
3) 175
4) 198
5) 28%
6) 2500
7) 325
8) 25%
9) 15
10) 100.8
11) 20%
12) 20
13) 18
14) 90%
15) 32
16) 62
17) 92%
18) 500
19) 42
20) 81.25%
21) 92% correct
22) 23 students
23) 3,077 people
24) 64% of his savings
25) 2,800 hairs
26) 68% of patients

Worksheet 3-4

1) $\frac{4}{25}$
2) $\frac{9}{20}$
3) $\frac{3}{50}$
4) 0.33
5) 0.045
6) 0.0294
7) 25%
8) 40%
9) 70%
10) 70%
11) 46.7%
12) 0.6%
13) 48
14) 6
15) 10%
16) 66.7%
17) 250
18) 27.3
19) 210
20) 5%
21) 7 instructors
22) 27 milligrams
23) 6.5 pounds
24) 24 patients
25) 2 holidays

Chapter 4 - 24 Hour Time, Exponents & Scientific Notation

Worksheet 4-1

1140 eleven forty hours 11:40 A.M.
1400 fourteen hundred hours 2:00 P.M.
2230 twenty-two thirty hours 10:30 P.M.
0030 zero zero thirty hours 12:30 A.M.
1050 ten fifty hours 10:50 A.M.
0615 zero six fifteen hours 6:15 A.M.
1050 ten fifty hours 10:50 A.M.
2145 twenty-one forty-five hours 9:45 P.M.
1315 thirteen fifteen hours 1:15 P.M.
0015 zero zero fifteen hours 12:15 A.M.
0005 zero zero zero five hours 12:05 A.M.
0930 zero nine thirty hours 9:30 A.M.
2230 twenty-two thirty hours 10:30 P.M.
0020 zero zero twenty hours 12:20 A.M.
1425 fourteen twenty-five hours 2:25 P.M.
0007 zero zero zero seven hours 12:07 A.M.
0100 zero one hundred hours 1:00 A.M.

Worksheet 4-2

1) 1
2) 2
3) 1
4) 4
5) 8
6) 9
7) 27
8) 81
9) 100
10) 216
11) 1

12) 25
13) 1
14) 81
15) 1
16) 20,736
17) 1,728
18) 10,000
19) 1
20) 343
21) 1×10^{3}
22) 1×10^{0}
23) 6.7×10^{7}
24) 1×10^{-1}
25) 3.06×10^{-3}
26) 1×10^{6}
27) 9.09×10^{5}
28) 1×10^{-3}
29) 2.8×10^{-4}
30) 6.14×10^{-7}
31) 0.000000614
32) 0.00306
33) 0.00028
34) 0.1
35) 0.001
36) 67,000,000
37) 909,000
38) 1
39) 1,000,000
40) 1,000

Worksheet 4-3

1040 ten forty hours 10:40 A.M.
2100 twenty-one hundred hours 9:00 P.M.
1230 twelve thirty hours 12:30 P.M.
0715 zero seven fifteen hours 7:15 A.M.
2245 twenty-two forty-five 10:45 P.M.
0830 zero eight thirty hours 8:30 A.M.
2215 twenty-two fifteen hours 10:15 P.M.
1020 ten twenty hours 10:20 A.M.
2200 twenty-two hundred hours 10:00 P.M.
1130 eleven thirty hours 11:30 A.M.
1915 nineteen fifteen hours 7:15 P.M.
1045 ten forty-five hours 10:45 A.M.
1230 twelve thirty hours 12:30 P.M.
2210 twenty-two ten hours 10:10 P.M.

1) 216
2) 25
3) 16,384
4) 81
5) 100
6) 36
7) 125
8) 1024
9) 27
10) 10,000
11) 1×10^{7}
12) 3.51×10^{2}
13) 7.1×10^{3}
14) 3.7×10^{-2}
15) 3.75×10^{-1}
16) 6.4×10^{-3}
17) 1×10^{6}
18) 3.5×10^{1}
19) 6.1×10^{3}
20) 3.7×10^{-1}
21) 3.75×10^{-1}
22) 5.6×10^{-3}

Worksheet 4-4

1) The medication should be scheduled for 1300, 1700, 2100, 0100, 0500, and 0900
2) 602,000,000,000,000,000,000,000
3) 81
4) 6×10^{12}
5) 5×10^{7}

Chapter 5 - Problem Solving Methods

Worksheet 5-1

1) 1:5
2) 2:11
3) 2:9
4) 15:23
5) 1:9
6) 5:1
7) 6:1
8) 8:3
9) 1:3
10) 1:10
11) 900 oz of sol; 3:900 or 1:300
12) 30:90 or 1:3
13) 60 oz of sol; 20:60 or 1:3
14) 50 g of sol; 10:50 or 1:5
15) 3:90 or 1:30
16) 2:40 or 1:20
17) 100 oz of sol; 5:100 or 1:20

Worksheet 5-2

1) 1:4
2) 3:17
3) 1:5
4) 2:7
5) 5:19
6) 23:25
7) 1:6
8) 5:1
9) 100:1
10) 1:4
11) 400 oz of sol; 5:400 or

1:80

12) 10 oz of sol; 2:10 or 1:5

13) 68:110 or 34:55

14) 744 oz of sol; 3:744 or 1:248

15) 1:10:20

Worksheet 5-3

1)
   a) The extremes are 5 and 40.
   b) The means are 8 and 25.
2) 6
3) 21
4) 2
5) 14
6) 12
7) 2
8) 1
9) 10
10) 3
11) 2
12) 5
13) 1
14) 6
15) 2
16) 60
17) 1
18) 1,000
19) 8

Worksheet 5-4

1) $\frac{3}{x}=\frac{6}{7}$
2) $\frac{5}{8}=\frac{N}{10}$
3) $\frac{K}{8}=\frac{4}{16}$
4) $\frac{7}{9}=\frac{14}{N}$
5) $\frac{2}{3}=\frac{x}{9}$
6) $\frac{7}{N}=\frac{14}{28}$
7) 3
8) 5
9) 9
10) 15
11) 3
12) 16
13) 1.25
14) 200
15) 100
16) 3.22 g
17) 0.175 oz of boric acid
18) 6 oz of magnesium sulfate
19) 135 g of salt
20) 0.16 oz of boric acid

Worksheet 5-5

1) $\frac{4}{X}=\frac{7}{9}$
2) $\frac{3}{5}=\frac{N}{8}$
3) $\frac{N}{72}=\frac{4.8}{12.0}$
4) $\frac{15.0}{N}=\frac{31.0}{1}$
5) $\frac{5}{17.1}=\frac{N}{2}$
6) $\frac{315}{32}=\frac{N}{35}$
7) 10
8) 10
9) 8.5
10) $5.\overline{5}$
11) 5
12) 0.6
13) 3.74
14) 60 mL
15) 3.15 L
16) 1.65
17) $49.84
18) 33.11 mL of colloids

Worksheet 5-6

1) $\frac{3\text{ days}}{1}\times\frac{2\text{ doses}}{\text{day}}\times\frac{1\text{ tab}}{\text{dose}}=6\text{ tablets}$
2) $\frac{7\text{ days}}{1}\times\frac{4\text{ doses}}{\text{day}}\times\frac{1\text{ cap}}{\text{dose}}=28\text{ capsules}$
3) $\frac{4\text{ days}}{1}\times\frac{4\text{ doses}}{\text{day}}\times\frac{1\text{ tab}}{\text{dose}}=16\text{ tablets}$
4) $\frac{10\text{ days}}{1}\times\frac{1\text{ doses}}{\text{day}}\times\frac{1\text{ tab}}{\text{dose}}=10\text{ tablets}$
5) $\frac{30\text{ days}}{1}\times\frac{2\text{ doses}}{\text{day}}\times\frac{1\text{ tab}}{\text{dose}}=60\text{ tablets}$
6) $\frac{2\text{ days}}{1}\times\frac{3\text{ doses}}{\text{day}}\times\frac{1\text{ tab}}{\text{dose}}=6\text{ tablets}$
7) $\frac{34\text{ days}}{1}\times\frac{4\text{ doses}}{\text{day}}\times\frac{1\text{ tab}}{\text{dose}}=136\text{ tablets}$
8) $\frac{4\text{ weeks}}{1}\times\frac{1\text{ capsule}}{\text{week}}=4\text{ capsules}$
9) $\frac{21\text{ days}}{1}\times\frac{3\text{ doses}}{\text{day}}\times\frac{1\text{ tab}}{\text{dose}}=63\text{ tablets}$
10) $\frac{72\text{ hr}}{1}\times\frac{1\text{ day}}{24\text{ hr}}\times\frac{4\text{ doses}}{\text{day}}\times\frac{1\text{ tab}}{\text{dose}}=12\text{ tablets}$

Worksheet 5-7

1) $\frac{7\text{ days}}{1}\times\frac{2\text{ doses}}{\text{day}}\times\frac{1\text{ tab}}{\text{dose}}=14\text{ tablets}$

2) $\frac{7\text{ days}}{1}\times\frac{3\text{ doses}}{\text{day}}\times\frac{1\text{ tab}}{\text{dose}}=21\text{ tablets}$

3) $\frac{90\text{ days}}{1}\times\frac{4\text{ doses}}{\text{day}}\times\frac{1\text{ tab}}{\text{dose}}=360\text{ tablets}$

4) $\frac{28\text{ days}}{1}\times\frac{4\text{ doses}}{\text{day}}\times\frac{1\text{ cap}}{\text{dose}}=112\text{ capsules}$

5) $\frac{4\text{ days}}{1}\times\frac{1\text{ dose}}{\text{day}}\times\frac{1\text{ tab}}{\text{dose}}=4\text{ tablets}$

6) $\frac{30\text{ days}}{1}\times\frac{3\text{ doses}}{\text{day}}\times\frac{1\text{ tab}}{\text{dose}}=90\text{ tablets}$

7) $\frac{90\text{ days}}{1}\times\frac{2\text{ doses}}{\text{day}}\times\frac{3\text{ caps}}{\text{dose}}=540\text{ capsules}$

8) $\frac{3\text{ days}}{1}\times\frac{1\text{ dose}}{\text{day}}\times\frac{1\text{ cap}}{\text{dose}}=3\text{ capsules}$

$\frac{5\text{ days}}{1}\times\frac{1\text{ dose}}{\text{day}}\times\frac{2\text{ cap}}{\text{dose}}=10\text{ capsules}$

$\frac{22\text{ days}}{1}\times\frac{1\text{ dose}}{\text{day}}\times\frac{4\text{ cap}}{\text{dose}}=88\text{ capsules}$

$3\text{ caps}+10\text{ caps}+88\text{ caps}=101\text{ capsules}$

Worksheet 5-8

1) $\frac{3\text{ days}}{1}\times\frac{3\text{ doses}}{\text{day}}\times\frac{50\text{ mg}}{\text{dose}}\times\frac{\text{cap}}{25\text{ mg}}=18\text{ tablets}$

2) $\frac{30\text{ days}}{1}\times\frac{1\text{ dose}}{\text{day}}\times\frac{1\text{ cap}}{\text{dose}}=30\text{ capsules}$

3) $\frac{3\text{ days}}{1}\times\frac{4\text{ doses}}{\text{day}}\times\frac{1\text{ tab}}{\text{dose}}=12\text{ tablets}$

4) $\frac{7\text{ days}}{1}\times\frac{1\text{ dose}}{\text{day}}\times\frac{1\text{ tab}}{\text{dose}}=7\text{ tablets}$

5) $\frac{90\text{ days}}{1}\times\frac{2\text{ doses}}{\text{day}}\times\frac{3\text{ caps}}{\text{dose}}=540\text{ capsules}$

6) $\frac{14\text{ days}}{1}\times\frac{4\text{ doses}}{\text{day}}\times\frac{1\text{ cap}}{\text{dose}}=56\text{ capsules}$

7) $\frac{30\text{ days}}{1}\times\frac{3\text{ doses}}{\text{day}}\times\frac{2\text{ caps}}{\text{dose}}=180\text{ capsules}$

8) $\frac{10\text{ days}}{1}\times\frac{2\text{ doses}}{\text{day}}\times\frac{1\text{ tab}}{\text{dose}}=20\text{ tablets}$

9) $\frac{5\text{ days}}{1}\times\frac{2\text{ doses}}{\text{day}}\times\frac{10\text{ mg}}{\text{dose}}\times\frac{1\text{ tab}}{10\text{ mg}}=10\text{ tablets}$

$\frac{4\text{ days}}{1}\times\frac{2\text{ doses}}{\text{day}}\times\frac{5\text{ mg}}{\text{dose}}\times\frac{1\text{ tab}}{10\text{ mg}}=4\text{ tablets}$

$\frac{2\text{ days}}{1}\times\frac{2\text{ doses}}{\text{day}}\times\frac{2.5\text{ mg}}{\text{dose}}\times\frac{1\text{ tab}}{10\text{ mg}}=1\text{ tablet}$

$10\text{ tabs}+4\text{ tabs}+1\text{ tab}=15\text{ tablets}$

Worksheet 5-9

1) 1:9
2) 1:6
3) 1:6
4) 20:100 or 1:5
5) 125 mL of a drug
6) 2.4 g of hydrocortisone
7) $\frac{5}{x}=\frac{10}{15}$
8) $\frac{6}{11}=\frac{Y}{12}$
9) $\frac{1}{3}=\frac{N}{18}$
10) 6
11) 8
12) 6
13) 40.5 g
14) 28.4 g
15) 18.3 g
16) 1 mL
17) $3.\overline{3}$ in.
18) 2 g

19) $\frac{7\text{ days}}{1}\times\frac{2\text{ doses}}{\text{day}}\times\frac{1\text{ tab}}{\text{dose}}=14\text{ tablets}$

20) $\frac{7\text{ days}}{1}\times\frac{3\text{ doses}}{\text{day}}\times\frac{1\text{ tab}}{\text{dose}}=21\text{ tablets}$

21) $\frac{30\text{ days}}{1}\times\frac{4\text{ doses}}{\text{day}}\times\frac{1\text{ tab}}{\text{dose}}=120\text{ tablets}$

22) $\frac{14\text{ days}}{1}\times\frac{4\text{ doses}}{\text{day}}\times\frac{1\text{ cap}}{\text{dose}}=56\text{ capsules}$

23) $\frac{14\text{ days}}{1}\times\frac{4\text{ doses}}{\text{day}}\times\frac{1\text{ cap}}{\text{dose}}=56\text{ capsules}$

24) $\frac{90\text{ days}}{1}\times\frac{3\text{ doses}}{\text{day}}\times\frac{1\text{ cap}}{\text{dose}}=270\text{ capsules}$

25) $\frac{3\text{ days}}{1}\times\frac{2\text{ doses}}{\text{day}}\times\frac{3\text{ caps}}{\text{dose}}=18\text{ capsules}$

26) $\frac{2\text{ days}}{1}\times\frac{3\text{ doses}}{\text{day}}\times\frac{1\text{ tab}}{\text{dose}}=6\text{ tablets}$

27) $\frac{7\text{ days}}{1}\times\frac{2\text{ tablets}}{\text{day}}\times\frac{30\text{ mg}}{\text{tab}}=420\text{ mg codeine}$

$\frac{7\text{ days}}{1}\times\frac{2\text{ tablets}}{\text{day}}\times\frac{300\text{ mg}}{\text{tab}}=4{,}200\text{ mg APAP}$

28) $\frac{0.5\text{ mL}}{1}\times\frac{480\text{ mcg}}{0.8\text{ mL}}=300\text{ mcg}$

Chapter 6 - Temperature Scale Conversions

Worksheet 6-1

1) 122° F
2) 116.6° F
3) 113° F
4) 104° F
5) 98.6° F
6) 89.6° F
7) 86° F
8) 77° F
9) 71.6° F
10) 68° F
11) 64.4° F
12) 59° F
13) 53.6° F
14) 50° F
15) 44.6° F
16) 41° F
17) 35.6° F
18) 32° F
19) 23° F
20) 14° F
21) -40° F
22) 37.8° C
23) 32.2° C
24) 31.7° C
25) 27.8° C
26) 26.7° C
27) 26.1° C
28) 23.9° C
29) 21.1° C
30) 17.2° C
31) 15.6° C
32) 14.4° C
33) 12.8° C
34) 7.2° C
35) 2.8° C
36) 0° C
37) -3.9° C
38) -5.6° C
39) -9.4° C
40) -11.1° C
41) -15° C
42) -17.8° C
43) -20.6° C
44) -25.6° C
45) -28.9° C

Worksheet 6-2

part I

- Cold: any temperature not exceeding 46° F
- Freezer: -13° to 14° F
- Refrigerator: 36° to 46° F
- Cool: 46° to 59° F
- Room temperature: the temperature prevailing in a working area
- Controlled room temperature: 59° to 86° F
- Warm: 86° to 104° F
- Excessive heat: any temperature above 104° F

part II

- Cold: any temperature not exceeding 8° C
- Freezer: -25° to -10° C
- Refrigerator: 2° to 8° C
- Cool: 8° to 15° C
- Room temperature: the temperature prevailing in a working area
- Controlled room temperature: 15° to 30° C
- Warm: 30° to 40° C
- Excessive heat: any temperature above 40° C

Worksheet 6-3

1) 266° F
2) No, it is too warm.
3)
   a) -13° to 14° F
   b) You will need to calculate this from when it is assigned.
4) 148.9° C
5) 121.1° C
6)
   a) 2.3° C
   b) 3.2° C
   c) 3.9° C
   d) 1.4° C*
   e) 2.7° C
   f) 2.6° C
   g) 3.8° C
   h) 6.6° C
   i) 9.3° C*

   1/28 & 2/2 fell outside the safe range
7)
   a) 4.8° F

b) 8.8° F
c) 8.2° F
d) 15.1° F*
e) 13.1° F
f) 7.7° F
g) 3.7° F
h) -2.0° F
i) -5.8° F

1/28 fell outside the safe range

Chapter 7 - Units Of Measurement

Worksheet 7-1

1) 1,000 mL
2) 10 mL
3) 0.697 L
4) 2,600 mL
5) 250 cc
6) 2,500 g
7) 0.415 kg
8) 2.16 g
9) 8,900 mg
10) 0.44 lb
11) 4.2 qt
12) 80.5 km
13) 197 in
14) 7.62 cm
15) 1.8 kg
16) 1.76 oz
17) 150 cc
18) 1,920 mL or 1,892 mL
19) 70.4 oz or 70.5 oz
20) 83 oz
21) 0.568 kg
22) 1.6 m
23) 1,986 to 1,989 g
24) 100 kg
25) 0.137 mg

Worksheet 7-2

1) 93 kg
2) 20 lb
3) 70 in = 178 cm = 1.78 m
4) 72 oz = 2,045 g = 2 kg
5) 32 fl oz = 2 pt
6) 2 qt
7) 15 mL = 1 Tbl = 0.5 *f*℥
8) 4 cups = 32 fl oz = 960 mL
9) 3 qt = 6 pt = 2,880 mL = 2.88 L
10) 3.35 kg
11) 1.88 m
12) 360 cc
13) 1.3 gal
14) 78.2 kg
15) 1.7 mL
16) 9.15 g
17)
   a) 1.5 tsp
   b) yes
18)
   a) 1 tsp
   b) 75 mL

Worksheet 7-3

1) 1 ℥ = 8ʒ
2) 1ʒ = 60 gr
3) 1 *f*℥ = 8 *f*ʒ
4) 1 fʒ = 60 gtt
5) 1 gtt = 1 ♍
6) 1 m = 100 cm
7) 1 m = 1000 mm
8) 1 kg = 1000 g
9) 1 g = 1000 mg
10) 1 mg = 1000 mcg
11) 1 mcg = 1000 ng
12) 1 L = 1000 mL
13) 1 L = 1000 cc
14) 1 cc = 1 mL
15) 1 ft = 12 in
16) 1 lb = 16 oz
17) 1 gal = 4 qt
18) 1 qt = 2 pt
19) 1 pt = 2 cups
20) 1 pt = 16 fl oz
21) 1 cup = 8 fl oz
22) 1 oz = 2 Tbsp
23) 1 Tbsp = 3 tsp
24) 1 gr = 60, 64.8, or 65 mg
25) 1 fl oz = 30 mL
26) 1 gtt = 0.06 mL
27) 1 in = 2.54 cm
28) 1 kg = 2.2 lb
29) 1 lb = 454 g
30) 1 oz = 28.4 g
31) 1 pt = 480 mL
32) 1 Tbsp = 3 tsp
33) 1 tsp = 5 mL
34) 3 tsp = 0.5 fl oz
35) 960 mL = 2 pt
36) 20 gtt = 1.2 mL
37) 6'1" = 185 cm
38) 165 lb = 75 kg
39) 8 lb 13 oz = 4 kg
40) 7.5 mL = 1.5 tsp
41) 4*f*ʒ = 14.4 mL to 20 mL

Worksheet 7-4

1) ʒ 0.27
2) ℥ 0.3125
3) 180, 194.4, or 195 mg
4) *f*℥ i
5) *f*ʒ iv
6) *f*℥ iii
7) 0.5 pt
8) 30 gtt

9) 10.8 to 15 mL
10) gr ii
11) 56.8 or 62.2 g
12) fʒ viii or fʒ vi
13) 120 mL
14) 170.4 g
15) gr iss
16) fʒ xvi or fʒ xii
17) f℥ iiss
18) ℥ ss
19) gr iii
20) f℥ iv
21) 3 or 4 tsp
22) 25 tsp
23) 24 tsp
24) f℥ viii
25) gr 0.23 or 0.25
26) 900 to 975 mg
27) gr .15 to .17
28) gr 38 to 42
29) 92.4 lb
30) 20 kg
31) 180 mL
32) 1.5 pt
33) 10 mL
34) 143 lb
35) 85 kg
36) 1.9 to 2.7 Tbsp
37) 960 mL
38) 180, 194.4, or 195 mg
39) 90, 97.2, or 95 mg
40) 22.5 mL

Worksheet 7-5

1) 100 cm
2) 10 mm
3) 1000 L
4) 1000 mm
5) 1 mL
6) 1000 g
7) fʒ viii
8) ♏ i
9) 3 tsp
10) 12 in
11) 3600 mm
12) 0.47 m
13) 94 mm
14) 0.482 L
15) 3900 mL
16) 3600 mg
17) 56.8 g
18) 227 g
19) 60 cc
20) 2 in
21) ʒ iss
22) 1 g
23) 6 Tbs
24) 25 mL
25) 1 Tbs
26) 54.5 kg
27) 42.3 kg
28) 1489 to 1491 g
29) 602 g
30) 1.8 m
31) 91.44 cm
32) gtt iii
33) 1.02 mL
34) a) Nitrostat 0.3 mg (nitroglycerin)
35) d) ferrous gluconate 325 mg tablets
36) 10 cups
37) 90 kg

Worksheet 7-6

1) Rx

Ipecacuanha	3.89 g
Codeine sulfate	0.648 g
Powdered digitalis	3.89 g
Honey	qs 15.55 g (*This will require 5 mL of honey.*)

2) Rx

Camphor	2.59 g
Eualyptol	0.648 g
Menthol	1.94 g
Petrolatum	qs 62.2 g

3) Rx

ASA	648 mg
Caffeine	32.4 mg
Salicylamide	194.4 mg

4) Rx

Magnesium Hydroxide Powder	38.88 g
Sodium Hypochlorite Powder	3.89 g
Purified Water	qs 480 mL

There are 2,430 mg of magnesium hydroxide powder are in 2 Tbs of this mixture.

5) Rx

Morphine Powder	60 mg
Cocaine HCl	60 mg
Simple Syrup	15 mL

90% Ethanol qs 60 mL

There are 5 mg of morphine and 5 mg of cocaine HCl are in a teaspoonful of this mixture.

Worksheet 7-7

1) 1000 mg
2) 100 cm
3) 1000 ml
4) 580 cm
5) 0.92 m
6) 3700 mL
7) 0.247 L
8) 4600 mL
9) 1.4 g
10) 4200 g
11) 55 in
12) 142 g
13) 272.4 g
14) 4.8 L
15) 90 cc
16) 3 in
17) 5.2 qt
18) 3 kg
19) 15.8 oz
20) 4500 mL
21) 18 cc
22) gr
23) oz, ℥
24) fl oz, *f*℥
25) pt
26) T, Tbs, Tbsp
27) teaspoon
28) drop
29) quart
30) dram
31) ounce
32) minim
33) fluid ounce
34) 30 mL
35) 0.06 mL
36) 28.4 or 31.1 g
37) 0.06 mL
38) 60, 64.8, or 65 mg
39) ♏ i
40) 5 mL
41) 3 tsp
42) ʒ iss
43) 30, 32.4, or 32.5 mg
44) ℥ ii
45) ʒ xxiv
46) *f*ʒ iiss
47) *f*℥ viii
48) 10 mL
49) 180, 194.4, 195 mg
50) 2.4 mL
51) gr 154 or 167
52) 3600, 3888, or 3900 mg

Chapter 8 - Working With Prescriptions

Worksheet 8-1

As the students are actually making up the prescription statements a good way to check this and create discussion is to have the students translate each others prescriptions.

Worksheet 8-2

1) This script for Okla Beaty is for one Flonase (fluticasone propionate) Nasal Spray. The instructions read, “Spray once in each nostril every morning.” The refills are written for as needed.
2) This script for Okla Beaty is for Nitrostat (nitroglycerin) 1/150 of a grain (0.4 mg) in a vial of 25 sublingual tablets with three refills The instructions read, “Place 1 tablet under the tongue every 5 minutes as needed for chest pain. May repeat 3 times.”
3) This script for Okla Beaty is for 30 Nitro-Dur (nitroglycerin) 0.4 mg patches with 3 refills. The instructions read, “Apply 1 patch at 8 A.M. and remove at 10 P.M. daily.” It is noteworthy that the patch is removed at night to provide a nitrate free interval.
4) This script for Okla Beaty is for a one month supply of Coumadin (warfarin) 5 mg tablets with no refills. The directions read, “Take ½ tablet on Sunday, Tuesday, Thursday, and Saturday, and take 1 tablet on Monday, Wednesday, and Friday.”
5) This prescription for Okla Beaty is for Spiriva (tiotropium) 30 capsules for inhalation with 3 refills. The instructions read, “Inhale 1 capsule by mouth daily.” It is noteworthy that this drug comes with a special inhaler that crushes the capsule so the patient can inhale it. The pharmacy may need to explain to the patient how to use the inhaler upon the initial fill.
6) This prescription for Patricia Pearson is for 90 Lipitor (atorvastatin) 10 mg tablets with no refills. The instructions read, “Take 1 tablet by mouth daily.”

7) This prescription for Patricia Pearson is for a 10 mL vial of Humulin R (regular insulin) with 2 refills. The directions read, “Inject 8 units subcutaneously before breakfast, 8 units before lunch, and 11 units before supper.
8) This prescription for Patricia Pearson is for a 10 mL vial of Novolin N (isophane – NPH – insulin) with 5 refills. The instructions read, “Inject 24 units subcutaneously every morning and 22 units subcutaneously every evening.”
9) This prescription for Patricia Pearson is for 90 capsules of Cardizem CD (extended release diltiazem) 240 mg with no refills. The instructions read, “Take 1 capsule by mouth daily.”
10) This prescription for Patrick Pearson is for a 1 month supply of Hytrin (terazosin) 1 mg capsules with 2 refills. The instructions read, “Take 1 capsule by mouth at bedtime for 3 days, then take 2 at bedtime for 5 days, then take 4 at bedtime thereafter.”
11) This prescription for Richard Stallman is for 30 tablets of Ambien (zolpidem tartrate) 5 mg tablets with 1 refill. The instructions read, “Take 1 tablet by mouth at bedtime as needed for sleep.” It is noteworthy that this is a prescription for a schedule 4 controlled substance and therefore Dr. Smith included his DEA number. If you check his DEA number you will find it does validate.
12) This prescription for Richard Stallman is for 30 capsules of Adderall XR (amphetamine and dextroamphetamine) 25 mg with no refills. The directions read, “Take 1 capsule by mouth daily.” It is noteworthy that this is a schedule 2 controlled substance and therefore no refills are allowed. Dr. Smiths DEA number does validate.
13) This prescription for Richard Stallman is for 100 mL of Augmentin (amoxicillin and clavulanic acid) 400 mg amoxicillin and 57 mg clavulanic acid per 5 mL with no refills. The directions read, “Take 1 teaspoonful by mouth every 12 hours for 10 days.”
14) This prescription for Richard Stallman is for Tobrex (tobramycin) ophthalmic drops with no refills. The instructions read, “ Place 2 drops into the left eye every 2 hours on day 1 and and place 2 drops every 4 hours on days 2 and 3. Call physician if eye infection persists.”
15) This prescription for Richard Stallman is for 180 tablets of Sinemet (carbidopa and levodopa) 25/100 with 5 refills. The instructions read, “Take 2 tablets by mouth three times a day.”
16) This prescription for Barbara Ericson is for 9 tablets of Imitrex (sumatriptan) 25 mg with 6 refills. “The instructions read, “Take 1 tablet every 6 hours as needed for migraine.”
17) This prescription for Kurt Thomas is for Trusopt (dorzolamide) 2% ophthalmic solution with 6 refills. The instructions are, “Place 1 drop in each eye three times a day.”
18) This Prescription for Tania Beltran is for 4 Fosamax (alendronate) 70 mg tablets with refills as needed. The directions read, “Take 1 tablet weekly.”

Worksheet 8-3

1) P
2) J
3) C
4) E
5) F
6) G
7) H
8) I
9) L
10) N
11) M
12) Q
13) K
14) D
15) B
16) A
17) O
18) U
19) R
20) S
21) T
22) A
23) L
24) B
25) C
26) D
27) E
28) H
29) I
30) G
31) J
32) K
33) F
34) T
35) S

36) X
37) C
38) D
39) E
40) F
41) H
42) I
43) K
44) J
45) G
46) N
47) O
48) P
49) Q
50) R
51) A
52) B
53) U
54) V
55) W
56) Y
57) L
58) M
59) E
60) N
61) B
62) Q
63) R
64) L
65) D
66) A
67) G
68) C
69) K
70) F
71) I
72) H
73) J
74) M
75) O
76) P
77) K
78) B
79) C
80) E
81) D
82) F
83) G
84) L
85) M
86) H
87) I
88) A
89) J
90) B
91) C
92) C
93) D
94) The abbreviation "U" can be misread as the number "0" or "4" or the cursive letters for "cc".
95) Apothecary symbols should be avoided because they can easily be misread as number or other symbols (i.e., the dram symbol looks like a 3, the abbreviation for grain can be misread as gram, and the minim symbol can be misread as milliliter).
96) A trailing zero should be avoided because if the decimal is missed it may result in a patient being overdosed.
97) Yes, you should place a leading zero before a number that is less than 1 as it will help to emphasize the decimal.

Chapter 9 - Basic Medication Calculations

Worksheet 9-1

1) 4 capsules/dose
2) 8 tablets/dose
3) 6 tablets/dose
4) 4 capsules/dose
5) 0.5 tablets/dose
6) 4 capsules/dose
7) 4 tablets/dose
8) 5 tablets/dose
9) 2 tablets/dose
10) 2 tablets/dose
11) 6 tablets/day
12) 24 tablets/day
13) 12 tablets/day
14) 8 capsules/day
15) 16 capsules/day
16) 48 tablets/day
17) 12 tablets/day
18) 1 tablet/day
19) 2 tablets/day
20) 6 tablets/day
21) 14 tablets
22) 14 tablets
23) 21 tablets
24) 30 tablets
25) 14 tablets
26) 12 capsules
27) 63 tablets
28) 84 tablets
29) 540 capsules

Worksheet 9-2

1) 0.75 cc
2) 0.4 cc
3) 0.5 cc
4) 0.4 cc
5) 1.5 cc
6) 1.875 cc
7) 1.6 cc
8) 0.8 cc
9) 1.5 mL
10) 21 mL
11) 6.5 mL
12) 1.2 cc
13) 2.5 cc
14) 1.5 cc

15) 0.5 cc
16) 0.15 mL
17) 0.28 mL
18) 2.4 mL
19) 2 cc
20) 1.5 cc
21) 0.5 cc
22) 3.75 cc
23) 0.75 cc
24) 8 mL
25) 22.4 mL

Worksheet 9-3
1) 1.5 g
2) 10%
3) 50 cc
4) 1.5 g
5) 10%
6) 150 cc
7) 1.6 g
8) 0.5 g
9) 20%
10) 1%
11) 200 cc
12) 300 mL
13) 50 cc
14) 4.5 g
15) 62.5%

Worksheet 9-4
1) 192.3 mL
2) 10 mL
3) 8 mL
4) 5 mL
5) 10 mL
6) 2.25 g
7) 1 g
8) 4 mL
9) 250 mL
10) ½ tablespoon

Worksheet 9-5
1) 2 tablets/dose
2) 3 tablets/dose
3) 2 tablets/dose
4) 1.5 tablets/dose
5) 1 capsule/dose
6) 8 tablets/day
7) 1 tablet/day
8) 8 capsules/day
9) 15 tablets/day
10) 2 capsules/day
11) 0.6 cc
12) 0.4 cc
13) 1.5 cc
14) 0.8 cc
15) 2.5 cc
16) 0.8 g
17) 3%
18) 90 mL
19) 0.5 g
20) 10%

Chapter 10 - Basic Infusion Calculations

Worksheet 10-1
1) 8 cc of 50% stock solution; 32 cc of diluent
2) 400 cc of 25% stock solution; 600 cc of diluent
3) 200 mL of 50% stock solution; 300 ml of diluent
4) 25 mL of 40% stock solution; 475 mL of diluent
5) 80 cc of 20% stock solution; 120 cc of diluent
6) 5 cc of 40% stock solution; 15 cc of diluent
7) 102.7 mL of 14.6% sodium chloride; 397.3 mL sterile water for injection
8) 100 mL of 50% mannitol; 150 mL of diluent
9) 24 mL of 10% povidone-iodine; 216 mL of diluent
10) 42 mL of 5.95% sodium hypochlorite solution; 958 mL of diluent
11) 0.61%
12) 167 mL of 70% dextrose solution; 833 mL of 40% dextrose solution

Worksheet 10-2
1) 31 gtt/min
2) 42 gtt/min
3) 21 gtt/min
4) 26 gtt/min
5) 150 mL/hr
6) 64.75 mL/hr
7) 25 minutes
8) 0300 on Wednesday
9) 10 mL/hr
10) approximately midnight (0000)
11) While technically either tubing set could be used, it would probably be easier to time 50 drops/minute with the microdrip tubing than 8 drops/minute with the other tubing.

Worksheet 10-3
1) 37.5 mL
2) 0.18 mL

3) 5.8 mL
4) 12.5 mL
5) 1.6 mL
6)
    a) 2850 mg
    b) 11,799 mg
7)
    a) 1400 mg
    b) 700 mg
8) 117,273 units

Worksheet 10-4
1) 470 mcg
2)
    a) 50 mg
    b) 12.5 mL
    c) 1 vial
3) 768 mL
4) 1 mL
5) 6.56 g

Worksheet 10-5
1)
    a) 12.7 mg
    b) 14.6 mg
    c) 2.3 mg
    d) 36.4 mg
2)
    a) 25 mg
    b) 20.8 mg
    c) 21.3 mg
    d) 36.4 mg
3)
    a) 125 mg
    b) 127.5 mg
    c) 146.7 mg
    d) 100 mg

Worksheet 10-6
1) 51.4 mL of 14.6% sodium chloride; 198.6 mL sterile water for injection
2) 12.5 mL of 2% stock solution
3) 1316 mL of 95% ethyl alcohol; 1184 mL of NS
4) 6 mL of 10% cyclosporine solution
5) 18 mL of 5% potassium permanganate stock solution; 162 mL of diluent
6) 56 gtt/min
7) 125 mL/hr
8) 31 gtt/min
9) 15 minutes
10) 21 gtt/min
11) 0.66 mg
12) 19.9 mg
13) 5.727 g
14) 45 mL
15) 9.4 mL
16) 117 mg
17) 2 mg
18)
    a) 1485 mg
    b) 29.7 mL
    c) 3 vials
19) 864 mL
20) 670 mg
21) Young's rule
22) 113.6 mg
23) 133.5 mg
24) 133.3 mg
25) 171.4 mg
26) 166.7 mg

Chapter 11 - Day's Supply

Worksheet 11-1
1) 30 days
2) 6 days
3) 10 days
4) 30 days
5) 50 days
6) 12 days
7) 90 days
8) 13 days
9) 30 days
10) 14 days
11) Many pharmacies would enter it as 5 days even though it works to $4.\overline{6}$ days.
12) 28 days
13) 30 days
14) 27 days
15) 15 days
16) 30 days
17) 5 days
18) 18 days
19) 25 days
20) 100 days
21) 60 days
22) 30 days
23) 6 days
24) 90 days
25) Even though there is enough gel for 7 days, the script is for 5 days.
26) Even though there is enough for 11 days, the script is for 10 days.
27) 14 days
28) 30 days
29) 30 days
30) 10 days
31) 30 days
32) 60 days
33) 16 days (many places automatically enter prn inhalers for 30 days)
34) 100 days
35) 84 days

36) 7 days
37) 90 days
38) 100 days
39) 30 days
40) 30 days
41) Even though there is enough cream for 9 days, the script is for 7 days.
42) 6 days
43) 12 days
44) 90 days
45) 12 days
46) 30 days
47) 10 days
48) 28 days
49) 5 days
50) 30 days

Worksheet 11-2

1) Dispense 1 inhaler at a time with 3 refills and no partials.
2) Dispense 5 boxes at a time with 5 refills and no partials.
3) Dispense 60 capsules at a time with 5 refills and no partials.
4) Dispense 30 tablets at a time with 11 refills and no partials.
5) Dispense 1 bottle at a time with 2 refills and no partials.
6) Dispense 4 tablets at a time with 2 refills and no partials.
7) Dispense 120 tablets at a time with 5 refills and no partials.
8) Dispense 1 vial at a time with 2 refills and no partials.
9) Dispense 30 tablets at a time with 9 refills and no partials.
10) Dispense 2 vials at a time with 2 refills and no partials.
11) Dispense 30 tablets at a time with 5 refills and a partial of 20 tablets.
12) Dispense 30 capsules at a time with 1 refill and no partials.
13) Dispense 30 tablets at a time with 5 refills and no partials.
14) Dispense 30 tablets at a time with 9 refills and no partials.
15) Dispense 60 capsules at a time with 2 refills and partial of 44 capsules.
16) Dispense 30 patches at a time with 2 refills and no partials.
17) Dispense 1 bottle at a time with 11 refills and no partials.
18) Dispense 1 bottle at a time with 2 refills and no partials.
19) Dispense 30 tablets at a time with 9 refills and no partials.
20) Dispense 90 tablets at a time with 5 refills and no partials.
21) Dispense 30 tablets at a time with 9 refills and no partials.
22) Dispense 30 tablets at a time with 11 refills and no partials.
23) Dispense 5 bottles at a time with 2 refills and a partial of 3 bottles.
24) Dispense 90 tablets at a time with 5 refills and no partials.
25) Dispense 60 tablets at a time with no refills and a partial of 40 tablets.

Chapter 12 - Compounding Math

Worksheet 12-1

1) Add 34 mL of water each time.
2) 4 mL clindamycin phosphate; 56 mL of Cetaphil Lotion
3) 40.8 mL diphenhydramine elixir; 12 mL lidocaine viscous; 48 mL nystatin suspension; 12.48 mL erythromycin ethyl succinate suspension; 6.72 mL cherry syrup
4) 8 capsules
5) 250 mL of each ingredient
6) 4.5 g ibuprofen powder
7) 5 capsules
8) 2.4 g testosterone; 5.2 g menthol; 112.4 g hydrophilic petrolatum
9) 3.4 g cholesterol; 3.4 g stearyl alcohol; 9.1 g white wax; 97.7 g white petrolatum *or* 3.6 g

cholesterol; 3.6 g stearyl alcohol; 9.6 g white wax; 103.2 g white petrolatum

10) 20 metfromin tablets; 100 mL Ora-Plus; 100 mL Ora-Sweet

Worksheet 12-2

1) 4% diclofenac sodium
2) 2.4 g glycopyrrolate; 2.3 mL benzyl alcohol; qs 240 mL purified water
3) 120 tablets
4) 25 g potassium bromide
5) 1% hydrocortisone
6) 9.6 mL promethazine; 125 mL codeine; 105.4 mL cherry syrup
7) 60 tablets
8) 4 tablets
9) 6 capsules
10) 160 mL nystatin; 16 mL gentamicin; 8 mL colistimethate
11) 12 acetazolamide tablets; 60 mL Ora-Plus; 60 mL Ora-Sweet
12) 24 amlodipine tablets; 60 mL Ora-Plus; 60 mL Ora-Sweet
13) 12 tablets
14) 120 tablets
15) 15 tablets

Worksheet 12-3

1) Add 30 mL of water initially to wet powder, then add another 60 mL
2) 20 mL Donnatal elixir; 20 mL lidocaine viscous; 80 mL Mylanta
3) 5.68 g ichthammol; 5.68 g lanolin; 45.44 g white petrolatum *or* 6 g ichthammol; 6 g lanolin; 48 g white petrolatum
4) 38.4 g zinc oxide; 1.2 g menthol; 2.4 g phenol; 220.8 mL calcium hydroxide solution; qs ad 480 mL olive oil
5) 4.8 g diclofenac sodium; 55.2 g Pentravan cream
6) 30 tetracycline capsules; 150 mL Ora-Plus; 150 mL Ora-Sweet
7) 120 mg benzocaine; 12 g powdered sugar; 840 mg acacia; 6 drops food coloring; 6 drops flavoring
8) 0.91 mL
9) 12.5 g erythromycin concentrate; 37.5 g ophthalmic base; 0.5% erythromycin
10) 12 clonazepam tablets; 60 mL Ora-Plus; 60 mL Ora-Sweet
11) 12 tablets
12) 30 bethanechol tablets; 30 mL Ora-Plus; 30 mL Ora-Sweet
13)
    a) 15 mL
    b) 240 mL
    c) 240 tablets
14) 40 diltiazem tablets; 150 mL Ora-Plus; 150 mL Ora-Sweet
15) 1.8 g gabapentin; 0.72 g xanthum gum; 1.35 g stevia; 1.35 g acesulfame; 0.18 g sodium saccharin; 0.36 mL magnasweet solution; 0.18 mL citric acid; 0.9 g sodium chloride; 1.8 mL bitter stopping agent flavor; 3.6 mL glycerin; 5.4 mL chicken flavor; qs 180 mL bacteriostatic water

Worksheet 12-4

1) 5.7%
2) crush and triturate 12 tablets then weigh out 3.71 grams
3) 2.4 g procaine; 25.24 g cocoa butter
4) size 3 capsule shell
5)
   a) 58.5%
   b) 5 mg morphine
   c) 5 mg cocaine
6) 30 levothyroxine tablets; 48 mL glycerin; qs 120 mL distilled water
7) 25 g bentonite; 475 mL purified water
8) crush and triturate 13 tablets then weigh out 7.94 grams
9)
   a) 2.4 g procaine
   b) 1.2 g hydrocortisone acetate
   c) 0.72 g witch hazel; 0.73 mL witch hazel
   d) 44.55 g cocoa butter
10) size 1 capsule shell

Worksheet 12-5

1) Add 30.5 mL of water each time.
2) Add 26 mL of water each time.
3) 30 mL lidocaine HCl; 90 mL Cetaphil Lotion
4) 6 g precipitated sulfur; 1.2 g salicylic acid; 52.8 g hydrophilic ointment
5) 45.4 g precipitated sulfur; 9.08 g salicylic acid; 399.52 g hydrophilic ointment
6) 160 mL of each ingredient
7)
   a) 368 mL of Mudd Mixture
   b) 320 mL nystatin; 32 mL gentamicin; 16 mL colistimethate
   c) 1 bottle of nystatin; 2 vials of gentamicin; 3 vials of colistimethate
8) 12 capsules
9)
   a) 0.568 g metronidazole; 0.568 g silver sulfadiazine; 2.84 g glycerin; qs 56.8 g hydrophylic ointment *or* 0.6 g metronidazole; 0.6 g silver sulfadiazine; 3 g glycerin; qs 60 g hydrophylic ointment
   b) 2.27 mL *or* 2.4 mL glycerin
10)
   a) 120 mg lidocaine HCl; 12 g powdered sugar; 840 mg acacia; 6 drops food coloring; 6 drops flavoring
   b) 3 mL lidocaine HCl
11) 5 tablets
12) crush and triturate 4 tablets then weigh out 0.3 grams
13) 0.6 g aminophylline; 12.47 g cocoa butter
14) size 0 capsule shell
15)
   a) 24 mL prednisone elixir
   b) 156 mL cherry syrup

Chapter 13 - Calculations for Billing Compounds

Worksheet 13-1
1) $19.84
2) $33.59
3) $33.80
4) $78.92
5) $38.53
6) $43.04
7) $30.91
8) $46.45
9) $20.52
10) $34.18

Worksheet 13-2
1) $83.21
2) $38.69
3) $29.95
4) $21.19
5) $25.03
6) $56.51
7) $19.92
8) $38.12
9) $27.41
10) $13.98

Worksheet 13-3
1) $34.49
2) $32.76
3) $14.55
4) $17.50
5) $96.92
6) $71.95
7) $39.71
8) $34.35
9) $21.76
10) $194.50

Worksheet 13-4
1) $36.23
2) $17.18 or $17.29
3) $110.15
4) $23.79
5) $14.03
6) $23.27
7) $20.83
8) $21.83
9) $11.23
10) $20.93

Chapter 14 - Pharmacy Business Math

Worksheet 14-1

Across
4) inventory
6) schedule II medications
7) inventory value
10) prime vendor purchasing
14) Material Safety Data Sheet
16) perpetual inventory
17) closed formulary
18) capitation fee
20) third party reimbursement
21) reorder point
22) depreciation

Down
1) Occupational Safety and Health Administration
2) net profit
3) direct purchasing
5) usual and customary price
8) open formulary
9) dispensing fee
11) wholesaler purchasing
12) average wholesale price
13) capital expenditures
15) gross profit
16) purchasing
19) formulary

Worksheet 14-2
1) 5 boxes
2) 0 bottles
3) 1 bottle
4) 41 bottles
5) 0 bottles
6) 5 bottles
7) 0 bottle
8) 3 bottles
9) 6 bottles
10) 7 bottles
11) 4 bottles
12) 6 bottles
13) 0 bottles
14) 6 bottles
15) 0 bottles
16) 11.5 turnovers annually
17) 12 turnovers annually
18) 12.5 turnovers annually
19) 31.65 days' supply inventory; $11,945.15 below budget
20) 30.33 days' supply inventory; $2,061.74 over budget
21) 29.12 days' supply inventory; $13,838.86 over budget
22) Most pharmacies use wholesalers to simplify their ordering process.
23) Schedule III – V medications may be ordered with the rest of the pharmacy's drug ordered.
24) Schedule II medications

are ordered via a DEA 222 order form using either a triplicate paper form, or through CSOS enabled software.

25) The pharmacist and the vendor should be notified immediately.
26) Pharmacists are required to check in controlled substances, although a technician may do so under the direct supervision of a pharmacist.

Worksheet 14-3

1) Environmental considerations include proper temperature, ventilation, humidity, light and sanitation.
2) Freezer: -25° to -10° C (-13° to 14° F); Refrigerator: 2° to 8° C (36° to 46° F); Controlled room temperature: 15° to 30° C (59° to 86° F)
3) More than 200 medications are considered light sensitive.
4) It breaks down into cyanide.
5) The state board of pharmacy sets sanitation standards.
6) A legend drug is a medication that requires a prescription and only "authorized personnel" should have access to it.
7) Schedule III – V medications must either be stored in a secured vault or be distributed throughout the pharmacy stock.
8) Schedule II medications must also either be stored in a secured vault or be distributed throughout the pharmacy stock; although, some states specifically require Schedule II medications to be stored in a secured vault.
9) The safety requirements include everything from the proper inventory rotation to avoid dispensing expired products, to material safety data sheets to provide the necessary information for safe clean up after accidental spills, to appropriate handling of oncology materials, and proper storage of chemicals and flammable items.
10) 10/31/2020

Worksheet 14-4

1) $14.97
2) $37.41
3) $105.60
4) $269.70
5) $361.41
6) $82.50
7) $205.50
8) $460.50
9) $201.36
10) $381.90
11) $44.63
12) $207.73
13) $341.60
14) $284.38
15) $330.66
16) $236.60
17) $348.97
18) $362.83
19) $370.39
20) $826.88
21) $18.50
22) $25.25
23) $18.50
24) $153.50
25) $9.50
26) $14.22
27) $35.54
28) $100.32
29) $256.22
30) $343.34
31) $78.38
32) $195.23
33) $437.48
34) $191.29
35) $362.81
36) $40.17
37) $186.96
38) $307.44
39) $255.94
40) $297.59
41) $212.94
42) $314.07
43) $326.55
44) $333.35
45) $744.19
46) 20%
47) $16.99
48) $3\frac{1}{3}$%
49) $5,225
50) $7.64
51) $16,200

Worksheet 14-5

1) $18.95 gross profit; $13.95net profit

2) $23.63 gross profit; $18.63 net profit
3) $6.45 gross profit; $1.45 net profit
4) $41.03 gross profit; $36.03 net profit
5) $8.83 gross profit; $3.83 net profit
6) $7.53 gross profit; $2.53 net profit
7) $24.01 gross profit; $19.01 net profit
8) $5.44 gross profit; $0.44 net profit
9) $80.33 gross profit; $75.33 net profit
10) $102.22 gross profit; $97.22 net profit
11) $103.34 gross profit; $98.34 net profit
12) $44.12 gross profit; $39.12 net profit
13) $29.15 gross profit; $24.15 net profit
14) $11.77 gross profit; $6.77 net profit
15) $106.06 gross profit; $101.06 net profit
16) $4.92 gross profit; $0.08 net loss
17) $9.43 gross profit; $4.43 net profit
18) $27.17 gross profit; $22.17 net profit
19) $4.26 gross profit; $0.74 net loss
20) $3.97 gross profit; $1.03 net loss
21) $18.95 gross profit; $13.95 net profit
22) $23.63 gross profit; $18.63 net profit
23) $6.45 gross profit; $2.95 net profit
24) $41.03 gross profit; $33.53 net profit
25) $8.83 gross profit; $5.33 net profit
26) $7.53 gross profit; $4.03 net profit
27) $24.01 gross profit; $19.01 net profit
28) $5.44 gross profit; $1.94 net profit
29) $80.33 gross profit; $72.83 net profit
30) $102.22 gross profit; $94.72 net profit
31) $103.34 gross profit; $95.84 net profit
32) $44.12 gross profit; $36.62 net profit
33) $29.15 gross profit; $24.15 net profit
34) $11.77 gross profit; $8.27 net profit
35) $106.06 gross profit; $98.56 net profit
36) $4.92 gross profit; $1.42 net profit
37) $9.43 gross profit; $5.93 net profit
38) $27.17 gross profit; $22.17 net profit
39) $4.26 gross profit; $0.76 net profit
40) $3.97 gross profit; $0.47 net profit

Worksheet 14-6

1) $30.47
2) $39.52
3) $7.50
4) $68.33
5) $7.50
6) $11.29
7) $26.24
8) $7.26
9) $144.30
10) $186.63
11) $188.80
12) $74.29
13) $11.91
14) $19.51
15) $194.05
16) $6.26
17) $14.95
18) $46.36
19) $5.06
20) $4.41
21) Fred's Pharmacy lost $121 last month on Senile Sally.
22)
   a) $2,750.00
   b) $847.37
   c) $1,902.63 profit

Worksheet 14-7

1)	Register 1	Register 2	Register 3	Total
+ Cash and Checks	513.12	---	300.44	*813.56*
+ Bank Charges	120.00	90.00	---	*210.00*
+ House Charges	---	---	52.02	*52.02*
+ Paid Outs	---	5.12	---	*5.12*
Total	*633.12*	*95.12*	*352.46*	*1,080.70*
+ Closing Reading	760.02	145.12	402.50	*1,307.64*
- Opening Reading	50.00	50.00	50.00	*150.00*
= Difference	*710.02*	*95.12*	*352.50*	*1,157.64*
- Coupons	---	---	---	*0.00*
- Discounts	8.90	---	---	*8.90*
- Voids	10.00	---	---	*10.00*
- Refunds	---	---	---	*0.00*
- Over-Rings	56.65	---	---	*56.65*
Total	*634.47*	*95.12*	*352.50*	*1,082.09*
Over + or Short -	*-1.35*	*0.00*	*-0.04*	*-1.39*

2)	Register 1	Register 2	Register 3	Total
+ Cash and Checks	3,897.65	12.00	1,533.12	*5,442.77*
+ Bank Charges	2,111.05	96.67	140.00	*2,347.72*
+ House Charges	25.00	---	---	*25.00*
+ Paid Outs	---	---	---	*0.00*
Total	*6,033.70*	*108.67*	*1,673.12*	*7,815.49*
+ Closing Reading	6,121.47	158.70	1,800.02	*8,080.19*
- Opening Reading	50.00	50.00	50.00	*150.00*
= Difference	*6,071.47*	*108.70*	*1,750.02*	*7,930.19*
- Coupons	14.25	---	1.10	*15.35*
- Discounts	23.47	---	8.90	*32.37*
- Voids	---	---	10.00	*10.00*
- Refunds	---	---	---	*0.00*
- Over-Rings	---	---	56.65	*56.65*
Total	*6,033.75*	*108.70*	*1,673.37*	*7,815.82*
Over + or Short -	*-0.05*	*-0.03*	*-0.25*	*-0.33*

3)	Register 1	Register 2	Register 3	Total
+ Cash and Checks	1,713.12	45.12	2,002.18	*3,760.42*
+ Bank Charges	240.00	20.00	350.44	*610.44*
+ House Charges	---	---	---	*0.00*
+ Paid Outs	---	---	10.00	*10.00*
Total	*1,953.12*	*65.12*	*2,362.62*	*4,380.86*
+ Closing Reading	2,172.67	115.12	2,429.66	*4,717.45*
- Opening Reading	50.00	50.00	50.00	*150.00*
= Difference	*2,122.67*	*65.12*	*2,379.66*	*4,567.45*
- Coupons	---	---	7.00	*7.00*
- Discounts	130.00	---	---	*130.00*
- Voids	27.50	---	---	*27.50*
- Refunds	12.00	---	---	*12.00*
- Over-Rings	---	---	10.00	*10.00*
Total	*1,953.17*	*65.12*	*2,362.66*	*4,380.95*
Over + or Short -	*-0.05*	*0.00*	*-0.04*	*-0.09*

4)	Register 1	Register 2	Register 3	Total
+ Cash and Checks	1234.56	789.01	234.56	*2,258.13*
+ Bank Charges	789.01	234.56	78.90	*1,102.47*
+ House Charges	---	---	25.00	*25.00*
+ Paid Outs	20.00	10.00	---	*30.00*
Total	*2,043.57*	*1,033.57*	*338.46*	*3,415.60*
+ Closing Reading	2,116.45	1,091.03	390.43	*3597.91*
- Opening Reading	50.00	50.00	50.00	*150.00*
= Difference	*2,066.45*	*1,041.03*	*340.43*	*3,447.91*
- Coupons	2.50	1.75	---	*4.25*
- Discounts	12.34	5.67	---	*18.01*
- Voids	---	---	---	*0.00*
- Refunds	8.00	---	---	*8.00*
- Over-Rings	---	---	---	*0.00*
Total	*2,043.61*	*1,033.61*	*340.43*	*3,417.65*
Over + or Short -	*-0.04*	*-0.04*	*-1.97*	*-2.05*

Worksheet 14-8

1) $4.13
2) $3.53
3) $1,026.00
4) $634.00

Worksheet 14-9

1) g
2) o
3) u
4) l
5) i
6) p
7) q
8) r
9) w
10) m
11) n
12) k
13) v
14) e
15) t
16) a
17) b
18) d
19) h
20) f
21) j
22) s
23) c
24) 1
25) 2
26) 0
27) 3
28) 0
29) 11.5 annual turnover rate
30) 12 annual turnover rate
31) 31.65 days; $10,750.65 under budget
32) 30.33 days; $1,855.55 over budget
33) DEA 222 form or CSOS
34) On the regular warehouse order form
35) Controlled substances are to be checked in by the pharmacist; a pharmacy technician may assist under the direct supervision of a pharmacist.
36) Under federal guidelines CIIs may be either locked in a secure vault or distributed throughout the inventory. CIII-V medications may also either be locked in a secure vault or distributed throughout the inventory.
37) Temperature, ventilation, humidity, light, safety, and sanitation.
38) Freezer -25° to -10° C (-13° to 14° F); refrigerated 2° to 8° C (36° to 46° F); controlled room temperature 15° to 30° C (59° to 86° F)
39) 02/28/2020 will expire first because 2020 is a leap year and therefore 02/2020 would not expire till the 29th of February that year.
40) 40% markup
41) $18.48
42) 3%
43) $6,175.00
44) $4.50
45) $17,820.00
46) $20.95 gross profit; $15.95 net profit
47) $25.63 gross profit; $20.63 net profit
48) $7.01 gross profit; $2.01 net profit
49) $44.03 gross profit; $39.03 net profit
50) $9.73 gross profit; $4.73 net profit
51) $8.13 gross profit; $3.13 net profit
52) $132.50
53) $39.52
54) $11.66
55) $68.12
56) $10.89
57) $21.25 profit
58) $1,138.25 annual depreciation

59)

	Register 1	Register 2	Register 3	Total
+ Cash and Checks	1264.56	790.01	233.56	*2,288.13*
+ Bank Charges	789.03	235.56	78.90	*1,103.49*
+ House Charges	---	---	25.00	*25.00*
+ Paid Outs	20.00	10.00	---	*30.00*
Total	*2,073.59*	*1,035.57*	*337.46*	*3,446.62*
+ Closing Reading	2,151.47	1,095.03	389.47	*3,635.97*
- Opening Reading	50.00	50.00	50.00	*150.00*
= Difference	*2,101.47*	*1,045.03*	*339.47*	*3,485.97*
- Coupons	7.50	1.75	---	*9.25*
- Discounts	12.34	7.67	---	*20.01*
- Voids	---	---	---	*0.00*
- Refunds	8.00	---	---	*8.00*
- Over-Rings	---	---	---	*0.00*
Total	*2,073.63*	*1,035.61*	*339.47*	*3,448.71*
Over + or Short -	*-0.04*	*-0.04*	*-2.01*	*-2.09*

60) $3.67

Chapter 15 - Parenteral Dosage Calculations

Worksheet 15-1

1) 24.75 mL
2) 1 mL
3) 50 mg
4) 0.12 mL
5) 0.8 mL
6) 2.5 mL
7) 12.5 mL
8) 14 mL
9) 1.6 mL
10) 4.7 mL
11) 31 mL
12) 70 mg
13) 0.5 mL
14) 4 mL
15) 7.5 mL
16) 12 mL
17) 6 mL

Worksheet 15-2

1) Parenteral means a route of administration other than the GI tract. Technically this includes everything from topical medications and inhalation therapies to ear drops and injections, but today the term parenteral is intended to mean various kinds of injections and infusions and generally excludes all other routes of administration.
2) d c p e i l k b a o n j h g m f
3) inhalation and ophthalmic
4) sterile, free of visible particulate material, pyrogen-free, stable for intended use, a pH similar to that of human blood, isotonic
5) 2.5 mL
6) 0.15 mL
7)
    a) 12.5 mL
    b) 1.5 mL
    c) 21 mL
    d) 31.25 mL
    e) 0.3 mL
    f) 0.5 mL
    g) 4 mL
8) The syringe should be marked at 2 mL
9) 5 mL
10) 3 mL
11) 100 mg

12) 250,000 u/mL = 26 mL; 500,000 u/mL = 13 mL; 1,000,000 u/m/L = 6.5 mL

13) Both syringes should be marked at 16 mL.

Chapter 16 - Insulin

Worksheet 16-1

1) A; Mark the syringe at 12 units.
2) B; Mark the syringe at 5 units.
3) D; Mark the syringe at 35 units.
4) E; Mark the syringe at 72 units.
5) I; Mark the syringe at 55 units.
6) C; Mark the syringe at 22 units.
7) F; Mark the syringe at 80 units.
8) H; Mark the syringe at 45 units.
9) B; Mark the syringe at 25 units.
10) G; Mark the syringe at 0.34 cc.
11) D; Mark the syringe at 40 units.
12) F; Mark the syringe at 77 units.
13) E; Mark the syringe at 21 units.
14) A; Mark the syringe at 0.5 cc.
15) The clear rapid or short-acting insulin is actually drawn into the syringe prior to adding the cloudy isophane (NPH) insulin.
16) A & I; First, calculate the total dose. (15 units + 30 units = 45 units) Then draw up 30 units of air and inject it into the Humulin N vial, but do not draw up any solution yet. Withdraw the needle from the vial. Next, draw up 15 units of air, inject it into the Humulin R, and draw 15 units of regular insulin. Next, insert the needle into the Humulin N vial and carefully invert the vial without injecting any solution into the isophane (NPH) insulin. Lastly, slowly withdraw insulin from the Humulin N vial until the vial contains a total of 45 units of insulin.
17) B & I; First, calculate the total dose. (20 units + 45 units = 65 units) Then draw up 45 units of air and inject it into the Humulin N vial, but do not draw up any solution yet. Withdraw the needle from the vial. Next, draw up 20 units of air, inject it into the Humalog, and draw 20 units of lispro insulin. Next, insert the needle into the Humulin N vial and carefully invert the vial without injecting any solution into the isophane (NPH) insulin. Lastly, slowly withdraw insulin from the Humulin N vial until the vial contains a total of 65 units of insulin.

Chapter 17 - Mmol, mEq, mCi, & IU

Worksheet 17-1

1)
   a) 12 carbon, 22 hydrogen, 1 calcium, 14 oxygen
   b) 1 sodium, 2 hydrogen, 1 phosphorous, 4 oxygen
   c) 2 carbon, 3 hydrogen, 2 oxygen, 1 potassium
   d) 2 carbon, 6 hydrogen, 1 oxygen
2)
   a) 1020 mg monobasic potassium phosphate (KH_2PO_4)
   b) 1305 mg dibasic potassium phosphate (K_2HPO_4)
3)
   a) 2.326 mMol calcium gluconate
   b) 4.65 mEq calcium
4) 1490 mg potassium chloride
5)
   a) 32.99 mCi
   b) 0.55 mL
6) 194.2 mCi
7) 10 mcg cholecalciferol
8) 10,000 IU ascorbic acid

Chapter 18 - Powder Volume Calculations

Worksheet 18-1

1) 8 mL
2) 92 mL
3)
   a) 0.8 mL
   b) ampicillin 250 mg/mL;

sulbactam 125 mg/mL
4) 27.7 mL
5) 0.4 mL
6) 500,000 units/mL
7)
   a) 0.1 mL
   b) 2080 mg
8)
   a) 500,000 units/mL
   b) 400,000 units/mL
   c) 250,000 units/mL
   d) 200,000 units/mL
9)
   a) 0.7 mL
   b) 200 mg/mL
   c) 1.25 mL
10) 5 mg/mL

Worksheet 18-2
1) 5 mL
2) 45 mL
3)
   a) 28 mL
   b) piperacillin 200 mg/mL; tazobactam 25 mg/mL
   c) 20 mL
4) 0.8 mL
5) 95.2 mg/mL
6)
   a) 10 mL
   b) 8 g
   c) 1.14
7)
   a) 1.17 mL
   b) 1.79
8)
   a) 0.4 mL
   b) 30 mL
9)
   a) 500,000 units/mL
   b) 400,000 units/mL
   c) 250,000 units/mL
   d) 200,000 units/mL
10)
   a) 4 mL
   b) 200 mg/mL
   c) 10 mL
11)
   a) 160 mg/mL
   b) 10 mL
12)
   a) $\frac{5\text{ g}}{100\text{ mL}}$
   b) 3.5 mL of powder volume in the 2.5 g vial; 7 mL of powder volume in the 5 g vial; 14 mL of powder volume in the 10 g vial
   c) 2 of the 10 g vials, 1 of the 5 g vials, and 1 of the 2.5 g vials
   d) reconstitute the 2.5 g vial with 21.5 mL of diluent; reconstitute the 5 g vial with 43 mL of diluent; reconstitute the 10 g vial with 86 mL of diluent
   e) 275 mL

Worksheet 18-3
1)
   a) 5 mL
   b) 105 mL
   c) 95.24 mg/mL
   d) 15.75 mL
   e) This bag needs to be infused via a central line due to its 6 mg/mL concentration.
2) 0.26 mL
3) 1.3 mL
4) 91.4 mL
5)
   a) 7 mL
   b) 10 mL
6)
   a) 43 mL
   b) 5 mL
7)
   a) 0.2 mL
   b) 1800 mg
8)
   a) 250,000 units/mL
   b) 14 mL
9)
   a) 1.4 mL
   b) 200 mg/mL
   c) 7.5 mL
10)
   a) 28 mL
   b) piperacillin 400 mg/mL; tazobactam 50 mg/mL
   c) 7.5 mL

Chapter 19 - Percentage Strength

Worksheet 19-1
1) 2.4 g
2) 25 L
3) 2 mL
4) 5 mL
5)
   a) 58.5 g
   b) 234 mg/mL
6) 12.5 mL
7) 5 mL
8) 908 mg
9) 17.5 mL
10) 26 g
11) 3000 mg
12) 4 capsules
13) 2.25 mg
14) No, there is not enough drug in stock to fill this

order.

15) 30%

16)

a) 2.5 mL

b) 0.075 mL

17) 2.5%

18) 4% sulfamethoxazole; 0.8% trimethoprim

19) 0.025 mL

20) 15 mg

Worksheet 19-2

1) 130 mg%
2) 150 mg%
3) 60 tablets
4) 2 mL
5) 6 capsules
6) 336 mL
7) 950 mcg
8) 18%
9) 16 mL
10) 48 mg
11) 1.8 g testosterone; 3.9 g menthol; 84.3 g hydrophylic petrolatum
12) 21.24 mL
13) Yes, there is enough drug in stock to fill this order.
14) The child has not reached a toxic level.
15) 25 mL
16) 25 mL
17) 8% sulfamethoxazole; 1.6% trimethoprim
18) 25% ampicillin; 12.5% sulbactam
19) 7 units
20) 7.5%

Chapter 20 - Ratio Strength

Worksheet 20-1

1) 1 g
2) 20 mg
3) 2.5%
4)
   a) 1:2000
   b) 500 mcg/mL
5) Yes, the pharmacy will be able to prepare this order.
6) No, the technician will not be able to fill this order.
7) 50 mcg
8) 450 mcg
9) 0.5 mL
10)
   a) 16 mL
   b) 4 ampules
11) 1 mL
12) 0.975 mL
13) 2%
14) $1.\overline{6}\%$
15) 1:1000
16)
   a) 0.01%
   b) 1:10,000
17)
   a) 5 g
   b) 1:12.5
18) 1:10
19) 25 L
20) 1:200,000

Worksheet 20-2

1) 1:20,000
2) 2.5 mg
3) 0.4%
4)
   a) 1:4000
   b) 250 mcg/mL
5) Yes, the pharmacy will be able to prepare this order.
6) No, the pharmacy does not have enough drug in stock to prepare both bags.
7) 75 mcg
8) 300 mcg
9) 1 mL
10)
   a) 16 mL
   b) 1 vial
   c) 1:15,625
11) 3 mL
12) 1.875 mL
13) 2.5%
14) 1.25%
15) 1:500
16)
   a) 0.001
   b) 1:100,000
17)
   a) 5 g
   b) 1:20
18) 1:10
19) 12.5 L
20) 1:500,000
21)
   a) 2 vials
   b) 248 mL
   c) 80 mcg/mL

Chapter 21 - PPM, and Reducing & Enlarging Formulas

Worksheet 21-1

1) 6 g precipitated sulfur; 1.2 g salicylic acid; 52.8 g hydrophilic ointment
2) 100 g precipitated sulfur; 20 g salicylic acid; 880 g

hydrophilic ointment
3) 15 g zinc oxide; 15 g starch; 30 g white petrolatum
4) 47.37 g talc; 3.16 g benzoic acid; 9.47 g bentonite
5) 0.001%
6) 750 g aspirin; 375 g phenacetin; 125 g caffeine
7) 13.2 kg yellow ointment
8) 52.5 g parachlorophenol; 97.5 g camphor
9) 5000 g lactose; 7500 g sucrose; 12,500 g starch, direct compressing formula; 25 g magnesium sulfate; yield 50,000 tablets
10) 38.4 g calamine; 38.4 g zinc oxide; 9.6 g glycerin; 120 mL bentonite magma; qs 480 mL calcium hydroxide
11) 3.028 mL trihalomethanes
12) 10 mcg proinsulin

Chapter 22 - Dosage Calculations Based on Body Weight

Worksheet 22-1

1) 750 mg
2) 12.5 mL
3) 18.75 mL
4) 9 tablets/day
5) 20 mg
6) 425 mg/day *or* 426 mg/day
7) The syringe should be marked for 1.95 mL
8)
   a) 250 mg/dose
   b) 1000 mg/day
   c) 1 mL/dose
9)
   a) 209 mg
   b) 157 mg
   c) 5.2 mL for the loading dose; 3.9 mL for each maintenance dose
10) 125 mL
11) 4.3 mL
12)
   a) Yes
   b) 1 mL
   c) The syringe should be marked for 1 mL.
13)
   a) Yes, this is within the dosage range for this patient.
   b) 4.7 mL
14) 11 mL
15)
   a) 7.5 mL/dose
   b) 225 mL
16)
   a) 2850 mg
   b) 22.8 mL
   c) 11,799 mg
   d) 94.4 mL methylprednisolone; 905.6 mL NS

Worksheet 22-2

1) 80 mg
2)
   a) 10,000 units
   b) 12,545 units
   c) 15,000 units
   d) 17,545 units
   e) 18,000 units
3) 15 hours 26 minutes
4) 3 hours 58 minutes
5)
   a) 1718 mg/day
   b) 286 mg/dose
   c) 2.86 mL
6)
   a) 6000 mg/day
   b) 1500 mg/dose
   c) 4 mL
7)
   a) 800 mg
   b) 800 mg
   c) 800 mg
   d) 800 mg
   e) 0 mg
   f) 409 mg
   g) 0 mg
   h) 409 mg
   i) 0 mg
   j) 409 mg
   k) 0 mg
   l) 409 mg
8)
   a) 16 mL
   b) 16 mL
   c) 16 mL
   d) 16 mL
   e) 0 mL
   f) 8.2 mL
   g) 0 mL
   h) 8.2 mL
   i) 0 mL
   j) 8.2 mL
   k) 0 mL
   l) 8.2 mL
9) 568 mg
10) 150 mg
11)
   a) 3 mg
   b) Dispense three of the 1 mg capsules every 12 hours.
12)
   a) 350 mg
   b) 70 mL
   c) 280 mL
13) 98,182 units
14) The bag needs 6.75 mL so

the syringe should be marked close to 6.8 mL
15) The patient was underdosed
16) 28 tablets
17)
    a) This is within the suggested dosing guidelines.
    b) 1 mL
18)
    a) 0.5 mg
    b) 1 mL
19)
    a) 750 mg
    b) 1125 mg
    c) 1500 mg
20)
    a) 2400 mg
    b) 19.2 mL
    c) 9936 mg
    d) 79.5 mL methylprednisolone; 920.5 mL NS

Chapter 23 - Dosage Calculations Based on BSA

Worksheet 23-1
1)
    a) 2 m^2
    b) 2 m^2
    c) 1.68 m^2
    d) 1.6 m^2
2)
    a) 0.61 m^2
    b) 0.79 m^2
    c) 0.86 m^2
    d) 1.27 m^2
3) 38 mg
4) 18.4 units
5)
    a) 2000 mg
    b) Two of the 1 g vials
    c) 52.6 mL
6)
    a) 54.6 mg
    b) 27.3 mL
7) 20.64 mg
8) 660 mg
9)
    a) 9 mL/dose
    b) 91 mL NS
    c) 2 mL (this is the maximum dose)
    d) 42 tablets
    e) 42 tablets
10)
    a) 99.5 mg doxorubicin; 2650 mg cyclophosphamide
    b) 424 mg Herceptin initially (loading dose); 212 mg Herceptin weekly (maintenance dose)

Worksheet 23-2
1) 2.11 m^2
2)
    a) 11 mg
    b) 154 mg
3)
    a) 75 kg
    b) 79.9 kg
    c) 109.375 mL/min
    d) 672 mg
    e) 1.98 m^2
    f) 198 mg
    g) 990 mg
4) 495 mcg/day
5) 3.2 g
6)
    a) 447.5 mcg/day
    b) 8.95 mcg/mL
    c) 50 mg
7) 216 mg
8) 11.7 mL
9) 2.8 mL
10)
    a) 1.72 m^2
    b) 1032 mg
    c) 50.1 kg
    d) 80.27 mL/min
    e) 421 mg
11) 18 mg/day
12) 1.6 mg vincristine; 64.44 mg doxorubicin; 480 mg dexamethasone
13)
    a) 1515 mg
    b) 101 mg
    c) 2 mg
    d) 2 tablets
    e) 757.5 mg
14)
    a) 250 mg
    b) 12.5 mL
    c) 1 g
    d) 20 mL
    e) 40 mg
    f) 4 mL
15)
    a) 989 mg
    b) 346 mg
    c) 509 mg

Chapter 24 - Drip Rates & Infusion Rates

Worksheet 24-1
1) 100 mL/hr
2) 104 mL/hr
3) 120 mL/hr
4) 120 mL/hr
5)

a) 750 mL/hr
b) 3 bags

6)
a) 23 hours 48 minutes
b) 42 gtt/min

7) 125 mL/hr
8) 175 mL
9)
a) 5 bags
b) bag 1 – 0900 on Saturday; bag 2 – 1400 on Saturday; bag 3 – 1900 on Saturday; bag 4 – 0000 on Sunday; bag 5 – 0500 on Sunday

10) 42 gtt/min
11) 1800
12)
a) 96 mL
b) 19.2 mL/hr
c) 8 hours

13)
a) 375 mg
b) 8 vials
c) 75 mL of amphotericin B lipid complex; 300 mL of D5W
d) 187.5 mL/hr
e) 2 hours
f) 63 gtt/min

14)
a) 3.15 mL/hr
b) 75.6 mL/day

15)
a) send 3 bags initially
b) 2 full bags and a partial bag with 2000 mL left
c) no additional bags need to be sent

Worksheet 24-2

1) 150 mL/hr
2)
a) 200 mL/hr
b) 33 gtt/min

3) 156.25 mL/hr
4) 90 mL/hr
5) 100 mL/hr
6) 20 gtt/min
7)
a) 8 hours 20 minutes
b) 30 gtt/min

8) 2.5 mEq KCl/hr
9) 280 mL
10) 972 mL/hr
11) 48 minutes
12)
a) 3 bags
b) bag 1 – 1100 on day 1; bag 2 – 1900 on day 1; bag 3 – 0300 on day 2
c) 21 gtt/min

13)
a) 2 bags
b) bag 1 – 0900 on day 1; bag 2 – 0140 on day 2
c) 20 gtt/min

14)
a) 130 mL/hr
b) 22 gtt/min

15) approximately 0000 (midnight)
16) 43.4 mL/day
17)
a) 140 g
b) 2800 mL
c) 48 minutes

18)
a) send 4 bags initially
b) 3 bags are left
c) 1 additional bag should be sent

Chapter 25 - Dilutions & Alligations

Worksheet 25-1

1) The problem can be solved using either the dilution formula or a series of ratio-proportions. 3%
2) The problem can be solved using either the dilution formula or a series of ratio-proportions. 0.5%
3) The problem can be solved using either the dilution formula or a series of ratio-proportions. 5%
4) The problem can be solved using either the dilution formula or a series of ratio-proportions. 1:3000
5) The problem can be solved using either the dilution formula, a series of ratio-proportions, or the alligation method. 5200 mL half-normal saline; 5100 mL diluent
6) The problem can be solved using either the dilution formula, a series of ratio-proportions, or the alligation method. 205.7 mL 70% isopropyl alcohol; 274.3 mL water
7) The problem can be solved using either the dilution formula, a ratio-proportion, or the alligation method. 4 mL 2.5% cortisone acetate suspension; 4 mL NS
8) The problem can be solved using a series of ratio-proportions. 31.9% alcohol
9) The problem can be solved using either the dilution formula or a series of ratio-proportions. 2.6% amino acid; 17.5%

dextrose

10) The problem can be solved using the alligation method. 192 mL of 1:10 solution; 808 mL of 1:1000 solution
11) The problem can be solved using either the dilution formula, a series of ratio-proportions, or the alligation method. 48 mL of 10% povidone-iodine
12) The problem can be solved using either the dilution formula, a series of ratio-proportions, or the alligation method. 1200 mL of 10% solution
13) The problem can be solved using a series of ratio-proportions. 50.65%
14) The problem can be solved using either the dilution formula, a series of ratio-proportions, or the alligation method. 60 mL micronutrients; 470.6 mL of 8.5% amino acids; 71.4 mL of 70% dextrose; 398 mL of sterile water.
15) The problem can be solved using a series of ratio-proportions. 71.7% alcohol

Worksheet 25-2

1) Mixing a substance (such as a medication) with a diluent/solvent to decrease its concentration is considered a dilution.
2) An alligation is when two items with different concentrations of the same substance are being mixed together and the final (or desired) concentration (which must be in-between the other two concentrations) is known.
3) A mixture involves mixing 2 or more ingredients of varying concentrations together.
4)
    a) 64.1 mL of 23.4% sodium chloride
    b) 435.9 mL of SWFI
5) 6.25 mL of 2% stock solution
6) 1579 mL of 95% ethyl alcohol; 1421 mL of normal saline
7) 60 mL of 6.5% concentrated quaternary ammonium compound; 3840 mL of SWFI
8)
    a) 200 mL
    b) 50%
9) 295.6 mL of alcohol
10) 3 g of 10% ointment; 57 g of petrolatum
11)
    a) 0.002%
    b) 1:50,000
    c) 20 mcg/mL
12)
    a) 72.5%
    b) 10.875 g
13) 0.46 mL of 14.6% sodium chloride; 2.54 mL of NS (0.9% sodium chloride)
14) 25 mL of 5% KMnO4 solution; 225 mL of water
15) 85.7 mL of 70% IPA; 34.3 mL of water
16) 4 mL of 0.3% tobramycin; 2 mL of NS
17) 1.95 mL of 80 mg/2 mL of tobramycin; 4.05 mL of 0.3% tobramycin
18) 150 mg of edetate; 0.25% tetracaine
19) 83.3 mL of 3% sodium hypochlorite; 916.7 mL of sterile water for irrigation
20) 25 mL of 3% sodium hypochlorite; 225 mL of sterile water for irrigation
21)
    a) 941.2 mL of 8.5% amino acid
    b) 542.9 mL of 70% dextrose
    c) 2.5% lipid emulsion
    d) 138.9 mL of sterile water
    e) 83 mL/hr
    f) 28 gtt/min

Chapter 26 - Parenteral Nutrition

Worksheet 26-1

1)
    a) 1000 mL of amino acid 10%; 1000 mL of dextrose 70%; 263.55 mL of SWFI; 20 mL sodium chloride; 30 mL potassium chloride; 5.9 mL magnesium sulfate; 10 mL potassium phosphate; 100 mL calcium gluconate; 10 mL MVI; 1 mL trace elements; 2 mL vitamin C; 0.55 mL regular insulin; 1 mL folic

acid; 40 mL sodium acetate; 14 mL potassium acetate;2 mL sodium phosphate

b) 104 mL/hr
c) 4% amino acid; 28% dextrose
d) As it comes out to 47.4 mEq/L it is likely to precipitate.

2)

a) 800 mL of amino acid 5.2%%; 700 mL of dextrose 70%; 360 mL of SWFI; 9.6 mL sodium chloride; 20 mL potassium chloride; 9.9 mL magnesium sulfate; 4 mL potassium phosphate; 14 mL calcium gluconate; 10 mL MVI; 1 mL trace elements; 4 mL vitamin C; 0.95 mL regular insulin; 1.5 mL folic acid; 35 mL sodium acetate; 20 mL potassium acetate;10 mL sodium phosphate
b) 83 mL/hr
c) 2.08% amino acid; 24.5% dextrose
d) As the number is 45.255 mEq/L this will probably not precipitate.

3)

a) 1000 mL of amino acid 8.5%; 1000 mL of dextrose 70%; 825.2 mL of SWFI; 24 mL sodium chloride; 20 mL potassium chloride; 8 mL magnesium sulfate; 10 mL potassium phosphate; 50 mL calcium gluconate; 10 mL MVI; 1 mL trace elements; 2 mL vitamin C; 0.4 mL regular insulin; 0.4 mL folic acid; 30 mL sodium acetate; 14 mL potassium acetate; 5 mL sodium phosphate
b) 125 mL/hr
c) 2.83% amino acid; 23.33% dextrose
d) As the number is 37.75 mEq/L this will probably not precipitate.

Worksheet 26-2

1) 59.3 kg
2) 1346.6 kcal
3) 2801 kcal
4) 77.1 g; 308.4 kcal
5) 840.3 kcal; 93.4 g
6) 1652.3 kcal 486 g
7) 2.34 m^2
8) 3510 mL
9) sodium chloride 87.75 mEq; potassium chloride 35.1 mEq; potassium acetate 70.2 mEq; calcium gluconate 15.8 mEq; potassium phosphate 52.7 mMol; magnesium sulfate 17.55 mEq
10) Amino Acid – 77.1 g – 907.1 mL; Dextrose – 486 g – 694.3 mL; Lipid Emulsion – 93.4 g – 467 mL; Sterile Water for Irrigation – q.s. 3510 mL – 1295.15 mL; sodium chloride – 87.75 mEq – 21.9 mL; potassium chloride – 35.1 mEq – 17.55 mL; potassium acetate – 70.2 mEq – 35.1 mL; calcium gluconate – 15.8 mEq – 34 mL; potassium phosphate – 52.7 mMol – 17.6 mL; magnesium sulfate – 17.55 mEq – 4.3 mL; folic acid – 5 mg – 1 mL; MVI – 1 vial – 10 mL; trace elements – 1 vial – 1 mL; famotidine – 40 mg – 4 mL
11) 146.25 mL/hr
12) 49 gtt/min
13) Since it comes out to 34.5 mEq/L it should not precipitate.
14) 13.8% dextrose; 2.2% amino acids
15) The estimated osmolarity is 910 mOsmol/L which should be able to be infused peripherally, but remember that this is only an estimated answer.

Worksheet 26-3

1) parenteral nutrition (PN), total parenteral nutrition (TPN), partial parenteral nutrition (PPN), hyperalimentation (HAL), intravenous hyperalimentation (IVH), total nutrient admixture (TNA), 3-in-1 admixture, all-in-one admixture, centrally infused parenteral nutrition (CPN), peripherally infused parenteral nutrition (PPN)
2) Lipids are not always mixed directly with parenteral nutrition admixtures.
3) Indications for parenteral nutrition may include diseases and conditions associated with a nonfunctional gastrointestinal tract (bowel obstruction, severe pancreatitis, severe

malabsorption), cancer therapy (radiation therapy, antineoplastic drugs, bone marrow transplantation), organ failure, hyperemesis during pregnancy, severe eating disorders (anorexia nervosa) when the patient cannot tolerate enteral nutrition, or failure when a patient attempted a trial of enteral nutrition.

4) The macronutrients in parenteral nutrition include water, amino acids, dextrose, and lipids. Their corresponding roles in enteral nutrition would be water, protein, carbohydrates, and fat.
5) Calcium ions and phosphate ions have a strong affinity for each other and have the potential to precipitate out of solution.
6) Parenteral nutrition is hypertonic.
7) A common practice is if a PN has an osmolarity less than a 1000 mOsmol/L it can be infused peripherally versus an osmolarity of 1000 mOsmol/L or more is infused through a central line (an IV line that feeds into the superior vena cava).
8)
    a) amino acid 941.2 mL, dextrose 542.9 mL, lipids 250 mL, SWFI 136.8 mL, sodium chloride 40 mL, potassium acetate 40 mL, calcium gluconate 20.2 mL, sodium phosphate 10 mL, magnesium sulfate 3.9 mL, MVI 10 mL, trace elements 1 mL, famotidine 4 mL
    b) 83.3 mL/hr
    c) As the concentration works out to 34.7 mEq/L it should not precipitate.
    d) The osmolarity is to high (estimated at 1350 mOsmol/L) to be infused peripherally. This admixture would need to be infused through a central line.
9)
    a) amino acid 470.6 mL, dextrose 71.4 mL, SWFI 388 mL, sodium chloride 20 mL, potassium acetate 20 mL, calcium gluconate 10 mL, sodium phosphate 5 mL, magnesium sulfate 4 mL, MVI 10 mL, trace elements 1 mL
    b) 125 mL/hr
    c) As the concentration works out to 34.65 mEq/L it should not precipitate.
    d) This bag could be infused peripherally as the estimated osmolarity is 650 mOsmol/L.
10)
    a) amino acid 598.9 mL, dextrose 266.6 mL, lipids 410.3 mL, SWFI q.s. 1909.1 mL – 514 mL, sodium chloride 26.7 mL, potassium acetate 33.4 mL, calcium gluconate 39.4 mL, potassium phosphate 4.3 mL, magnesium sulfate 3.8 mL, folic acid 0.34 mL, MVI 10 mL, trace elements 1 mL, regular insulin 0.40 mL
    b) 79.5 mL/hr
    c) As the concentration works out to 23 mEq/L it should not precipitate.
    d) This bag could be infused peripherally as the estimated osmolarity is 755.5 mOsmol/L.

Chapter 27 - Aliquot

Worksheet 27-1

1) 150 mg
2) 1 mL
3) Mix 1 mL of phytonadione with 9 mL of diluent. The ordered dose will require 1 mL of this mixture.
4)
    a) If you use the conversion 1 grain = 60 mg, then you will need 120 mg of codeine phosphate and 840 mg of lactose. A dose would require 120 mg of this mixture.

or

If you use the conversion 1 grain = 64.8 mg, then you will need 129.6 mg of codeine phosphate and 907.2 mg of lactose. A dose would require 129.6 mg of this mixture.

or

If you use the conversion 1 grain = 65 mg, then you will need 130 mg of codeine phosphate and 910 mg of lactose. A dose would require 130 mg of this mixture.

b) Use a size 5 capsule shell for this

5) Mix 200 mg of chlorpheniramine powder with 200 mg of diluent and then the ordered dose will require 200 mg of this mixture.
6) Mix 0.2 mL of the substance with 0.6 mL of SWFI and then you would require 0.2 mL of the mixture to provide the appropriate dose.
7) Dose 1 requires 2 mL of Dilution 2 and 98 mL of D5W.
   Dose 2 requires 4 mL of Dilution 2 and 96 mL of D5W.
   Dose 3 requires 8 mL of Dilution 2 and 92 mL of D5W.
   Dose 4 requires 16 mL of Dilution 2 and 84 mL of D5W.
   Dose 5 requires 32 mL of Dilution 2 and 68 mL of D5W.
   Dose 6 requires 60 mL of Dilution 2 and 40 mL of D5W.
   Dose 7 requires 12 mL of Dilution 1 and 88 mL of D5W.
   Dose 8 requires 24 mL of Dilution 1 and 76 mL of D5W.
   Dose 9 requires 50 mL of Dilution 1 and 50 mL of D5W.
   Dose 10 requires 10 mL of the Initial Bag and 90 mL of D5W.
   Dose 11 requires 20 mL of the Initial Bag and 80 mL of D5W.
   Dose 12 requires 40 mL of the Initial Bag and 60 mL of D5W.
   Dose 13 requires 80 mL of the Initial Bag and 20 mL of D5W.
8) Mixture 1 requires 80 mg of aspirin and 63.92 g of powdered sugar.
   - Dose 1 requires 80 mg of Mixture 1.
   - Dose 2 requires 240 mg of Mixture 1.
   - Dose 3 requires 800 mg of Mixture 1.
   - Dose 4 requires 2.4 g of Mixture 1.
   - Dose 5 requires 8 g of Mixture 1.

   Mixture 2 requires 90 mg of aspirin and 180 mg of powdered sugar.
   - Dose 6 requires 90 mg of Mixture 2.
   - Dose 7 requires 120 mg of Mixture 2.

INDEX